ANHARMONIC LATTICES, STRUCTURAL TRANSITIONS AND MELTING

NATO ADVANCED STUDY INSTITUTES SERIES

Proceedings of the Advanced Study Institute Programme, which aims at the dissemination of advanced knowledge and the formation of contacts among scientists from different countries.

Applied Sciences No. 1

Vh

ANHARMONIC LATTICES, STRUCTURAL TRANSITIONS AND MELTING

edited by

T. RISTE

NOORDHOFF - LEIDEN - 1974

ISBN 90 286 0533 9

Printed in the Netherlands.

PREFACE

This book contains the papers presented at the NATO Advanced Study Institute held at Ustaoset Høyfjellshotel, near Geilo, Norway, 24th April - 1st May, 1973. The members of the Programme Committee were

J. D. Axe	Brookhaven National Laboratory, USA
J. Feder	University of Oslo, Norway
T. Riste	Institutt for Atomenergi, Norway (Chairman)

The title of the institute, "Anharmonic Lattices, Structural Transitions and Melting", gives an idea of the programme and of the content of this book. The first part is devoted to the so-called "central mode" at 2nd order structural phase transitions, a phenomenon which was first reported in 1971 at a NATO Advanced Study Institute held at Geilo. (Several references to the proceedings of that institute, entitled "Structural Phase Transitions and Soft Modes", are found in the present volume). From 2nd order transitions the emphasis of the programme moves gradually to 1st order transitions by covering such topics as order-disorder structural transitions in simple molecular crystals and tricritical phenomena. The last part is devoted to melting, including discussions on the dynamics of solids and liquids near their melting points.

The lectures are divided almost equally between review talks and research reports dealing with subjects closely connected with the topics under review. With a few exceptions the lectures are printed in the sequence of presentation at the study institute.

The total number of participants was limited to 71, their names are given at the end of the book. They all survived the skiing between the lectures.

The publication of these lectures could be realized through the kind cooperation of the lecturers. I am indebted to Mrs. Gerd Jarrett and to my colleagues at Institutt for Atomenergi for their help in planning and arranging the study institute. The generous support from the NATO Scientific Affairs Division is gratefully acknowledged.

Institutt for Atomenergi
Kjeller, September 1973

Tormod Riste

C O N T E N T S

PHASE TRANSITIONS, ANHARMONIC SOLIDS AND LIQUIDS

G. Niklasson and A. Sjölander

Institute of Theoretical Physics, Göteborg, Sweden

I. INTRODUCTION

Intensive experimental and theoretical studies of liquids in the last ten or fifteen years have revealed new and rather detailed information on the motions of the individual atoms as well as on the collective character of the atomic motions. Both for gases and for crystalline solids one has for a long time had standard methods for calculating various quantities of interest. For instance, calculations of the transport properties of dilute gases are based on Boltzmann's equation and on the methods for solving this equation. For crystalline solids, the harmonic approximation has been the starting point, and anharmonicity has been included as a perturbation. The situation for the liquid phase is considerably more complicated and the search for reasonably simple models, which still give the essential physics accurately enough, has met with great difficulties. At the same time Raman scattering, Brillouin scattering, neutron scattering and other experimental techniques, and also big computer calculations have revealed new detailed structure in the atomic motions and these results have to be explained quantitatively. In the last few years, considerable progress has been made, particularly on the theory for atomic motions in simple liquids like Argon. In calculating thermodynamic properties of these liquids the main uncertainty seems now to lie in our knowledge of the interatomic potential (1,2). Also, in calculations of the dynamic structure factor $S(\vec{q}\omega)$, which is directly measured in neutron scattering experiments, one has been very successful in using Mori's memory function approach or analogous methods (3-6).

In these lectures, we shall mainly concentrate on the dynamical aspects and direct our attention to some questions of great interest at present. Computer calculations on two- and three-dimensional hard core gases, carried out a few years ago by Alder and Wainwright (7), revealed some very long relaxation times, which could not be explained within the conventional treatment of the Boltzmann equation. This discovery leads to the conclusion that Navier-Stoke's equation should not be valid for a two-dimensional gas. It also means that the conventional way of deriving the hydrodynamic equations for three-dimensional gases and liquids by expanding in higher and higher spatial derivatives of the local density, current and temperature can be seriously questioned. When approaching the gas-liquid critical point the ordinary hydrodynamics ceases to be valid due to the large critical fluctuations (8-10). At present we do not know the form of the proper equation of motion within the critical regime. Concerning crystalline solids it has been shown that low order perturbation theory cannot give the proper equations of motion in the hydrodynamic regime (11,12). Starting from the harmonic theory and including anharmonic terms to finite order would not bring in entropy as a proper thermodynamic mode. This mode shows up as a quasi-elastic peak in $S(\vec{q}\omega)$. After it was established that many structural transitions in solids are caused by a certain phonon mode becoming soft (13-15), lattice dynamics has attracted new attention and somewhat new aspects were brought in. Experimentally one has found that $S(\vec{q}\omega)$ contains not only the soft phonon peak but also in some cases a central peak when approaching the transition temperature (16-18). This happens even when symmetry arguments tell that the particular soft phonon mode cannot couple to entropy fluctuations and, therefore, should not give a heat diffusion peak in $S(\vec{q}\omega)$.

In the following we shall try to bring some order into the various observations mentioned above and we shall try to get a somewhat unified picture of what is happening. We will, therefore be concerned mainly with the general behaviour and less with those properties which are specific for only one class of systems.

After making some general comments on phase transitions we will first discuss the ordinary hydrodynamic equations for a liquid and a solid. We will briefly discuss the conditions under which these equations are valid. In the next section we very briefly summarize the results from neutron scattering experiments for those wave-lengths and frequencies for which hydrodynamics is not valid. It is possible to carry out a detailed derivation of the hydrodynamic equations for a weakly anharmonic solid, and we shall in the following sections give the main steps in such a derivation. This will show how hydrodynamic modes enter the theory, and it will bring in the concept of mode-mode coupling. In the last part, we shall return to liquids and discuss the long relaxation times, found by Alder and Wainwright, and also the idea of mode-mode coupling due to Fixman, Kawasaki, Kadanoff and Swift, and others (19-21).

II. PHASE TRANSITIONS

Most of the phase transitions we know are of first order and are characterized by a discontinuous change in various thermodynamic quantities; for instance, in a crystal changing from a closed packed cubic to a closed packed hexagonal structure, the energy, the entropy, the lattice parameter etc. are all changing discontinuously. The quantity which controls the phase transition is Gibb's free energy

$$G = U + pV - TS \qquad \text{(II.1)}$$

where U = total energy, V = total volume, S = total entropy, p = pressure, and T = temperature. In considering structural transitions in solids one often ignores the term pV and then discusses the Helmholtz free energy instead. The stable phase is the one with the lowest free energy and a transition occurs when two or more phases happen to have equal value of G for the same pressure and temperature. In order to calculate the position of the transition point or to determine which phase is thermodynamically stable one calculates as accurately as possible the free energy for each phase and compares the values. In metals for instance, the difference in G for some crystal structures is found to be very small and this means that very accurate calculations are required in order to determine the stable phase conclusively.

In a second order phase transition, all of the quantities mentioned above change continuously at the transition point. This implies that no abrupt change in symmetry can occur. A typical example of a second order structural phase transition occurs in $SrTiO_3$ at about 105^oK. Above this temperature the stable phase has cubic structure and this continuously changes into a tetragonal structure when the temperature is lowered. Similarly a simple cubic structure could change continuously into a hexagonal one by changing the ratio c/a from unity. In such second order transitions, the higher order thermodynamic derivatives, such as compressibility, specific heat etc. change abruptly at the transition point and many of them become singular.

At the gas-liquid critical point, the sound velocity becomes zero and this means the system has a tendency to be unstable against long wave-length density fluctuations. In many second order structural transitions in solids one has found that a certain optical mode becomes soft. This means that the crystal becomes unstable against certain particular distortions. Under such circumstances it is not surprising that we have large, slowly varying, fluctuations of certain thermodynamic variables, the so-called critical fluctuations. The system is facing a situation like a person who has to decide between two equally attractive bids. The search for a proper description of these critical fluctuations is still a very challenging problem.

III. THE HYDRODYNAMIC REGIME

Away from a phase transition point, the long wave-length fluctuations are governed by the ordinary hydrodynamic equations. In liquids, it means that the only slowly varying quantities are the particle density $n(\vec{x}t)$, the particle current density $\vec{j}(\vec{x}t)$, the local temperature $T(\vec{x}t)$ and quantities related to these. All other modes are varying more rapidly and they are assumed to average out within the time scale considered here. Therefore, we are able to get a closed set of equations for these macroscopic variables. After linearization, we get the following well-known equations (22,23):

$$\frac{\partial}{\partial t} n + \vec{\nabla}\cdot\vec{j}_l = 0,$$

$$\frac{\partial}{\partial t} \vec{j}_l + c_{is}^2 \vec{\nabla} n + n_o \alpha c_{is}^2 \vec{\nabla} T - \frac{\zeta + \frac{4}{3}\eta}{mn_o} \vec{\nabla} (\vec{\nabla}\cdot\vec{j}_l) = 0, \tag{III.1}$$

$$\frac{\partial}{\partial t} \vec{j}_{tr} - \frac{\eta}{mn_o} \nabla^2 \vec{j}_{tr} = 0,$$

$$\frac{\partial}{\partial t} T - (\lambda/C_V)\, \nabla^2 T - \frac{1}{n_o\alpha} \left(\frac{C_p}{C_V} - 1\right) \frac{\partial}{\partial t} n = 0,$$

where $c_{is} = (mn_o\chi_{is})^{-1/2}$ = isothermal velocity (χ_{is} is the compressibility)

mn_o = equilibrium mass density

α = thermal expansion coefficient

ζ, η = bulk and shear viscosity coefficients

λ = thermal conductivity

C_p, C_V = specific heat per unit volume at constant pressure and volume, respectively.

The current has been split into its transverse and longitudinal parts and this means

$$\vec{\nabla}\cdot\vec{j}_{tr} = 0, \; \vec{\nabla}\times\vec{j}_l = 0. \tag{III.2}$$

We notice that the density, the longitudinal current and the temperature are all coupled. By going over to the Fourier space and further eliminating the current and temperature we get

$$[-\omega^2 + c_{is}^2\, q^2 + \omega q^2\Gamma(\vec{q}\omega)\,]\; n(\vec{q}\omega) = 0, \tag{III.3}$$

where

$$\Gamma(\vec{q}\omega) = -\, i\, \frac{\zeta + \frac{4}{3}\eta}{mn_o} - i(\frac{C_p}{C_V} - 1)\, c_{is}^2\, \frac{1}{-i\omega + (\lambda q^2/C_V)}\,. \tag{III.4}$$

This leads to the following expression for the dynamic structure factor:

$$S(\vec{q}\omega) = -\,\frac{2k_BT}{m}\, \frac{q^4\Gamma''(\vec{q}\omega)}{[\omega^2 - c_{is}^2 q^2 - \omega q^2\Gamma'(\vec{q}\omega)]^2 + [\omega q^2\Gamma''(\vec{q}\omega)]^2} \tag{III.5}$$

where Γ' and Γ'' are the real and imaginary parts of $\Gamma(\vec{q}\omega)$.

For solids we obtain very similar results. The slow hydrodynamic variables are the displacements, $\vec{u}(\vec{x}t)$, from the true equilibrium positions and the temperature, and they are governed by the following hydrodynamic equations (24):

$$\frac{\partial^2}{\partial t^2} u^\alpha - \frac{1}{mn_o} \sum_{\beta\gamma\delta} C_{is}^{\alpha\beta;\gamma\delta}\, \nabla^\beta\nabla^\gamma\, u^\delta - \frac{1}{mn_o} \sum_{\beta\gamma\delta} \eta^{\alpha\beta;\gamma\delta}$$

$$\times\, \nabla^\beta\nabla^\gamma\, \frac{\partial}{\partial t} u^\delta + (\alpha/mn_o\chi_{is})\nabla^\alpha T = 0, \tag{III.6}$$

$$\frac{\partial}{\partial t} T - (\lambda/C_V)\, \nabla^2 T + \frac{1}{\alpha}\, (\frac{C_p}{C_V} - 1)\, \frac{\partial}{\partial t}\, (\vec{\nabla}\cdot\vec{u}) = 0,$$

where α denote the Cartesian components and $C_{is}^{\alpha\beta;\gamma\delta}$ is the isothermal elastic tensor and $\eta^{\alpha\beta;\gamma\delta}$ is the viscosity tensor.

As in the liquid case, we find that only the longitudinal

waves are coupled to the temperature. The quantity of particular interest is the displacement correlation function $\langle u^{\alpha}(\vec{x}t)u^{\beta}(oo)\rangle$, whose Fourier transform is directly measured in Brillouin scattering and in neutron scattering. It has a structure very similar to that of $S(\vec{q}\omega)$ above.

For small values of the wave vector $S(\vec{q}\omega)$ splits into three sharp peaks as shown in Fig. 1.

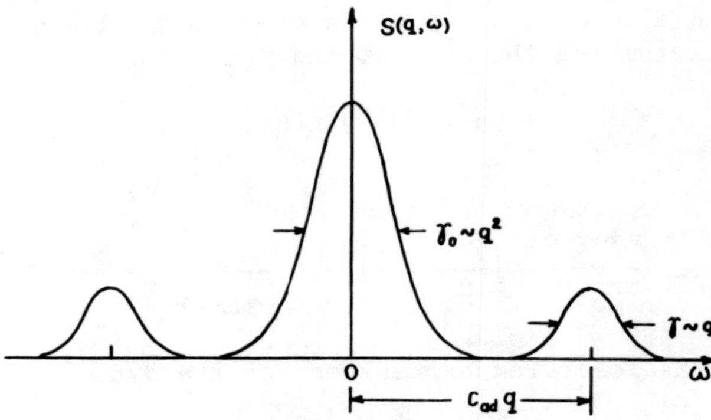

Figure 1. Dynamic structure factor for a liquid in the hydrodynamic regime.

The two Brillouin components are located at

$$\omega = \pm\, c_{ad}\, q, \tag{III.7}$$

where the adiabatic sound velocity is given by

$$c_{ad}^2 = c_{is}^2 + \left(\frac{C_p}{C_V} - 1\right) c_{is}^2 = \frac{C_p}{C_V} c_{is}^2 . \tag{III.8}$$

The full width of the two peaks is determined by

$$\gamma = \left[\frac{\zeta + \frac{4}{3}\eta}{mn_o} + \frac{\lambda}{C_p}\left(\frac{C_p}{C_V} - 1\right)\right] q^2 . \tag{III.9}$$

The third peak is located around $\omega = o$ and has the width

$$\gamma_o = \frac{2\lambda}{C_p} q^2 \tag{III.10}$$

We notice that the width of the peaks vary as q^2. From Eqs. (III.3,4) we find that the coupling to the heat diffusion mode is proportional to $(C_p/C_V - 1)\omega q^2$. The intensity of the central peak relative to the two displaced peaks is given by

$$I_o/2I = \frac{C_p}{C_V} - 1 . \tag{III.11}$$

As we go to larger values of the wave vector, the peak widths increase and the three peaks begin to overlap and finally go over into a single broad peak around $\omega = o$. Another conclusion we can draw from the hydrodynamic equations is that as we approach the critical point, where χ_{is} and C_p/C_V tend to infinity, the two sound wave peaks move towards the central peak. This is in qualitative agreement with observations. However, a closer comparison with experiments has shown that hydrodynamics does not give the correct predictions for the critical fluctuations and we have reason to ask what the conditions are for using the hydrodynamic equations (8-10).

In the derivation one is assuming that only certain modes vary slowly. All other modes of importance are assumed to decay rapidly, and for these one may introduce a typical mean free path l and relaxation time τ. Hydrodynamics is then valid for wave vectors and frequencies such that

$$ql \ll 1, \quad \omega\tau \ll 1. \tag{III.12}$$

This is the hydrodynamic regime. In the opposite case, when

$$ql \gg 1, \quad \omega\tau \gg 1 \tag{III.13}$$

one talks about the collisionless regime or sometimes the zero-sound regime.

We have situations where other hydrodynamic modes appear. In superfluid He^4 and in superconductors, the velocity of the condensate is an independent macroscopic variable and has to be included among the hydrodynamic variables. A similar situation occurs in a soft mode transition. When the frequency of an optic mode becomes very small, the corresponding displacement has to be considered as a hydrodynamic mode and the conventional theory has to be modified accordingly.

IV. NON-HYDRODYNAMIC REGIME

In this section we shall only briefly summarize what happens when we move out of the hydrodynamic regime, as observed by neutron scattering in liquids and solids.

In solids one finds, even for large wave vectors, sharp resonances in the scattering cross section and these correspond to the ordinary phonons, which are the elementary excitations in a crystal. No central heat diffusion peak is found and should not appear, according to elementary theory. In liquids no sharp resonances are found, but in liquid metals and also recently in liquid hydrogen one has seen rather broad peaks, corresponding to the Brillouin components above (25,26). In liquid argon, for instance, one observes a broad central peak (27). However, this is not due to heat diffusion. The width of the peak is instead determined by the diffusion of individual atoms.

For simple liquids, like argon, one has been very successful in explaining quantitatively the neutron scattering data, using approaches introduced by Kadanoff and Martin and by Mori (28,29). It means, in practice, using frequency and wave vector dependent transport coefficients in the hydrodynamic equations. The result for $S(\vec{q}\omega)$ is the same as in Eq. (III.5) with the difference that $\Gamma(\vec{q}\omega)$ is a more complicated function. Also c_{is}^2 is changed to

$$c_{is}^2 \to \frac{k_B T}{m} \frac{1}{S(\vec{q})}, \qquad \text{(IV.1)}$$

where $S(\vec{q})$ is the static structure factor. For small values of $\vec{q}$ and ω the result goes over to that in the previous section. For large values of ω but arbitrary $\vec{q}$ certain sum rules for $S(\vec{q}\omega)$ determine $\Gamma(\vec{q}\omega)$ in terms of the interparticle potential and the static structure factor. One has then used some extrapolation between the two extreme cases and has been remarkably successful in doing this (3-6).

From observations we have found that the dynamics can change quite drastically when moving out of the hydrodynamic regime. Certain hydrodynamic modes may disappear when we go to shorter wavelength, whereas other hydrodynamic modes may remain, although somewhat modified. The elastic waves in a solid go over into ordinary phonons, whereas the temperature mode disappears. This brings us to a fundamental question:

(i) How do the hydrodynamic modes enter into the picture, when they do not appear in the basic microscopic equation of motion for the system.

Particularly in connection with critical fluctuations we may have to consider situations, where there is no sharp distinction in relaxation times for the strictly hydrodynamic modes and the non-hydrodynamic ones. If so, we may ask the question:

(ii) Would the hydrodynamic equations be modified if some of the non-hydrodynamic modes also have long relaxation times.

Observations on some structural phase transitions have shown a central peak in $S(\vec{q}\omega)$ in spite of the fact that symmetry arguments rule out the possibility of coupling to the temperature. Therefore, we may ask:

(iii) Can a central peak in $S(\vec{q}\omega)$ arise from other sources than temperature fluctuations.

For a weakly anharmonic crystal it is possible to get a deeper understanding of these questions, starting from the microscopic equation of motion for the atoms. In the next section we shall outline the main steps in such an analysis.

V. ANHARMONIC SOLIDS

One way of getting information on the dynamics of a system is to disturb it by some external means and study how the system responds. Let us for illustration consider a harmonic linear chain on which is applied a weak external force field, varying in space and time. The equation of motion for this system is

$$\frac{d^2}{dt^2} \langle u(Rt)\rangle + \sum_{R'} \phi(R-R') \langle u(R't)\rangle = J(Rt), \qquad (V.1)$$

where $J(Rt)$ is the external force devided by the atomic mass m, $\phi(R)$ are the harmonic force constants and $\langle u(Rt)\rangle$ are the induced mean displacements of the atoms. Written in Fourier space we get

$$\langle u(q\omega)\rangle = D(q\omega)\, J(q\omega), \qquad (V.2)$$

where the response function is

$$D(q\omega) = \frac{-1}{\omega^2-\omega^2(q)}, \qquad (V.3)$$

$\omega(q)$ being the harmonic phonon frequency. We find that the response function has a sharp resonance corresponding to the elementary excitations in the system. In a real system anharmonic effects would lead to modifications of our results, but we still expect to see resonances as long as the anharmonicity is reasonably small. In general we may write the response function as

$$D(q\omega) = \frac{-1}{\omega^2-\omega^2(q)-M(q\omega)} , \qquad (V.4)$$

where all effects of anharmonicity are included in the "self-energy" $M(q\omega)$. This is in general a complex quantity, the real part giving a shift and the imaginary part a broadening of the resonance peak. $M(q\omega)$ may itself contain a resonance corresponding to new collective modes entering through the interaction between the phonons. In order to see how this can occur, we shall modify Eq. (V.1) to include third order anharmonic terms. We write

$$\frac{d^2}{dt^2} \langle u(Rt)\rangle + \sum_{R'} \phi(R-R')\langle u(R't)\rangle + \sum_{R'R''} \phi(R,R',R'') \qquad (V.5)$$

$$\times \langle u(R't)u(R''t)\rangle = J(Rt),$$

where $\phi(R,R',R'')$ are the anharmonic force constants. We notice that the equal time displacement correlation function enters, and we have to know how this changes under the external force.

In equilibrium we can write generally

$$\langle u(R't)u(R''t)\rangle = \frac{\hbar}{m} \int \frac{dq}{a} \int \frac{d\omega}{\omega} [1 + n(\omega)]\, A(q\omega)\, e^{iq(R'-R'')}, \qquad (V.6)$$

where $A(q\omega)$ is the phonon spectral function and $n(\omega)$ is the ordinary Bose-Einstein distribution or phonon occupations number. The q-integration goes over the first Brillouin zone with length a. In the harmonic approximation

$$A(q\omega) = \frac{1}{2} [\delta(\omega-\omega(q)) + \delta(\omega+\omega(q))] , \qquad (V.7)$$

and it gives information on the phonon frequencies. Due to anharmonicity the δ-functions are broadened and the widths of the peaks tell us what the phonon life times are. The Bose-Einstein distribution is

$$n(\omega) = [\exp(\hbar\omega/k_B T) - 1]^{-1}, \qquad (V.8)$$

where T is the equilibrium temperature. It gives information on the number of phonons present with the energy $\hbar\omega$.

The external forces will lead to local distortions of the lattice, described by $\langle u(Rt)\rangle$. Particularly, if the distortion varies slowly in space and time we can consider this as macroscopic and it leads to a local change in the phonon frequencies, entering through a change of $A(q\omega)$, and a local change in the number of phonons, entering through a change of the temperature in $n(\omega)$. Using many-body theory one can define formally a non-equilibrium spectral function $A(q\omega|Rt)$ and phonon number density $n(q\omega|Rt)$ and derive a set of coupled equations for these quantities and $\langle u(Rt)\rangle$. For further details we refer to a series of papers by Niklasson and Sjölander on this subject (12, 30-32). It was found that $A(q\omega|Rt)$ adjusts itself rather quickly to the external disturbance with a characteristic relaxation time of the order inverse Debye frequency. The generalized phonon number density satisfies a modified Boltzmann equation, which we write schematically

$$\left(\frac{\partial}{\partial t} + c(q)\frac{\partial}{\partial R} + L_{op}\right) n(q\omega|Rt) = \gamma\langle u(Rt)\rangle, \qquad \text{(V.9)}$$

where c is the phonon group velocity and γ gives the coupling to the lattice distortion. The collision operator L_{op} is responsible for bringing the system to local equilibrium. It has the energy and momentum conservation built in and this means that the local temperature or entropy appears as a hydrodynamic mode. Mathematically it appears as an eigen state of L_{op} with zero eigenvalue; i.e. infinite life time. If Umklapp processes were absent, the phonon current would also be a hydrodynamic mode. All other modes will decay more rapidly with a relaxation time of the order thermal phonon life time τ. This is normally much longer than the inverse Debye frequency. We have therefore, to consider three quite different relaxation times; the <u>inverse Debye frequency</u> $1/\omega_D$, the <u>thermal phonon life time</u> τ and the <u>hydrodynamic relaxation time</u>, which is proportional to $1/q^2$. <u>Fig. 2</u> illustrates the situation. The time-dependent phonon self-energy $M(qt)$ may for a qualitative argument be considered as a superposition of three different relaxation functions, each one decaying exponentially.

If the lattice distortion is slow enough so that

$$ql \ll 1, \quad \omega\tau \ll 1, \qquad \text{(V.10)}$$

where $l = c\tau$ is the phonon mean free path, the system has time to relax close to local equilibrium. The non-equilibrium value of $\langle u(R't)\, u(R''t)\rangle$ in Eq. (V.5) is then governed by the local temperature, which satisfies the ordinary heat diffusion equation. The heat diffusion mode shows up as a resonance in the phonon self-energy. After carrying out the detailed calculations one obtains the following expression for $D(q\omega)$, valid in the hydrodynamic region

$$D(q\omega) = -\left[\omega^2 - c_{is}^2 q^2 + i\frac{\eta\omega}{mn_o} q^2 + i\left(\frac{C_p}{C_V} - 1\right) c_{is}^2 \frac{\omega q^2}{-i\omega + (\lambda q^2/C_V)}\right]^{-1} \quad \text{(V.11)}$$

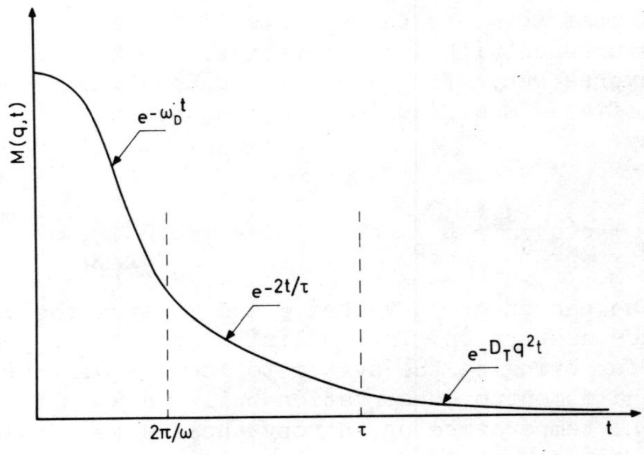

Figure 2. The figure shows the qualitative behaviour of M(qt) with three different relaxation regions.

The notations are the same as in Eq. (III.4). In Brillouin and neutron scattering one essentially measures [Im $D(q\omega)/\omega$] . The contribution from the spectral function as well as from the non-hydrodynamic modes in $n(q\omega|Rt)$ enter into the numerical values of c_{is} and η. As already mentioned in Chapter III, the heat diffusion mode gives a three peak structure to [Im $D(q\omega)/\omega$] with a central diffusion peak, whose width is $(\lambda q^2/C_p)$. If no Umklapp processes were present, one would find a propagating heat mode, second sound, instead of ordinary diffusion and this would in Im $D(q\omega)$ show up as two peaks for $\omega \neq o$.

For frequencies such that

$$1 < \omega\tau < \omega_D\tau, \quad \text{(V.12)}$$

no heat diffusion mode would appear and the dynamics would be governed by processes where $n(q\omega|Rt)$ is approaching local equilibrium. The corresponding relaxation time, being of the order thermal phonon life time, can be quite long. Even in this case we may expect to get some kind of resonance in $M(q\omega)$, now with the width $1/\tau$. The sound velocity and the damping should then show some changes within the frequency range $1/\tau$ and one may call this the transition region from the collisionless to the hydrodynamic regime.

The above considerations are essentially unchanged for a three dimensional lattice, except for mathematical complications associated with the geometry of the lattice and with the different polarization directions. The coupling between the long wavelength purely transverse acoustic modes and the thermodynamic heat mode vanishes and in the corresponding response function no heat diffusion peak appears. However, the transverse sound velocity and its damping show changes within the transition region, where transverse sound velocities change from their hydrodynamical values towards their "zero-sound" values. Results from calculations of the sound velocities in solid argon in this transition region is shown in Fig. 3.

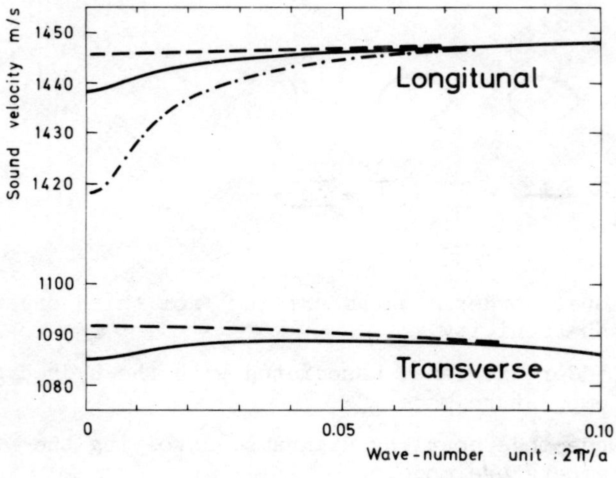

Figure 3. Calculated sound velocities in the (1,0,0)-direction for solid argon at 20 °K (32).

Dashed curves: Lowest order perturbation, neglecting effects of the finite phonon life-times.

Dot-dashed curves: Lowest order perturbation, but including effects of the finite phonon life-times. For $\vec{q} = 0$, one gets the isothermal sound velocities.

Full curves: Higher order perturbation, including terms associated with heat diffusion. For the transverse modes these higher order terms do not contribute.

It is conventional to discuss the phonon self-energy in terms of equilibrium Green functions and in terms of diagrams. For this reason we shall briefly comment on how the various effects that we have discussed above enter in such a description. The lowest order term of interest for us is represented by the diagram in Fig. 4a. It corresponds to an incoming phonon with polarization index j interacting with two other phonons of polarization j_1 and j_2, respectively.

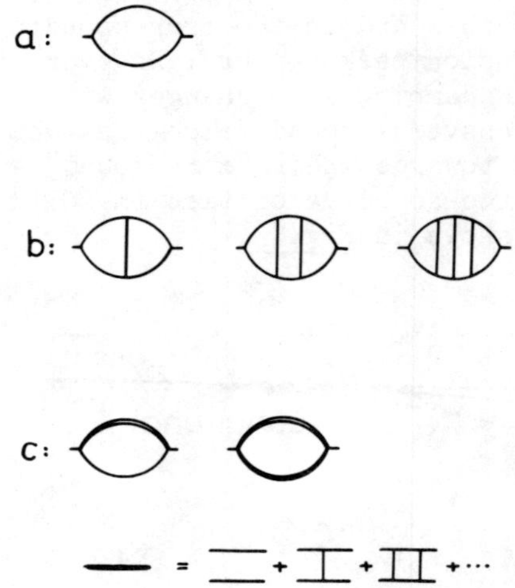

Figure 4. a: Lowest order diagram arising from third order anharmonicity.

b: Ladder diagrams, associated with the hydrodynamic temperature mode.

c: Mode-mode coupling diagrams, involving the temperature mode.

This diagram leads to a contribution to the phonon self-energy of the form

$$M_j(\vec{q}\omega) = \int d\vec{q}\,' \frac{|V_{jj_1j_2}(\vec{q}\,',\vec{q}-\vec{q}\,')|^2}{\omega \pm \omega_{j_1}(\vec{q}\,') \pm \omega_{j_2}(\vec{q}-\vec{q}\,') + \frac{i}{\tau_1} + \frac{i}{\tau_2}} , \qquad (V.13)$$

where $V_{jj_1j_2}(\vec{q}\,',\vec{q}-\vec{q}\,')$ is essentially the third order anharmonic force constant and τ_1 and τ_2 are the life times of the two phonons j_1 and j_2. These terms can be devided into two classes:

(a) terms, where either $\omega_{j_1}(\vec{q}')$ and $\omega_{j_2}(\vec{q}-\vec{q}')$ enter with the same sign or with $j_1 \neq j_2$, lead to a frequency dependence of $M_j(\vec{q}\omega)$, which corresponds to a relaxation time of order $1/\omega_D$. They are associated with the response of the spectral function in the previous discussion,

(b) terms, where both $j_1 = j_2$ and $\omega_{j_1}(\vec{q}')$ and $\omega_{j_2}(\vec{q}-\vec{q}')$ enter with opposite sign, have the common feature that for $\vec{q} \to o$ the two phonon frequencies cancel each other in the denominator for all values of $\vec{q}'$. It leads to a relaxation time of the order thermal phonon life time. Such terms are associated with the non-hydrodynamic part of $n(\vec{q}\omega|\vec{R}t)$.

The hydrodynamic part of $n(\vec{q}\omega|\vec{R}t)$ is not contained in this lowest order diagram. It has been shown that terms giving the hydrodynamic heat mode are represented by the infinite set of ladder diagrams in Fig. 4b (11,12). By neglecting these, but at the same time including phonon life time effects in Eq. (V.13), one is actually violating energy-conservation. This is disastrous for obtaining proper hydrodynamics. However, introducing phonon life times in Eq. (V.13) has the effect of making the resulting phonon widths quadratic in $\vec{q}$ instead of linear in the long wave length limit.

As an illustration we consider a long wavelength transverse acoustic phonon. Since it does not couple to the temperature we need only consider the diagrams in Fig. 4a. With certain approximations we can write the resulting response function in the following form (32).

$$D(\vec{q}\omega) = \left[\omega^2 - c_{is}^2 q^2 - \gamma^2 \omega q^2 \sum_j \int \frac{d\vec{q}'}{\omega - \vec{c}_j(\vec{q}')\cdot\vec{q} + \frac{2i}{\tau}} \right]^{-1} , \quad (V.14)$$

where γ is proportional to the third order anharmonic force constants. $\vec{c}_j(\vec{q})$ is the phonon group velocity. Assuming that the dominant contribution to the integral comes from phonons with small group velocities, such as optical phonons or short wavelength acoustic phonons, we may ignore the term $\vec{c}_j(\vec{q}')\cdot\vec{q}$. This gives

$$D(\vec{q}\omega) = \left[\omega^2 - c_{is}^2 q^2 - \frac{\gamma'^2 \omega q^2}{\omega + \frac{2i}{\tau}} \right]^{-1} , \quad (V.15)$$

where γ' is constant. An expression of similar form was used by Shirane and Axe to analyse some recent neutron scattering results on Nb_3Sn near its phase transition at 45 $^{\circ}K$ (17). The transition is caused by a soft transverse acoustic mode, which cannot couple to the thermodynamic heat diffusion mode.

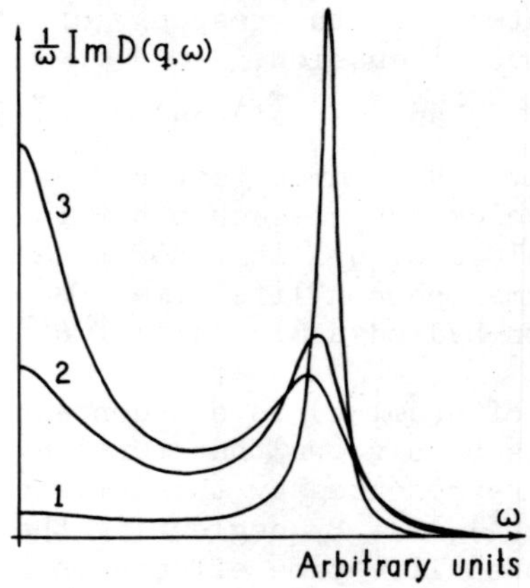

Figure 5. Dynamic structure factor $S(\vec{q}\omega) \propto \mathrm{Im}\ [D(\vec{q}\omega)/\omega]$, calculated from Eq. (V.15) for $\omega_\infty\tau = 4$, and $\omega_0/\omega_\infty =$ = 0.9 (<u>curve 1</u>), 0.7 (<u>curve 2</u>) and 0.6 (<u>curve 3</u>), with $\omega_0^2 = c_{is}^2 q^2$ and $\omega_\infty^2 = (c_{is}^2 + \gamma'^2)q^2$.

<u>Fig. 5</u> shows the shape of $\mathrm{Im}\ D(\vec{q}\omega)/\omega$ for three different parameter choices. It seems to us that a central peak structure should be a rather common feature of a soft-mode transition. It follows from the fact that certain processes entering the self-energy relax with a time of the order of the thermal phonon life time. This conclusion is supported by recent theoretical work (33-37). The details of the structure depends on the particular soft mode involved and the characteristics of the coupling to other modes. The situation here seems in one respect to differ significantly from second order transitions in liquids and in magnetism, for instance. There all short wavelength fluctuations decay rapidly and we have, therefore, no reason to distinguish between two time scales τ and $1/\omega_D$ as was done above.

The diagram in <u>Fig. 4a</u> represents a mode-mode coupling analogous to what is generally considered near a second order phase transition (20,21). If we would be fully self-consistent in our calculations, the phonon lines in the diagrams should for long wavelengths go over to the proper hydrodynamic modes. Furthermore one should introduce some vertex corrections.If only long wave-length modes contribute, such vertex corrections can be calculated through purely thermodynamic arguments. A sound wave can also interact with two hydrodynamic modes, where one or both is a heat diffusion mode. Such processes are not included in the diagrams considered above but would be included in the diagrams in Fig. 4c.

VI. MODE-MODE COUPLING IN LIQUIDS

In the introduction we referred to computer calculations on two- and three dimensional hard core gases and we mentioned that they revealed some unexpected long relaxation times. A series of papers have appeared explaining these results (38-42). For our discussion here a paper of Ernst and Dorfman (43) is of particular interest and their results will be briefly summarized.

The Boltzmann equation gives a closed expression for the one-particle phase space distribution. The linearization of the equation, which is conventionally done, results in the loss of one interesting aspect of the dynamics. This can be illustrated in the following way. Consider a single atom moving in a medium of surrounding atoms and interacting with these. The first atom disturbs its surroundings and the slowly decaying part of the disturbance are the hydrodynamic modes of the surrounding medium. The disturbance reacts back on the first atom and hence influences its motion. In particular, the slowly decaying hydrodynamic modes will react back for a long time and give rise to long memory effects. They give rise to low frequency contributions to the transport coefficients. In a dilute gas, considered by Ernst and Dorfman, the effect was found to be numerically very small but it is of great conceptual interest. It may be of more importance in liquids, and near a second order phase transition it is of major importance and corresponds to the mode-mode coupling introduced to explain critical fluctuations (19-21).

A hydrodynamic mode can decay into two or possibly more other hydrodynamic modes. The smallness of the available phase space would limit the importance of these processes, but on the other hand they give long time effects and would increase in importance after time integration. These processes are not included in the conventional hydrodynamics. They are erased by assuming a finite relaxation time for all non-hydrodynamic modes, and modes of pairs of hydrodynamic modes are assumed to decay rapidly compared to a single hydrodynamic mode. Near a critical point one is facing a situation where the most important processes have a relaxation time proportional to $1/q^2$. This implies that the ordinary hydrodynamics is reached only for wave vectors small compared with a certain inverse correlation distance and the latter becomes infinite at the phase transition point. The mode-mode terms give to the self-energy contributions of the form.

$$M(\vec{q}\omega) = \sum_{r,s} \int \frac{d\vec{q}'}{(2\pi)^3} \frac{|A_{rs}(\vec{q}',\vec{q}-\vec{q}')|^2}{\omega-\omega_r(\vec{q}')-\omega_s(\vec{q}-\vec{q}')} , \qquad (VI.1)$$

where $\omega_{r,s}(\vec{q})$ are the complex frequencies of the various modes and $A_{rs}(\vec{q}',\vec{q}-\vec{q}')$ is a coupling parameter, whose form depends on which modes are involved. For a sound wave mode $\omega(\vec{q}) = cq - i\gamma q^2$ and for a diffusive mode $\omega(\vec{q}) = -iDq^2$. Ernst and Dorfman find by evaluating the integrals above the following sound wave dispersion

$$\omega_{sound} = cq - i\gamma q^2 + \delta q^{5/2} + \ldots , \qquad (VI.2)$$

where δ is a complex number. For the shear mode and the heat diffusion mode they get

$$\begin{aligned} \omega_{shear} &= - i(\eta/mn_o)\, q^2 + i\delta_1\, q^{5/2} + \ldots , \\ \omega_{heat} &= - i(\lambda/C_p)\, q^2 + i\delta_2\, q^{5/2} + \ldots , \end{aligned} \qquad (VI.3)$$

Contrary to what has been expected the dispersion law cannot be expanded in integer powers of the wave number. For a two-dimensional system it leads to disastrous results. The ordinary transport coefficients become infinite for all temperatures (7). Expressed as a function of time it means that

$$\begin{aligned} M(\vec{q},t) &\propto t^{-3/2} \qquad (3\text{-dim}), \\ M(\vec{q},t) &\propto t^{-1} \qquad (2\text{-dim}), \end{aligned} \qquad (VI.4)$$

and hence we find that the memory decays extremely slowly.

Let us close the discussion by briefly stating how Ernst and Dorfman's treatment is connected to the by now conventional mode-mode coupling theory for critical fluctuations. We shall follow closely the presentation by Kadanoff and Swift (21). In principle they solve the full Liouville equation and they extract the purely hydrodynamic modes, which represent the most low lying eigenstates of the Liouville operator. They derive formally the hydrodynamic equations, which take the form

$$-i\omega\, a_l(\vec{q}\omega) - \sum_{l'} L_{ll'}(\vec{q})\, a_{l'}(\vec{q}\omega) - \sum_{l'} U_{ll'}(\vec{q}\omega)\, a_{l'}(\vec{q}\omega) = 0. \tag{VI.5}$$

These are a set of coupled equations for the five hydrodynamic variables, denoted by a_l. Ignoring the last term we would get the equations for the ideal fluid with no diffusion and viscosity. These latter effects arise from states representing deviation from equilibrium. Kadanoff and Swift consider non-hydrodynamic states containing coupling of two or three hydrodynamic modes. They are able to show how these contribute to $U_{ll'}(\vec{q}\omega)$. The processes they consider are the same as Ernst and Dorfman did. From $U_{ll'}(\vec{q}\omega)$ one extracts the various transport coefficients, and Kadanoff and Swift find, for instance, that the thermal conductivity goes as

$$\lambda \sim |T - T_c|^{-2/3} \tag{VI.6}$$

due to coupling of a heat mode to another heat mode + a viscous flow mode. A coupling to a heat mode + sound wave gives no singularity. Similarly they find for the bulk viscosity

$$\zeta \sim |T - T_c|^{-2} \tag{VI.7}$$

from coupling to two heat modes, and

$$\zeta \sim |T - T_c|^{-2/3} \tag{VI.8}$$

from coupling to a heat mode + a sound wave.

V. CONCLUSION

In Section IV we posed three rather general questions and we are now able to answer these.

(i) Well defined collective modes of the elementary excitations can appear in the hydrodynamic regime. The ordinary heat diffusion in solids is an example of this, and it represents the evolution of the number of phonons per unit volume. In contrast to this, the propagating hydrodynamic waves are slightly renormalized phonons.

(ii) The hydrodynamic equations would be modified, if a significant number of the non-hydrodynamic modes have relaxation times of the same order as the hydrodynamic modes. The mode-mode coupling terms, representing anharmonicity in the hydrodynamic equations, cannot then be ignored.

(iii) It was illustrated on Nb_3Sn that a central peak in $S(\vec{q}\omega)$ can appear if the relaxation towards local thermal equilibrium is slow compared with the inverse frequency of the soft mode. A basic reason for this is that, as the soft mode frequency decreases, the anharmonicity plays an increasing role for this mode and it leads to a drastic change in the shape of the corresponding spectral function. For other phonons the anharmonicity plays a much smaller role, and one finds for these only the ordinary phonon peak in $S(\vec{q}\omega)$. The argument, given in these lectures, for the appearance of a central peak is too qualitative to actually specify which physical mechanism is responsible for the long relaxation time. It is, therefore necessary to carry out detailed numerical calculations of the contributions from the various terms in the diagram expansion and to compare with the experimental results. It is also necessary to do such calculations in order to find how close to the transition point one has to come, before proper critical fluctuations enter in a significant way.

1. H.C. Andersen, D. Chandler and J.D. Weeks, J. Chem. Phys. 56, 3812 (1971), 57, 2626 (1972).

2. L. Verlet and J.J. Weis, Phys. Rev. A5, 939 (1972).

3. J. Kurkijärvi, Ann. Acad. Sci. Fenn. A346, 1 (1970).

4. N.K. Ailawadi, A. Rahman and R. Zwanzig, Phys. Rev. A4, 1616 (1971).

5. K. Kim and M. Nelkin, Phys. Rev. A4, 2065 (1971).

6. K.N. Pathak and K.S. Singwi, Phys. Rev. A2, 2427 (1970).

7. B.J. Alder and T.E. Wainwright, Phys. Rev. A1, 18 (1970), See also: T.E. Wainwright, B.J. Alder and D.M. Gass, Phys. Rev. A4, 233 (1971).

8. H.L. Swinney and H.Z. Cummins, Phys. Rev. 171, 152 (1968).

9. C.W. Garland, D. Eden and L. Mistura, Phys. Rev. Letters 25, 1161 (1970).

10. D. Eden, C.W. Garland and J. Thoen, Phys. Rev. Letters 28, 726 (1972).

11. L.J. Sham, Phys. Rev. 156, 494 (1967).

12. G. Niklasson and A. Sjölander, Ann. Phys. (N.Y.) 49, 249 (1968).

13. R.A. Cowley, Phys. Rev. 134, A981 (1964).

14. P.A. Fleury and J.M. Worlock, Phys. Rev. Letters 18, 665 (1967), 19, 1176 (1967).

15. See also K. Gesi, J.D. Axe and G. Shirane, Phys. Rev. B5, 1933 (1972).

16. T. Riste, E.J. Samuelsen, K. Otnes and J. Feder, Solid State Comm. 9, 1455 (1971).

17. G. Shirane and J.D. Axe, Phys. Rev. Letters 27, 1803 (1971).

18. S.M. Shapiro, J.D. Axe, G. Shirane and T. Riste, Phys. Rev. B6, 4332 (1972).

19. M. Fixman, J. Chem. Phys. 36, 310 (1961).

20. K. Kawasaki, Phys. Rev. 150, 291 (1966).

21. L. Kadanoff and J. Swift, Phys. Rev. 166, 89 (1968).

22. See for instance, L.D. Landau and E.M. Lifshitz, Fluid Mechanics, Pergamon (London 1959).

23. R.D. Mountain, Rev. Mod. Phys. 38, 205 (1966).

24. See for instance, L.D. Landau and E.M. Lifshitz, Theory of Elasticity, Pergamon (London 1959).

25. K.E. Larsson, Proc. Int. Symposium on Inelastic Neutron Scattering, p. 397, IAEA, Copenhagen 1968.

26. K. Carneiro, M. Nielsen and J.P. Mc Tague, Phys. Rev. Letters 30, 481 (1972).

27. K. Sköld, J.M. Rowe, G. Ostrowski and P.D. Randolph, Phys. Rev. A6, 1107 (1972).

28. L. Kadanoff and P.C. Martin, Ann. Phys. (N.Y.) 24, 419 (1963).

29. H. Mori, Progr. Theor. Phys. 33, 423 (1965).

30. G. Niklasson, Fortschr. Phys. 17, 235 (1969).

31. G. Niklasson, Ann. Phys. (N.Y.) 59,263 (1970).

32. G. Niklasson, Phys. Kondens. Materie 14, 138 (1972).

33. R.A. Cowley, J. Phys. Soc. Japan (Suppl) 28, 239 (1970).

34. R. Silberglitt, Solid State Comm. 11, 247 (1972).

35. C.P. Enz, Phys. Rev. B6, 4695 (1972).

36. T. Schneider, Phys. Rev. B7, 201 (1973).

37. G.J. Coombs and R.A. Cowley, J. Phys. C 6, 121 (1973), 6, 143 (1973).

38. R. Zwanzig and M. Bixon, Phys. Rev. A2, 2005 (1970).

39. J.R. Dorfman and E.G.D. Cohen, Phys. Rev. Letters 25, 1257 (1970), Phys. Rev. A6, 776 (1972).

40. M.H. Ernst, E.H. Hauge and J.M.J. van Leeuwen, Phys. Rev. A4, 2055 (1971).

41. K. Kawasaki, Progr. Theor. Phys. 45, 1691 (1971).

42. E.H. Hauge, Phys. Rev. Letters 28, 1501 (1972).

43. M.H. Ernst and J.R. Dorfman, Physica 61, 157 (1972).

NEUTRON SCATTERING STUDIES OF SOFT MODE DYNAMICS*

J. D. Axe, S. M. Shapiro, and G. Shirane

Brookhaven National Laboratory

Upton, New York 11973 U. S. A.

T. Riste

Research Establishment, Kjeller, Norway

At the NATO Advanced Study Institute held at nearby Geilo in April 1971, Riste, Samuelsen, and Otnes[1] reported on the critical behavior of $SrTiO_3$ near the 105°K structural phase transformation as studied by neutron scattering. They observed that in addition to the expected condensing soft mode phonon side bands that there was in addition a very narrow central component. This paper is in the nature of a progress report on the neutron scattering work of the last two years on the "central mode problem."

β-TUNGSTEN SUPERCONDUCTORS

The first thing we learned was that the central mode was not an isolated phenomenon occurring only in $SrTiO_3$. At about the same time, Shirane and I were studying the structural transformation in Nb_3Sn.[2] This structural transformation, which occurs in many binary or pseudobinary compounds with the same β-tungsten structure, is characterized by a drastic softening of the acoustic shear mode with propagation vector $\vec{q}||[110]$ and displacement vector $\vec{e}||[1\bar{1}0]$.[3] At the transition temperature T_M = 46°K, the crystal structure changes from cubic to a slightly distorted tetragonal structure.[4] Fig. 1 summarizes the frequency shifts seen

*Work performed under the auspices of the U. S. Atomic Energy Commission and NATO Research Grant.

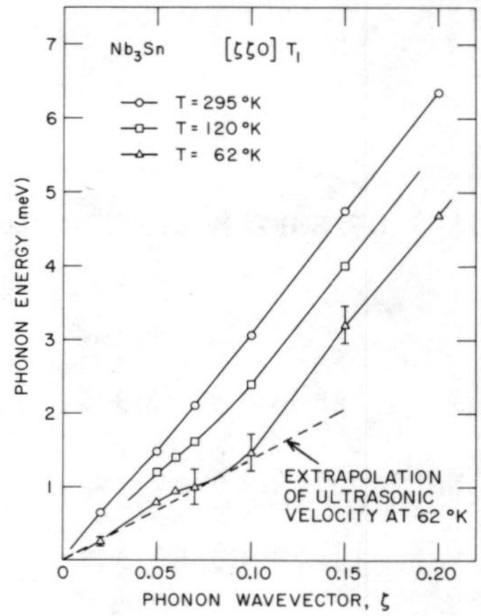

Figure 1. Temperature dependent transverse acoustic phonon dispersion at small wave vectors propagating in the "soft" [110] direction in Nb_3Sn.

in these soft [ζζ0] TA phonons as the temperature is lowered toward T_M. It is probable that the change in slope which occurs at q ∿ 0.1 is related to the change in screening which occurs as the phonon wave vector passes through parallel edges of the Fermi surface. Most interesting for the purposes of the present discussion are the changes in the power spectrum of the soft phonons in the range of wave vector shown in Fig. 1 as the temperature is lowered further. Typical of these observations are the data in Fig. 2 showing that as the temperature is lowered there is a gradual evolution of a central component in the scattering spectrum in addition to the familiar "phonon-like" sidebands. Although the sideband structure continues to move to lower frequencies as the temperature is lowered, far more dramatic (note the logarithmic scale) is the growth of intensity of the central component which completely dominates the fluctuation spectrum near T_M. The apparent width of the central component can be essentially accounted for by the resolution of the instrument alone. Thus the intrinsic width which adds in quadrature to the instrumental width is small, 1/3 or less of the observed width. Fig. 3 demonstrates that this central peak intensity maximizes at or very near T_M and thus represents the major contribution to the critical scattering associated with the structural transformation. An additional qualitative observation is that although this additional central component can be observed at least out to q = 0.1,its

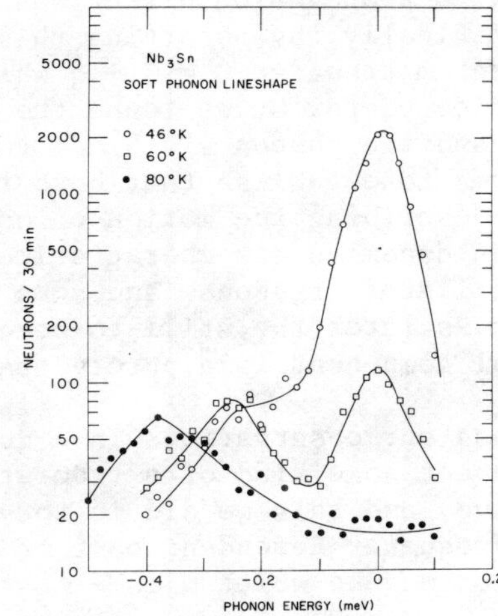

Figure 2. Observed spectral profiles of [ζζ0] T_1 phonon mode in Nb_3Sn with $\zeta = 0.02a^*$ at several temperatures above T_M. Only the phonon annihilation portion of the spectrum is shown. The data were taken with an incoming neutron energy of 5 meV and the energy resolution is $\sim$ 0.1 meV.

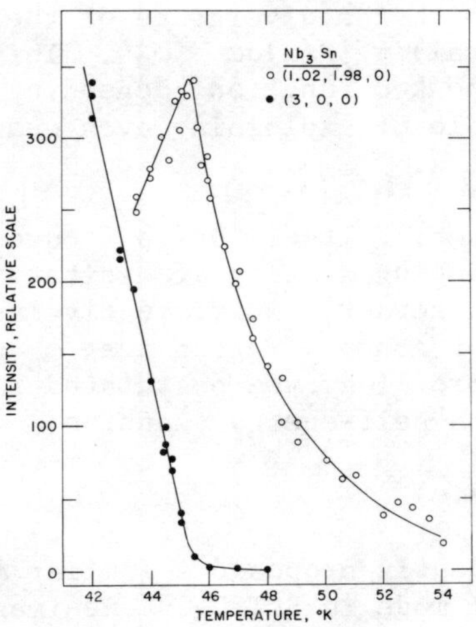

Figure 3. The closed circles show the onset of the structural phase transformation as monitored by the "forbidden" (300) Bragg reflection. The open circles represent the temperature dependence of the central component in the neutron critical scattering spectrum of Nb_3Sn.

presence is restricted to propagation directions nearly along the soft [110] direction. Finally, by observing this feature for several values of momentum transfer $\vec{Q} = \vec{G} + \vec{q}$ which differed by only a reciprocal lattice vector $\vec{G}$, we found the intensities of the central component and the phonon sidebands to be in a constant ratio. This seems to establish that both the central peak and the sidebands are describing the motion of one and the same mode, but that the mode dynamics are characterized by both the normal phonon-like oscillatory response and some slower response as well. (I will discuss later the still-to-be-eliminated possibility that the central component is a purely static phenomenon.)

In order to discuss our observations in a quantitative way, it was necessary to invent some kind of a theoretical construction, however tentative, and this we did in the following way. Quite generally, the frequency dependent part of the one-phonon scattering cross section may be written as

$$S(\omega) = \left(\frac{k_B T}{\hbar\omega}\right) \mathrm{Im}\{\omega_0^2 - \omega^2 - i\omega\Gamma(\omega)\}^{-1} \tag{1}$$

where ω_0 is a temperature-dependent quasiharmonic frequency and we assume $k_B T >> \hbar\omega$ for the frequency range of interest. Soft-mode line shapes have previously been discussed in the "viscous damping" approximation in which Γ is taken to be a frequency-independent constant. The condition for dynamical instability, $\omega_0 \to 0$, is connected with the divergence of the integrated scattering intensity $I(\text{total}) = \int S(\omega)d\omega \propto \omega_0^{-2}$. This form leads to either a two- or one-peaked function, depending upon the ratio Γ/ω_0, and is not capable of explaining even qualitatively the profiles shown in Fig. 2.

In general, however, Γ itself has a frequency dependence which reflects the changing density of excitations with which the one-phonon state can interact. Sufficiently large changes in $\Gamma(\omega)$ in the important frequency region near ω_0 can produce more complicated spectral profiles. We postulated the following simple form for the phonon self-energy function

$$\Gamma(\omega) = \Gamma_0 + \delta^2/(\gamma - i\omega). \tag{2}$$

Schwabl later independently proposed a similar Ansatz for $\Gamma(\omega)$ in discussing the central mode in $SrTiO_3$.[5] We imagined the first term in Eq. (1) to represent the normal damping due to phonon-phonon scattering and the second term as due to coupling with another (as yet unspecified) fluctuation with a Debye relaxation spectrum. Eq. (1) has been criticized on the grounds that it violates certain moment sum rules. This is related to the failure of $\Gamma(\omega) \to 0$ as $\omega \to \infty$. Γ_0 should be thought of as the low frequency approximation to a function $\Gamma_o(\omega)$ which varies slowly over the region $\omega \approx \omega_o \ll \omega_D$, where ω_D is the Debye frequency of the solid. Thus in systems studied thus far it is mathematically correct but physically improper to omit Γ_o in Eq. (1) on the basis that the remaining term satisfies higher moment sum rules.

Eq. (1) seems now to be established as the canonical line shape in the central mode problem. One would do well to remember however that it has a very slender foundation in microscopic theory. It has survived rather well whatever experimental tests we have subjected it to, but these have not been stringent enough to allow us to say that a correct theory must yield this exact analytical form.

Inserting Eq. (2) into Eq. (1) gives

$$S(\omega) = \frac{\left(\frac{k_B T}{\hbar}\right)\left[\Gamma_o + \frac{\delta^2\gamma}{\omega^2+\gamma^2}\right]}{\left[\left(\omega_\infty^2 - \frac{\delta^2\gamma^2}{\omega^2+\gamma^2} - \omega^2\right)^2 + \omega^2\left(\Gamma_o + \frac{\delta^2\gamma^2}{\omega^2+\gamma^2}\right)^2\right]} \tag{3}$$

where $\omega_\infty^2 = \omega_o^2 + \delta^2$. This formula has the general qualitative features of Fig. 2. In the limit $\omega_o^2 \gg \delta^2$, it shows three distinct peaks with side bands at $\pm\omega_\infty$. In the other limit $\omega_o \to 0$ (i.e. $\omega_\infty \to |\delta|$), Eq. (3) shows a profile with shoulders similar to that observed in Fig. 2 at 46°K.

For a more quantitative comparison with experiment, we can conveniently divide the cross section into $S(\omega)_{total} = S(\omega)_{central} + S(\omega)_{sideband}$ with

$$S(\omega)_{central} = \left(\frac{k_B T}{\hbar}\right)\left(\frac{\delta^2}{\omega_o^2\omega_\infty^2}\right)\left(\frac{\gamma'}{\omega^2+\gamma'^2}\right) \tag{4}$$

where $\gamma' = (\omega_o^2/\omega_\infty^2)\gamma$. This formula is valid for the range of parameters of interest, $\omega_\infty \gg \gamma$ and $\Gamma_o \ll (\delta^2/\gamma)$. As with the simpler damping, $I_{total} \propto \omega_o^{-2}$ and dynamical instability occurs as $\omega_o \to 0$. The fractional integrated central peak intensity is simply

$$\frac{I(\text{central})}{I(\text{total})} = \frac{\delta^2}{\omega_\infty^2} . \tag{5}$$

From data of the type shown in Fig. 1, it is straightforward to obtain both the ratio I(central)/I(total) and the value of ω_∞ (essentially the peak of the phonon sideband), so that Eq. (5)

can be used to deduce a value of $|\delta|$. Investigation for various wave vectors establishes that $|\delta|$ varies approximately linearly with q, $|\delta| = \lambda q$. The principal prediction of this phenomenological theory is that the central mode intensity grows not in proportion to the increasing intensity of the soft phonon sidebands ($\sim T/\omega_\infty^2$) but rather at the much faster rate ($\sim T\delta^2/\omega_\infty^2\omega_o^2$). The observed temperature dependence is consistent with this relation if δ is taken as temperature independent.

There is one additional initially puzzling feature of our data which is resolved by our present understanding of the significance of the central mode. At room temperature the value of the $(\zeta\zeta 0)T_1$ phonon velocity derived from neutron measurements is essentially equal to that obtained by ultrasonic techniques[3]. However, as shown in Fig. 4 there is an increasing systematic discrepancy between the two types of measurements. We may suppose that the ultrasonic frequency (40 MHZ in this instance) is negligibly small compared with the inverse relaxation time γ, in which case one measures a low frequency sound velocity v_o. On the other hand, phonon frequencies from which the neutron determined velocity is derived is much greater than γ, in which case the observed phonon velocity can be shown from Eq. (3) to be

$$v_\infty^2 = (\omega_\infty^2/q^2) = v_o^2 + \lambda^2 \qquad (6)$$

Plotted in Fig. 4 are the ultrasonic velocities v_o and the values of v_∞ calculated from Eq. (6), and using the value $\lambda = 6.42 \times 10^4$ cm/sec obtained from the above analysis of the central peak amplitude. v_∞ agrees very well with the velocities obtained from the direct analysis of the sideband frequencies obtained from the

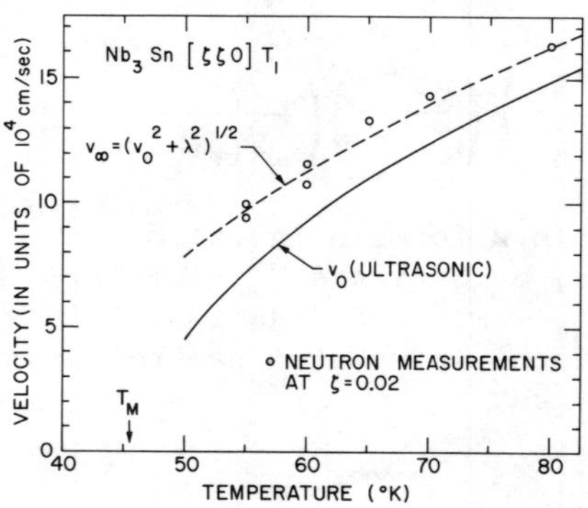

Figure 4. The $(\zeta\zeta 0)T_1$ phonon velocity in Nb_3Sn determined by neutron scattering differs from ultrasonic velocities by an amount which is predictable from the amplitude of the central component.

neutron experiments.

$SrTiO_3$

As is now well known, the structural phase transformation in $SrTiO_3$ at $\sim$105°K results from the instability of a zone boundary phonon.[7-9] In the cubic phase the displacements of the soft mode transform according to the triply degenerate R_{25} representation of the group of R and correspond to rotation of the oxygen tetrahedra about the <100> axes. At T_c there is a doubling of the unit cell and the zone boundary of the cubic phase now becomes the zone center of the tetragonal phase. The triply degenerate zone-boundary R_{25} mode of the cubic phase splits into two zone-center modes: a doublet with rotation axes perpendicular to the fourfold axis and a singlet with rotation axis parallel to the fourfold axis. Subsequent to the discovery of the "central mode" in $SrTiO_3$ by Riste et al further experiments were carried out in collaboration with Riste at the HFBR at Brookhaven.[10]

In the study of Nb_3Sn the small sample size (0.05 cm^3) severely limited the scope of the study, but fortunately the extension of the central component in reciprocal space was much larger than the corresponding q-width of the instrumental resolution. While the observed line shapes were still subject to considerable correction for finite energy resolution, to within a reasonable first approximation the relative integrated intensities of the central and sideband components were given directly from the experimental data. In the case of $SrTiO_3$ (as well as $KMnF_3$ and $LaAlO_3$, to be discussed later) the cross section was observed to vary rapidly over a range of q comparable to the instrumental width. This necessitated resolution corrections, the importance of which can be visualized with the aid of Fig. 5, which schematically shows a cross section corresponding to a phonon dispersion surface, some additional cross section around ω=0, and the resolution ellipse drawn to scale for a typical high resolution experiment. A constant Q scan moves the resolution function parallel to the energy axis through the scattering cross section. For a particular setting $(\vec{Q}_s,\omega_s)$ of the spectrometer the observed neutron intensity is

$$I(\vec{Q}_s,\omega_s) = \int d\vec{Q}d\omega R(\vec{Q}-\vec{Q}_s,\omega-\omega_s)\sigma(\vec{Q},\omega) \quad (7)$$

and

$$\sigma(\vec{Q},\omega) = \frac{k_F}{k_I} |F_{in}(\vec{Q})|^2 S_q(\omega) \quad (8)$$

where k_F and k_I are the final and initial neutron momenta and

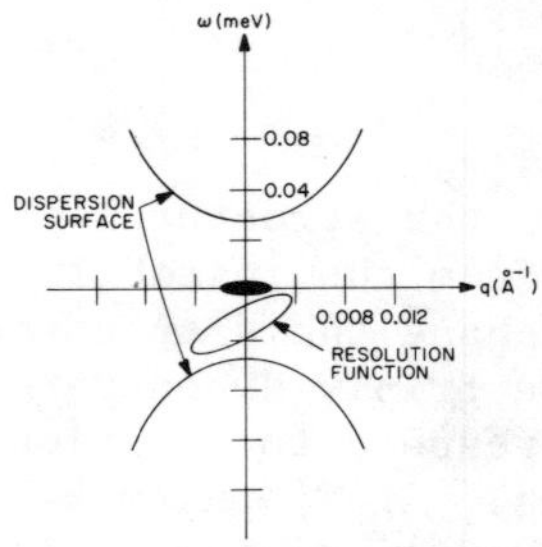

Figure 5. Schematic representation of a cross section of the soft phonon dispersion near the R-point in $SrTiO_3$, with additional scattering near $\omega=0$. The resolution ellipse is drawn to scale for 5-meV incident neutron energy, 20' horizontal collimation.

$F_{in}(\vec{Q})$ is the inelastic structure factor for the mode with a reduced momentum $\vec{q} = \vec{Q} - \vec{G}$.

The intensity of the phonon-like sidebands changes with $\vec{Q}$ much less rapidly than does that of the central component, the two components are therefore weighted differently by the spectrometer. The proper resolution correction can be applied only if we know the $\vec{Q}$ dependence of $\sigma(\vec{Q},\omega)$. If the dynamical matrix is expanded about $\vec{q}_R = (1/2,1/2,1/2)$, the behavior of the "bare" phonon modes near $\vec{q}_R$ are determined by a truncated dynamical matrix[11]

$$\underline{\underline{C}}_T(\vec{q}) = \omega_\infty^2(\vec{q}_R)\underline{\underline{1}} + (\vec{q}-\vec{q}_R)\cdot \underline{\underline{\lambda}}\cdot (\vec{q}-\vec{q}_R) \qquad (9)$$

where $\underline{\underline{1}}$ is a 3x3 unit matrix and $\underline{\underline{\lambda}}$ contains the 3 independent constants which can be adjusted to agree with the measured frequencies $\omega_\infty^2(\vec{q})$. The inelastic structure factors $F_{in}(\vec{Q})$ are easily evaluated from the eigenvalues of Eq. (9). Although we can safely neglect the q-dependence of Γ_o, our phenomenological derivation of $S_q(\omega)$ gives us no guide for predicting $\delta(\vec{q})$ and $\gamma(\vec{q})$. Our analysis will be independent of any detailed assumptions about $\gamma(\vec{q})$, but in setting $\delta(\vec{q})$ = constant, we are in effect averaging whatever $\vec{q}$-dependence there is in this quantity over the sampled region of momentum transfer.

Fig. 6 is an example of the observed line shape at about ten degrees above T_c. Very high resolution was obtained by using low energy (4.9 meV) incident neutrons and higher energy contamination was reduced to a negligible level by a Be filter. The

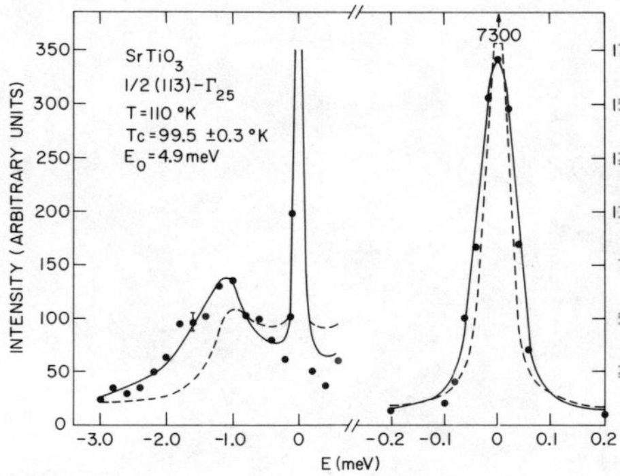

Figure 6. Scattered-neutron spectrum of $SrTiO_3$ at T = 110°K. The circles are the observed data, the full line represents the fit of the data with Eq. (3) folded with the instrumental resolution; Γ_0 = 0.88 ± 0.1 meV, δ^2 = 0.63± 0.1 meV^2,ω^2_∞ = 1.27 ± 0.1 meV^2, λ_1 = 500 meV^{2}A^2, λ_2 = 1000 meV^{2}A^2. The dashed curve is a plot of S(ω) with the above parameters.

incoherent scattering (a small correction at this temperature)was measured separately and has been subtracted. The importance of the resolution corrections, seen by comparing the solid and dotted curves is, as expected, most pronounced on the central peak.

Since the observed linewidth of the central peak was always resolution limited we chose a nominal value of γ' such that it was always smaller than the energy resolution of the spectrometer. After a good fit was obtained γ' was allowed to vary. The fit was insensitive to γ' provided it was smaller than the instrumental resolution and larger than the mesh size required for the numerical folding. As can be seen most easily by the approximate Eq. (4), δ^2 can still be determined accurately under these conditions from the integrated central peak intensity.

Fig. 7 shows the temperature dependence of δ^2, $\omega^2_\infty(\vec{q}_R)$ and $\omega^2_0(\vec{q}_R)$ deduced by fitting many experimental results to Eq. (3). Several internal checks, such as the consistency of parameters obtained with different spectrometer resolution as well as the overall goodness of fit, convince us that the form of the parameterized cross section we have chosen is not grossly inappropriate. There are at least two features here worthy of comment. δ^2 is apparently constant at $\sim$ 0.9 meV^2 for T - T_c > 25°K, decreasing to about 0.3(±0.1)meV^2 at T_c. It is possible, however, that all or part of this apparent T-dependence results from our rather ad hoc assumption that δ^2 is q-independent.

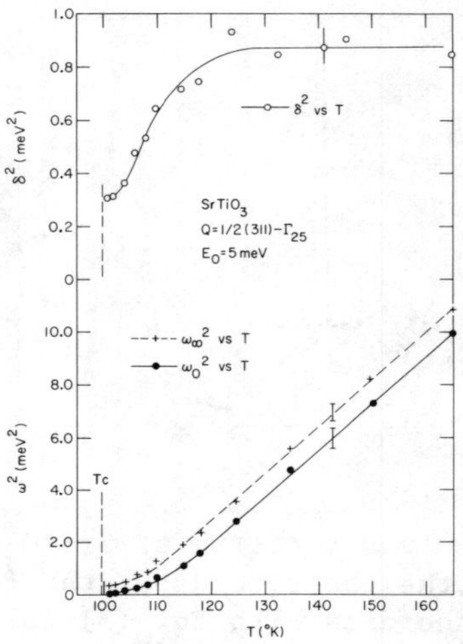

Figure 7. δ^2, ω_∞^2, ω_o^2 vs. T for $SrTiO_3$.

In agreement with previous studies[1,12] ω_∞^2 is essentially linear with temperature in the range $10°K < T - T_c < 180°K$, but a deviation from this simple mean field behavior is clearly evident at lower temperatures for both ω_∞^2 and ω_o^2. Since ω_o^2 is essentially the inverse static susceptibility an attempt was made to deduce a critical exponent, γ, using the relation $\omega_o^2(T \to T_c) \sim (T-T_c)^\gamma$ but the results were not satisfactory, perhaps because of the limited temperature range of our observations. However, our measurements strongly suggest that if such a limiting form is applicable, $1.5 < \gamma < 2.5$. It was not possible to make meaningful measurements nearer T_c because of our inability to determine T_c to better than $\sim \pm 0.5°$. This determination of T_c was made by observing the intensity change of the cubic (222) reflection caused by extinction relief in the strained tetragonal crystal.

$KMnF_3$

$KMnF_3$ exhibits the same cubic-tetragonal transformation that exists in $SrTiO_3$ with $T_c = 186°K$.[13] Gesi et al[14] studied the temperature dependence of the soft mode line shape with moderately high resolution and were able to describe their observations with a simple damped harmonic oscillator response, but found an unexpected temperature dependence of the damping parameter Γ_o. This led us to a reinvestigation of $KMnF_3$ with higher resolution and the observation of a narrow central component in addition to a broader overdamped phonon peak.[10]

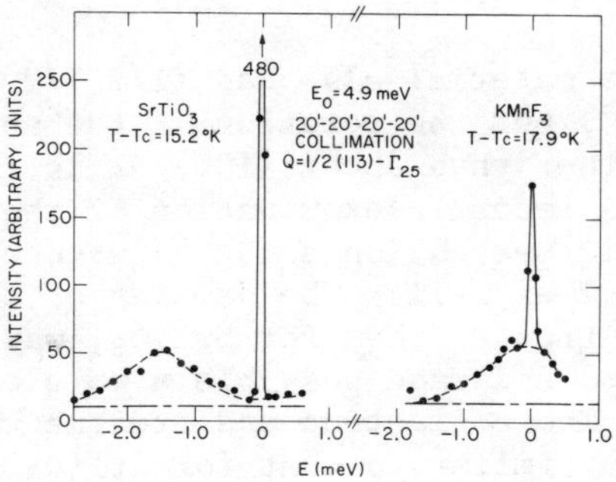

Figure 8. Scattered-neutron spectra of $SrTiO_3$ and $KMnF_3$ at $T-T_c = 15.2$ and 17.9°K, respectively. The dotted line corresponds to the phonon peak and the solid line to the central component. The dashed line corresponds to level of room background. All incoherent scattering has been subtracted.

Fig. 8 compares the soft phonon line shapes of $SrTiO_3$ and $KMnF_3$ under identical high resolution and at nearly identical values of $T-T_c$. It shows unambiguously the existence of the central peak in $KMnF_3$ as well as in $SrTiO_3$. In the latter, the phonon peak (dotted line) is underdamped, whereas in $KMnF_3$ the Phonon is well overdamped at this value of $T-T_c$ and appears as a broad line centered around $\omega=0$. Sitting on top of this broad line is the narrow central peak. In addition, the linewidth of the central peak in both systems is that of the instrumental resolution. This sets an upper limit on the value of the central component: $\gamma < 0.02$ meV ($=0.16$ cm^{-1} $= 4.8 \times 10^9$ Hz).

Although there is no difficulty in fitting the line shape observed in $KMnF_3$ to Eq. (3), the results are considerably less reliable because severe correlations develop between the fitting parameters, particularly ω_∞^2 and δ^2. If, however, we use the values of $\omega_\infty^2(T)$ given by Gesi et al (which should be reasonably reliable for T not too close to T_c) values of $\delta^2 \sim 0.3 \pm 0.1$ meV^2 can be estimated.

If, in accordance with our assumption that $\delta(\vec{q})$ can be regarded as $\vec{q}$-independent, the central mode intensity depends upon $\vec{q}$ only through $\omega_\infty(\vec{q})$, and from Eq. (4), it is given approximately by $[\omega_\infty^2(\vec{q})\omega_0^2(\vec{q})]^{-1}$. $\omega_\infty^2(\vec{q})$ is much more anisotropic in $KMnF_3$, the soft branch being especially low along the Brillouin zone boundary from R(1/2 1/2 1/2) to M(1/2 1/2 0). We should therefore expect the central mode intensity to be more diffuse in $KMnF_3$, extending especially in the direction R → M. This indeed appears to be the case.

LaAlO$_3$

This perovskite material also has (1/2 1/2 1/2) wave vector phonon instability[14], but the rotation of the anion octahedra is about (111) axes rather than about (100) as is the case for $SrTiO_3$ and $KMnF_3$. A recent reexamination of the soft mode line shape by Kjems et al[15] establishes the existence of a central peak in this material as well. The damping of the phonon-like sidebands, although greater than for $SrTiO_3$, was not found to be as large as in $KMnF_3$, and made possible a more extensive quantitative discussion. The collection and treatment of the experimental data was very similar to that for $SrTiO_3$. The energy width was resolution limited, with an upper limit of $\gamma' \leq 0.03$meV. Eq. (3) adequately represented the observed scattering and for $T-T_c \geq 15°C$, $\delta^2 = 0.38 \pm 0.1$ meV^2 and was nearly temperature independent. As was the case in $SrTiO_3$, ω_0^2 and ω_∞^2 exhibited linear temperature dependence well away from T_c but the extrapolated values go to zero well above the observed transformation temperature, and the observed frequencies deviate considerably from the mean field behavior near T_c. The behavior of both $\omega_\infty^2(T)$ and $\delta^2(T)$ are shown in Fig. 9. The central component showed a clearly resolvable q-width which agreed well with that predicted by the q-dependence of ω_∞^2.

DISCUSSION

From the outset, we have assumed that the scattering associated with the central mode is a part of the one-phonon response and is dynamical in nature, but with a narrow frequency response. However, in view of the fact that we have not succeeded in demonstrating experimentally an energy width, it is important to

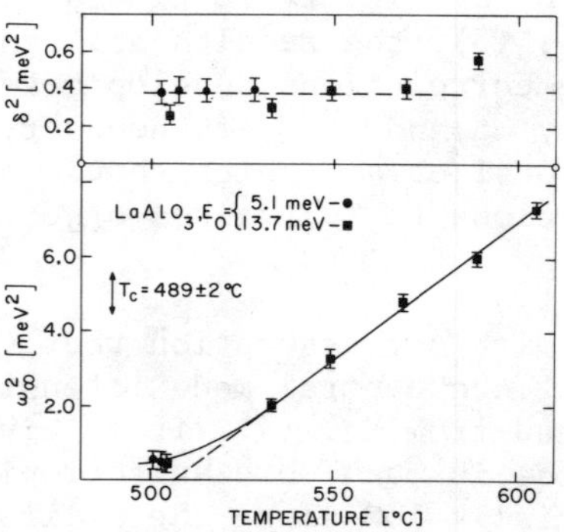

Figure 9. δ^2 and ω_∞^2 vs T for $LaAlO_3$.

consider whether an alternative static mechanism could explain our observed central component. While it is true that the intensity of the central component near T_M strongly suggests its dynamical origin, it is possible to imagine a plausible and somewhat trivial nondynamical mechanism involving scattering from static strain fields, which is at least in qualitative agreement with many of our observations. It is well known that point defects in a lattice will in general give rise to displacements of neighboring atoms from their equilibrium positions in the homogeneous impurity free crystal. These displacement fields cause diffuse scattering of x-rays or neutrons sometimes known as Huang scattering. The magnitude of the displacement field about an impurity is calculated as a linear response to a force field $\vec{F}(\vec{r})$ which the impurity exerts on the undisplaced lattice. For our purposes it is convenient to Fourier transform the resulting displacements and express them in terms of a linear combination of phonon modes with wave vector q and branch index j. The amplitude of this impurity induced set of static phonon-like displacements is easily shown in the harmonic approximation to be given by

$$\langle Q_{qj}\rangle_{\text{static}} = F_{qj}/\omega^2_{qj}$$

where

$$F_{qj} = -N^{-\frac{1}{2}} \sum_{\ell k} M_k^{-\frac{1}{2}} \vec{F}_{\ell k}\cdot\vec{e}^{\,*}_k(qj)\ \exp(-i\vec{q}\cdot\vec{\ell})$$

is the projection of $\vec{F}(\vec{r}) \equiv \vec{F}_{\ell k}$ upon the phonon eigenvector $\vec{e}(q,j)$. The intensity of this static diffuse scattering can be calculated from the corresponding expression for the integrated phonon scattering under the same conditions by simply replacing $\langle Q^2_{qj}\rangle_{\text{thermal}}$ by $|\langle Q_{qj}\rangle_{\text{static}}|^2$.

$$I_{qj}(\vec{Q})_{\text{static impurity}} \propto |\langle Q_{qj}\rangle_{\text{static}}|^2 = \frac{|F_{qj}|^2}{\omega^4_{qj}} \quad . \qquad (7)$$

In normal materials the major contributions come from long wavelength acoustic modes (because of the weighting by ω_{qj}^{-4}) and the effect of impurity concentrations of $\sim 10^{-2}$ can easily be detected and studied against the thermal diffuse background by x-ray scattering. However, Eq. (7) also suggests that this same factor of ω_{qj}^{-4} would greatly enhance the contribution of any phonon mode whose frequency becomes anomalously small near a structural transformation. If impurities are of the proper symmetry to couple to the soft mode, there will be a central component, static in origin, whose intensity grows more rapidly ($\omega^{-4}_{\text{soft}}$) than that of the collapsing phonon sidebands ($\omega^{-2}_{\text{soft}}$).

In spite of these obvious similarities, we do not believe that this impurity mechanism provides a satisfactory explanation of our observations. Although we have not until now made the distinction, it is clear that it is the low frequency stiffness ω_o^2 which goes into Eq. (7) not ω_∞^2, if there is a difference between the two quantities. However, the static impurity mechanism acting alone provides no frequency dependent terms to the phonon self-energy, so that $\omega_o^2 = \omega_\infty^2$ and $I(Q)$static $\propto 1/\omega_\infty^4$. For Nb_3Sn our measurements closely follow $(1/\omega_o^2\omega_\infty^2)$, and there is a substantial difference between the two predictions especially near T_M. Simply put, ω_∞^2 (as obtained from the phonon sidebands) saturates near T_M while the central intensity continues to increase. Also the observed agreement with the discrepancy between the extrapolated long wavelength acoustic velocities and our measurements and the magnitude of the central component would be entirely fortuitous for this (or for that matter any other) static description of the central component.

We believe that it is unlikely that the impurity effect outlined above is the dominant one in the observations described here. On the other hand, it is certainly a plausible mechanism for producing unusual line shape effects near phonon instabilities and as such deserves further consideration. It is obvious from these comments that it is most important to try to characterize, directly if possible, the energy width of the central component. If the estimate of $\gamma \gtrsim 10^{10}$Hz($\sim$0.4 meV) obtained[16] indirectly from ESR data in $SrTiO_3$ is correct, these widths are at the limit of conventional neutron resolution, and it is possible that light scattering is the more appropriate tool.

REFERENCES

1. T. Riste, E. J. Samuelsen, and K. Otnes in "Structural Phase Transitions and Soft Modes," ed. E. J. Samuelsen, E. Andersen, and J. Feder (Universtetsforlaget, Oslo, Norway, 1971). See also, T. Riste, E. J. Samuelsen, K. Otnes, and J. Feder, Solid State Commun. 9, 1455 (1971).

2. G. Shirane and J. D. Axe, Phys. Rev. Letters 27, 1803(1971).

3. W. Rehwald, M. Rayl, R. W. Cohen, and G. D. Cody, Phys. Rev. (to be published). This contains references to earlier work.

4. R. Mailfort, B. W. Batterman, and J. J. Hanak, Phys. Lett. 24A, 315 (1967); L. J. Vieland, R. W. Cohen, and W. Rehwald, Phys. Rev. Letters 26, 373 (1971).

5. F. Schwabl, Phys. Rev. Letters 28, 500 (1972).

6. T. Schneider, to be published.

7. H. Unoki and T. Sakudo, J. Phys. Soc. Japan 23, 546 (1969).

8. P. A. Fleury, J. F. Scott, and J. M. Worlock, Phys. Rev. Letters 21, 15 (1968).

9. G. Shirane and Y. Yamada, Phys. Rev. 177, 858 (1969).

10. S. M. Shapiro, J. D. Axe, G. Shirane, and T. Riste, Phys. Rev. B 6, 4332 (1972).

11. K. Gesi, J. D. Axe, and G. Shirane, Phys. Rev. B 5, 1933 (1972).

12. K. Otnes, T. Riste, G. Shirane, and J. Feder, Solid State Commun. 9, 1103 (1971).

13. V. J. Minkiewicz and G. Shirane, J. Phys. Soc. Japan 26, 674 (1969); G. Shirane, V. J. Minkiewicz, and A. Linz, Solid State Commun. 8, 1941 (1970).

14. W. Cochran and A. Zia, Phys. Stat. Sol. 25, 273 (1968); G. Shirane, and K. A. Müller, Phys. Rev. 183, 820 (1969).

15. J. K. Kjems, G. Shirane, K. A. Müller, and H. J. Scheel, to be published.

16. Th. von Waldkirch, K. A. Müller, and W. Berlinger, Phys. Rev. B 7, 1052 (1973).

SLOW LOCAL FLUCTUATIONS NEAR T_c IN $SrTiO_3$ STUDIED BY EPR

K.A. Müller

IBM Zurich Research Laboratory
8803 Rüschlikon, Switzerland

Contents

I. INTRODUCTION

In the present two lectures a survey is given of recent paramagnetic resonance (EPR) linewidth, shape and intensity studies. With them slow local fluctuations of the order parameter $\varphi_\ell(t)$ near the phase transition in $SrTiO_3$ were observed. Local is understood on an atomic scale, and slow is meant with respect to underdamped soft-mode frequencies. The information obtained is new and novel in the field of phase transitions: the mean frequency distribution width $\Delta\omega$ of slow local fluctuations φ_ℓ could be determined and close to T_c the probability distribution $P(\varphi_\ell)$ of φ_ℓ measured. These two quantities are of value for the following reasons: a) From $\Delta\omega$ an estimate of the frequency width of the central peak at $\vec{q} = \vec{q}_R$ is possible, which has so far not been the case using scattering techniques. b) Near and at T_c the probability $P(\varphi_\ell)$ is found experimentally to differ markedly from a Gaussian distribution occurring for statistical independence. This deviation results mainly from the long-range correlations of particles near the transition and was first observed in $SrTiO_3$. It should be present near all displacive second-order phase transitions.

The temperature-dependent spread $\Delta\omega(T)$ could be found from the behavior of two paramagnetic centers which couple in an entirely different way to fluctuations $\varphi_\ell(t)$. This is discussed in section III. Very close to T_c the fluctuations become so slow that the EPR line shapes are directly proportional to $P(\varphi_\ell)$. Proof of this "slow-motion regime" is given in IV. There the anisotropy of fluctuations in a sample which transforms into a monodomain determined by EPR are compared to recent birefringence measurements. Finally, in section V the very recent EPR line-shape analyses above and below T_c in the slow-motion regime are given. Below T_c, where the symmetry is broken, $P(\varphi_\ell)$ is asymmetric. The asymmetry is a critical quantity. We start in section II by recapitulating the most important features of the $SrTiO_3$ phase transition, and discuss the EPR information on the time dependence of the order-parameter fluctuations.

II. OCTAHEDRAL TIME-DEPENDENT ROTATIONS IN $SrTiO_3$

In the Geilo Spring 1971 School the use of paramagnetic resonance (EPR) to study structural "distortive" (1) phase transitions was reviewed (2). Especially emphasized were the cubic-to-tetragonal transitions in $SrTiO_3$ ($O_h^1 \rightarrow I4/mcm$) and the cubic-to-trigonal transition in $LaAlO_3$ $(O_h^1 \rightarrow R\bar{3}c)$. In both ABO_3 perovskite-type crystals antiferrodistortive transitions occur when an alternate static rotation of BO_6 octahedra sets in around the cubic <100> or <111> crystal axes, respectively. EPR of Fe^{3+} ions substitutional on B sites yielded the temperature dependence of the rotational order parameter $\langle\varphi\rangle$ with an accuracy unattained till now by other methods. This high accuracy allowed for the first time the observation of static critical behavior in the temperature dependence of $\varphi \propto |T - T_c|^{\beta}$, with $\beta = 0.33 \pm 0.01$, near T_c in a distortive transition (3). This occurs when the correlation length ξ exceeds the range of pair forces λ. Recently, Steigmeier and Auderset measured with similar high accuracy the soft-mode frequency ω_A in $SrTiO_3$ below T_c with Raman scattering (4). They found ω_A to be proportional to $\langle\varphi\rangle$ from inside the critical region near T_c $(\beta = 1/3)$ through the changeover region up to where $\langle\varphi\rangle$ shows Landau behavior $(\beta = 1/2)$ down to $T/T_c = 0.7$.

The accurate values of $\langle\varphi\rangle$ were obtained by measuring EPR line positions. The EPR of Fe^{3+} ions substitutional on B sites reflects the local orientation of BO_6 octahedra (see reference (2) for a detailed discussion). For a certain fixed direction of the magnetic field $\vec{H}$ with respect to the crystallographic axes (direction cosini 1, m, n) the resonance field H_r is given by

$$H_r = H_o(1, m, n; \nu) + A(1, m, n; \nu) \cdot \varphi_\ell^\alpha . \tag{1}$$

In equation (1) ν is the frequency of the applied microwave field, φ_ℓ^α is a specific component $\alpha = [x,y,z]$ of the local rotation of a BO_6 octahedron, for example, $\alpha = [001]$ or $[1,1,1]$. H_o is the resonance field in the absence of rotation, and A is a proportionality constant which is practically temperature independent. It is the derivative of H_r with respect to φ_ℓ^α. H_o and A depend on 1, m, n and ν. It could be shown that for certain directions of $\vec{H}$, i.e., values of 1, m, n, the second-order term, proportional to φ_ℓ^2 in an expansion-like equation (1), becomes negligibly small. The constant A is largest for the Fe^{3+}-V_O pair center due to the large anisotropy of H_o along the pair axis. The Fe^{3+}-V_O center consists of a trivalent iron impurity on a B site with a nearest-neighbor oxygen vacancy (2). The resonance pattern of the latter has been analyzed above and below T_c (5), (6). For H ‖ [110] and K-band, A = 26 Gauss/degree for an $\alpha = [001]$ octahedral rotation. The local random fluctuations φ_ℓ are of the order of a degree, whereas the background linewidth is about 3 Gauss. Thus, owing to the high sensitivity of this center the stochastic variation $\delta\varphi_\ell(t) = \varphi_\ell(t) - \langle\varphi\rangle$ could be observed. The time-dependent departure $\delta H(t)$ from $H_o + A \langle\varphi\rangle$ is a random function proportional to $\delta\varphi_\ell(t)$ from equation (1). The situation can be characterized in an analogous way to the NMR linewidth and shape in the presence of nuclear motion (7). A measure of the time dependence is the autocorrelation function for octahedra with center at $\vec{R}$: $\langle\delta\varphi^\alpha(\vec{R},t)\ \delta\varphi^\alpha(\vec{R},0)\rangle$. The local spectral density $J(\omega)$ is the frequency Fourier transform of this autocorrelation function and is independent of $\vec{R}$,

$$J^{\alpha\alpha}(\omega) = \frac{1}{\pi}\int_{-\infty}^{+\infty} \langle\delta\varphi^\alpha(\vec{R},t)\ \delta\varphi^\alpha(\vec{R},0)\rangle\ e^{i\omega t}\ dt\ . \tag{2}$$

The information $J(\omega)$ from resonance experiments may be related to the one gained by scattering techniques. There the wave vector $(\vec{q})$-frequency (ω) domain is probed yielding the well-known structure factor $S^{\alpha\alpha}(\vec{q},\omega)$. Measuring the motion at one particular coordinate $\vec{R}$ means looking at the spatial Fourier transform of $S^{\alpha\alpha}(\vec{q},\omega)$, then taking $\vec{R}$ at the origin:

$$J^{\alpha\alpha}(\vec{R},\omega) = \sum_{\vec{q}} S^{\alpha\alpha}(\vec{q},\omega)\ e^{-i\vec{q}\vec{R}}\ ,\quad J^{\alpha\alpha}(\omega) \equiv J^{\alpha\alpha}(0,\omega) = \sum_{\vec{q}} S^{\alpha\alpha}(\vec{q},\omega)\ . \tag{3}$$

In the high-temperature approximation where $S^{\alpha\alpha}(\vec{q},\omega) = S^{\alpha\alpha}(\vec{q},-\omega)$, and one has $J(\omega) = J(-\omega)$ as also from equation (2), in magnetic resonance $J(\omega)$ is usually only defined for positive ω.

In magnetic resonance until recently the characteristic measuring time $\tau_m = 2\pi/\omega_m$ was so long, or ω_m so small, that one probed $J(\omega = 0)$. This statement applies to all EPR and NMR work in magnetic phase transitions. It is hereafter called the "fast-motion regime", the fluctuation time τ being much faster than τ_m. On approaching T_c the fluctuations become longer, then $J(\omega)$ narrows and thus $J(0)$ increases. An especially good example is the recent NMR linewidth study of Gottlieb and Heller (8) above T_c in FeF_2. There, from the linewidth $\Delta H \propto J(0)$, the critical exponent ν of the correlation length $\xi = \xi_o\ [(T - T_c)/T_c]^{-\nu}$ was determined and dynamic scaling verified. The analogous experiment for the antiferrodistortive phase transition in $SrTiO_3$ was carried out by von Waldkirch et al. (9), (10). It also yielded the critical exponent ν and the anisotropy Δ of spatial correlations introduced by Schwabl in his theory of the linewidth broadening in the fast-motion regime (11). In $SrTiO_3$ magnetic-resonance experiments, close to T_c, for the first time the critical part of the frequency components of $J(\omega)$ fell below the measuring frequency ω_m (9) (see figure 1). This

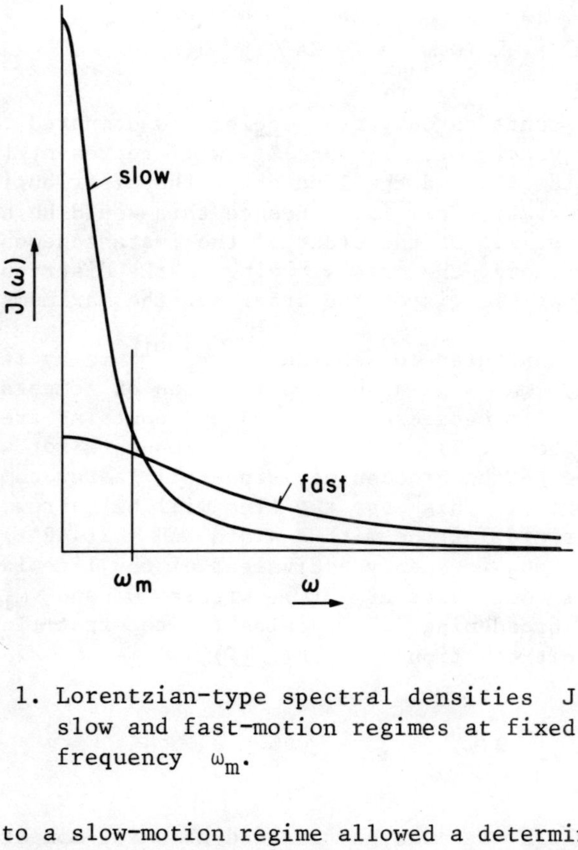

Figure 1. Lorentzian-type spectral densities $J(\omega)$ for slow and fast-motion regimes at fixed measuring frequency ω_m.

changeover to a slow-motion regime allowed a determination of the mean-frequency distribution $\Delta\omega$ of $J(\omega)$. This is discussed in part A of section III. Because the critical part of $S^{\alpha\alpha}(\vec{q},\omega)$, the so-called central peak discovered by Riste et al. (12), extends over a small volume in wave-vector space $\vec{q}$ (13) an estimate of the mean width of $S^{\alpha\alpha}(\vec{q} = \vec{q}_R,\omega)$ from $\Delta\omega$ is possible.

In the slow-motion regime, where the major part in the spectral density $J(\omega)$ occurs below ω_m, the observed linewidth is easily related to $J(\omega)$. Consider the definition of the second moment of the line. It is given for $\alpha = [100]$ by

$$\langle \delta H^2 \rangle = A^2 \langle (\varphi_\ell^{[001]})^2 \rangle = A^2 \langle \varphi_\ell^{[001]}(t)\, \varphi_\ell^{[001]}(t) \rangle \qquad (4)$$

and is proportional to the temperature-dependent mean-square value of the fluctuations. From the Fourier back transform of equation (2) we then have, setting $t = 0$,

$$\langle \delta H^2 \rangle = A^2 \int_{-\infty}^{+\infty} J^{\alpha\alpha}(0,\omega)\, d\omega = 2A^2 \int_0^{\infty} J(\omega)\, d\omega \quad . \qquad (5)$$

Because $J(\omega)$ contains only frequencies low compared to the characteristic frequency ω_m, $2(\langle \delta H^2 \rangle)^{1/2}$ will represent the experimental linewidth ΔH, and the line shape the distribution of $\varphi_\ell^{[001]}$. In the case of statistical independence this would be a Gaussian. The frequency ω_m is of the order of the instantaneous spread in Larmor frequency over the crystal owing to the distribution of $\varphi_\ell^{[001]}$, which at T_c is of the order of the maximum linewidth (7).

Since the instantaneous spread in $\varphi_\ell^{[001]}$ is by itself temperature dependent, ω_m will also be a function of temperature. With increasing $|\varepsilon|$ it decreases. Then $J(\omega)$ contains frequencies large compared to $\omega_m(T)$, and only the components for which $\omega \leq \omega_m$ will contribute to the broadening, since all faster components are averaged to zero. In this case the line will be narrower than in the case of a rigid lattice with a width $\Delta H \propto 2(\langle \delta H^2 \rangle)^{1/2}$. We denote by $\omega_1(T)$ the frequency equivalent of the linewidth $\Delta H(T)$, $g\beta\Delta H = \hbar\omega_1$. All components of $J(\omega)$ within 0 and ω_1 will contribute to the broadening. This yields for the fast-fluctuation case the implicit equation for ΔH_f (7),

$$\Delta H_f^2 = 2A^2 \int_0^{\omega_1} J(\omega)\, d\omega \; . \qquad (6)$$

In the limit where ω_1 can be regarded as small compared to the frequency dependence of $J(\omega)$ between $\omega = 0$ and ω_1, equation (6) can be rewritten using the linear relation between ω_1 and $\Delta H(T)$,

$$\Delta H_f = (4\pi g\beta/h)\, A^2\, J(0) \; . \qquad (7)$$

The detailed analysis of Abragam (7) gives in this case of fast fluctuations a Lorentzian line shape. The implications of the changeover from the fast to the slow-motion regime have been discussed recently by Rigamonti (14). The main result for the linewidth analysis is that in the extreme of slow motion the square of ΔH_s is proportional to the integral over $J(\omega)$, whereas in the fast-motion regime ΔH_f varies linearly with $J(0)$.

III. MEAN-FREQUENCY WIDTH OF SLOW FLUCTUATIONS NEAR T_c

A) Determination from the Changeover of Slow to Fast Motion of the Fe^{3+}-V_O Center

1. Experimental

In Geilo 1971 the first report on temperature-dependent EPR linewidth $\Delta H(T)$ broadening of the Fe^{3+}-V_O center near T_c was given (2). It results from the local fluctuations $\varphi_\ell{}^\alpha$ in equation (1). For these experiments the external magnetic field $\vec{H}$ was in a (001) plane nearly aligned along a [110] direction of the $SrTiO_3$ crystal.

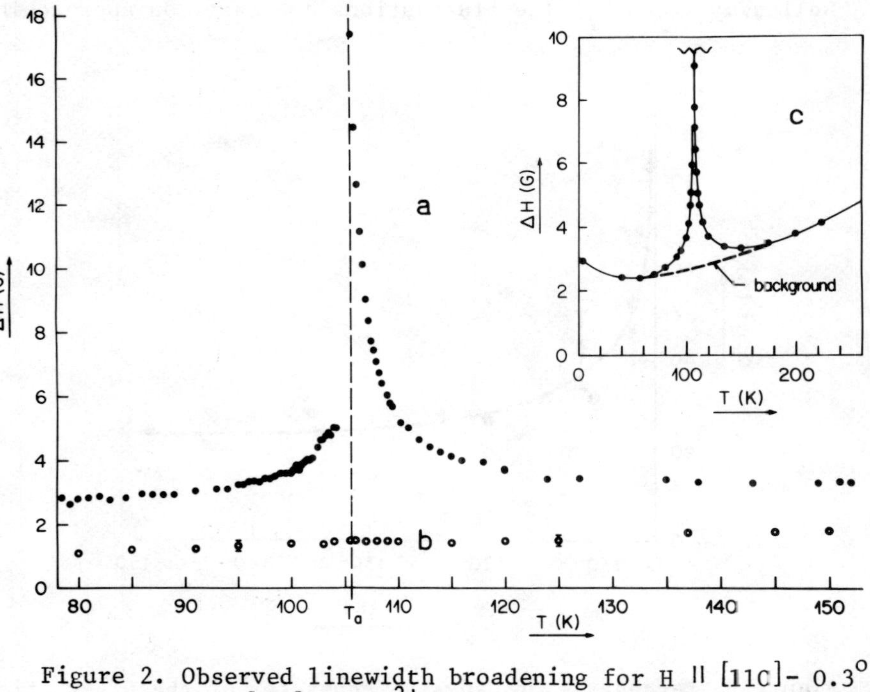

Figure 2. Observed linewidth broadening for H ∥ [110]- 0.3°, c ∥ [001] of Fe^{3+}-V_O center at the SPT of $SrTiO_3$ at 19.2 GHz. (a) For high-field line. (b) For low-field g ∿ 6 line. (c) Determination of the background width.

With this geometry φ_ℓ [001] rotations are probed by the Fe^{3+}-V_O centers lying either along [100] or [010] (6) ($g \simeq 4.5$). To avoid distrubing influences of different domains for the measurements below T_c, plate-shaped crystals were used which transform into a monodomain sample (15), Their domain axis was oriented parallel to the [001] axis. Figure 2(a) shows recent high-accuracy measurements of $\Delta H(T)$ at K-band (10). The broadening towards T_c from above and below is quite marked. Below and near T_c, ΔH could not be determined due to the asymmetry in the line shapes discussed in section V.A). The width reaches the same maximum value at T_c from both sides. Figure 2(b) exhibits the broadening of the low-field ($g \sim 6$) resonance which for this direction is independent of φ_ℓ (6) and therefore shows no additional broadening near T_c. To isolate the influence of the critical rotational fluctuations on the measured linewidth the background width resulting from different mode fluctuations must be subtracted. This background is by itself temperature-dependent and has been interpolated as indicated in figure 2(c). It has also been shown that the critical line broadening is due to rotations alone and not to spin-lattice relaxation (10).

Well away from T_c the fluctuations are fast. On approaching

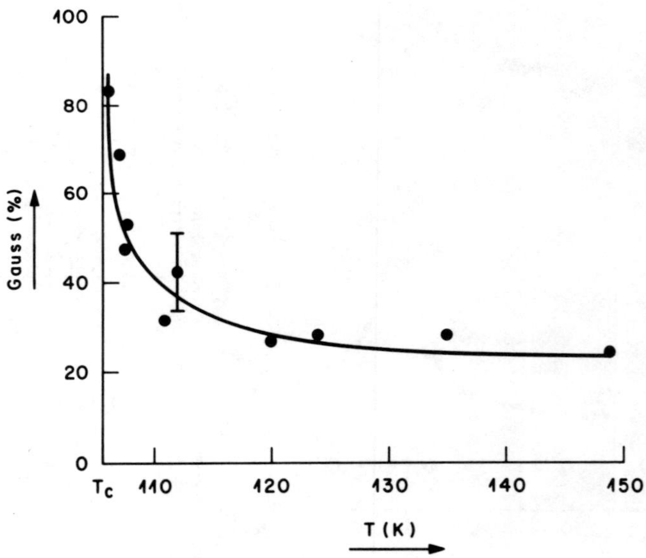

Figure 3. Percentage of Gaussian character of the fluctuation broadened lines of figure 2(a). From the thesis of Th. von Waldkirch (1972) unpublished.

T_c, the fluctuations slow down critically. Their correlation time and J(0) increase as shown schematically in figure 1. In the region where the halfwidth $\Delta\omega$ of $J(\omega)$ is still broader than ω_m, the resonance lines are near Lorentzian as shown in figure 3 for $T > T_c$; there $\Delta H \propto J(0)$ from equation (7).

2. Fast-motion regime

The analysis of $\Delta H(T)$ in this regime is first reviewed. It is based on an expression of J(0) obtained by Schwabl (11).

The central mode in neutron scattering has been analyzed by a dynamic form factor with three peaks $S(\vec{q},\omega)$ at $\omega = \pm\,\omega_\infty$ and $\omega = 0$ in $SrTiO_3$ (12), $KMnF_3$ (13), and now also in $LaAlO_3$ (16) (see the lecture notes by J. Axe and F. Schwabl). From this form of $S(\vec{q},\omega)$, and $\omega_\infty^2 = \omega_0^2 + \delta^2$

$$\begin{aligned} S(\vec{q},\ \omega{=}0) &= (\tfrac{1}{\pi})\ k_B T\ \chi(\vec{q},\varepsilon)\ \omega_0^{-2}\ \Gamma'(\vec{q},\omega = 0) \\ &= C \cdot \Gamma'(\vec{q},\omega = 0)\ T\chi^2(\vec{q},\varepsilon) \end{aligned} \tag{8}$$

is obtained if ω_0 is related to the static susceptibility $\chi^{\alpha\alpha}(\vec{q},\varepsilon)$ by the expression

$$\omega_0^2 = [I^\alpha\ \chi^{\alpha\alpha}(\vec{q},\varepsilon)]^{-1} \quad .$$

$\chi^{\alpha\alpha}(\vec{q},\varepsilon)$ is the static susceptibility of φ_q^α which is the Fourier transform of the order parameter φ_ℓ^α of the octahedra around the α axis, while $\varepsilon = (T - T_c)/T_c$ and I^α is the moment of inertia of the octahedra with respect to the α axis. C is a temperature-independent constant, and $\Gamma(\vec{q},\omega) = \Gamma'(\vec{q},\omega) + i\Gamma''(\vec{q},\omega)$ is a complex damping function. Due to anharmonic coupling one uses the Ansatz

$$\Gamma(\vec{q},\omega) = \delta^2(\vec{q})/[\lambda(\vec{q}) + i\omega] + \sigma \quad . \tag{9}$$

In this expression the parameters $\delta(\vec{q})$, $\lambda(\vec{q})$, and σ are taken to be uncritical near T_c. From equation (8) it is seen that the specific temperature behavior of $S(\vec{q},\omega = 0)$ near T_c is given by the divergence of $\chi(\vec{q},\varepsilon)$. Schwabl (11) assumed an anisotropic Ornstein-Zernike type susceptibility for rotations of the oxygen octahedra around α = [100], [010], or [001]:

$$\chi(\vec{q},\varepsilon) = \chi_0 [\vec{q}^2 - (1 - \Delta)\, q_\alpha^2 + \kappa^2]^{-1 + \eta/2} \quad . \qquad (10)$$

Here χ_0 contains the single-particle susceptibility, and $\kappa(\varepsilon) = \kappa_0 \varepsilon^\nu$ is the inverse of the correlation length $\xi(\varepsilon)$. ν and η are critical exponents (17), the latter describing the deviation from classical Ornstein-Zernike behavior on $\vec{q}$. The parameter Δ takes into account the possible anisotropy of the dispersion of $\chi(\vec{q},\varepsilon)$ near $\vec{q}_R$. It can vary between 0 and 1 characterizing pure two or three-dimensional correlations, respectively. This is an important refinement since for rotations around a certain axis α, the in-plane coupling of the oxygen octahedra by common corners is large compared to the interplane coupling. Using equations (8) and (10) Schwabl obtained the linewidth broadening for $T \rightarrow T_c^+$ for fast fluctuations,

$$J(0) \propto \Delta H(\varepsilon) = \frac{D}{T} \; \frac{\kappa(\varepsilon)^{-(1-2\eta)}}{\sqrt{\Delta}} \; \text{arc tg} \, \frac{\pi \sqrt{\Delta}}{\kappa(\varepsilon) \cdot a} \quad , \qquad (11)$$

where a is the lattice-constant. In equation (11) the single-particle susceptibility $\chi_0 \propto 1/T$ was taken into account. Equation (11) differs by this factor from Schwabl's result (11) who assumed it to be near constant.

The experimental $\Delta H(\varepsilon)$ points could be fitted to equation (11) above and below T_c with the same constants D, $\nu = 0.65 \pm 0.05$, $\Delta \simeq 1/40$ and $\kappa_0 = 0.23$ Å^{-1} but taking into account that the lattice constant $a = 3.9$ Å for $T \geq T_c$ is doubled for $T < T_c$ (10). η was set to zero as it is of the order of the experimental error. κ_0 is further mentioned in IV.A). Figure 4 shows the result above the phase transition occurring at $T_c = 105.6$ K. A good fit is obtained between T = 106.5 K and 152 K. Below T_c the measurements were also reproduced well analytically between 103.5 and 91 K (see figure 7 of reference (10)).

The value of the anisotropy parameter Δ means that about 40 octahedral units in a (001) plane are correlated when one in a next (001) plane is correlated to them in the sense of the I4/mcm structure. This is understood qualitatively by the strong in-plane coupling between the alternately rotating oxygen octahedra through the oxygen ions between the planes. Δ has been related to the measured anisotropy in dispersion of the soft R_{25} mode (10). The "pancake"-type correlations found here in the cubic phase may, to

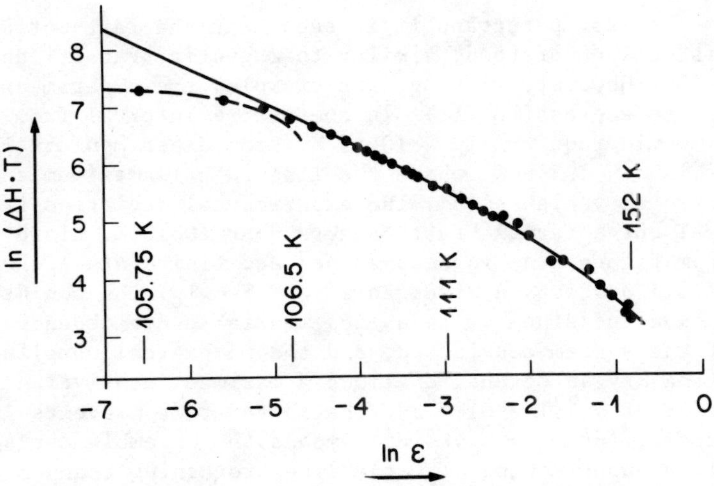

Figure 4. Comparison between experimental and theoretical φ_c-broadened linewidths for the fast- and slow-motion regimes for $T \geq T_c$. For the latter, the theoretical curve shows $(\langle \overline{\Delta H}^2_{crit} \rangle)^{1/2}$; •, experimental data; solid line, theory fast; dashed line, theory slow.

a certain extent, be compared to the behavior of the XY Heisenberg model as suggested by Stanley (17). This model yields critical exponents $\beta = 1/3$, $\gamma = 4/3$ and $\nu = 2\beta = 2/3$ with $\eta = 0$. This is near the experimental values $\beta = 0.33 \pm 0.01$ from static EPR experiments (3) and $\nu = 0.65 \pm 0.05$ obtained here, as well as $\gamma = 1.29 \pm 0.10$ calculated by Schneider and Stoll (18) using the observed changeover from $\beta \sim 1/2$ outside the critical region, to 1/3 near T_c. The comparison with the XY model is limited as this model allows a free rotation of its spins in the xy plane, whereas the octahedral rotations are bounded to within a few degrees. Furthermore, "pancake"-type correlations occur along all three equivalent <100> directions for $T > T_c$. A monodomain sample showing above T_c only one type of "pancake" is therefore more closely related to the XY model. This question is discussed further in section IV.A) with other models arrived at recently by theory.

Schwabl (11) also obtained from his theory an expression for the ultrasound absorption near T_c. In the region of quasi-two-dimensional correlations the attenuation of ultrasound in a [100] direction is proportional to $|T_c - T|^{-\rho}$ with $\rho = \nu(2-\eta) = \gamma$. Careful measurements by Fossheim and Berre (19) and Rehwald (20) for the propagation of ultrasound along a [100] direction, yield $\rho = 2\nu = 1.25 \pm 0.05$ $(\eta = 0)$ in agreement with the result of ν and γ obtained by EPR.

3. Changeover to slow-motion regime

From the last paragraph it is seen that the fast-motion regime as such is well understood, similar to magnetic transitions studied with NMR (8). However, fitting, for example, the experimental points above T_c to expression (11) in the entire interval from 105.73 K to 152 K would require $T_c \simeq 104.5$ K. From experiment it is found that T_c is at 105.6 K where the linewidth is maximum and the order parameter vanishes (3). The experimental deviation from the theoretical curve termed "fast" occurs near 106.3 K. Close to T_c, $\Delta H(T)$ from figure 2(a) is cusp-shaped and finite at T_c, whereas equation (11) predicts a divergence for $T \to T_c$. The non-divergent width is expected since it is a local variable not depending on the volume of the system considered, and the mechanical coupling between the rotating oxygen octahedra allows a maximum local value of the order of $\varphi_\ell^\alpha(T = 0) \sim 2^o$. The latter fact prevents $<\varphi_\ell^2>$ from diverging for $T \to T_c$. $<\varphi_\ell^2>$ means the ensemble average of (φ_ℓ^α) (2). On approaching T_c the line broadening changes over to a different regime. This is also evident from the line-shape analysis in figure 3. The percentage of Gaussian contribution increases markedly on approaching T_c. Although the data scatter appreciably, the 50:50 % Lorentzian-to-Gaussian relation occurs roughly at the same temperature where the deviation of the $\Delta H(T)$ points is found.

An expression for the linewidth in the slow-motion regime is easily derived (9), (10) using equation (5). This allows to approach the changeover region from below. Inserting equation (3) into equation (5) and employing the fluctuation dissipation theorem

$$\int_{-\infty}^{+\infty} S(\vec{q},\omega)\, d\omega = k_B T\, \chi(\vec{q},\varepsilon)$$

gives

$$\Delta H_s^2 = 4 <\delta H^2> = 4 A^2 k_B T \int_0^{\pi/a} \chi(\vec{q},\varepsilon)\, d^3q \,. \qquad (12)$$

Inserting expression (10) for $\chi(\vec{q},\varepsilon)$ into equation (12) with $\eta = 0$ yields for the critical contribution $(\kappa \to 0)$, using cylinder coordinates:

$$\Delta H_{crit}^2(\varepsilon) \propto \text{const} - [\kappa(\varepsilon)/\sqrt{\Delta}]\, \arctan\, [\pi\sqrt{\Delta}/\kappa(\varepsilon)\, a] \quad . \qquad (13)$$

The second term which for $\kappa \to 0$ becomes linear in κ, vanishes at $T = T_c$ so that the first term describes the finite maximum width at the phase transition. Close to T_c, where $(\pi\sqrt{\Delta})/[\kappa(\varepsilon)\ a] >> 1$, equation (13) is rewritten

$$\Delta H^2_{crit}(\varepsilon) = \Delta H^2_{max}\ (1 - C'\ \varepsilon^{\nu})\ , \tag{14}$$

where C' is a proportionality constant. Equation (14) does indeed give a cusp behavior of $\Delta H(\varepsilon)$ near T_c in accordance with the experimental observations.

A fit using $\Delta H_{max} = 19.7$ G and $C' = 18.6$ is shown in figure 4. The experimental uncertainties of T_c and ΔH^{expt}_{max} do not allow an accurate independent determination of the exponent ν. Therefore, $\nu = 0.65$ has been taken from the fast-motion analysis. Following the slow-motion theoretical curve one sees that the experimental, measured points start deviating at the same temperature as those following the fast-motion curve. Actually, both curves nearly touch at $T_{ch} = T_c + (0.8 \pm 0.3)$K. This is the changeover temperature.

To characterize the changeover let us assume that $J(\omega)$ for not too large ω can be approximated by a Lorentzian form

$$J(\omega) = \frac{1}{\pi}\ \frac{\Delta\omega}{\Delta\omega^2 + \omega^2}\ \sum_{\vec{q}} <\varphi^2_{\vec{q}}>\ ,$$

the half-width $\Delta\omega(\varepsilon)$ being strongly temperature dependent near T_c. In the slow-motion regime we have $\omega_1(\varepsilon) = \omega_m(\varepsilon) >> \Delta\omega(\varepsilon)$. For increasing $|\varepsilon|$, $\omega_1(\varepsilon)$ and $\omega_m(\varepsilon)$ decrease and $\Delta\omega(\varepsilon)$ increases. For the changeover region $\omega_1 \simeq \Delta\omega$. For higher $|\varepsilon|$ the line becomes motionally narrowed. This region is characterized by $\omega_1 < \omega_m << \Delta\omega(\varepsilon)$. Since $\omega_1(\varepsilon)$ decreases and $\Delta\omega$ increases for growing $|\varepsilon|$, the temperature region covered by the changeover must be fairly small. As pointed out by Rigamonti (14), the changeover thus arises at that temperature where the half-width in $J(\omega)$ becomes of the order of the instantaneous spread in Larmor frequency $\omega_m(T)$. This interpretation allows a direct estimate of the half-width $\Delta\omega$ from $\Delta H(T)$ to be of the order of 70 MHz at $T_{ch} \simeq T_c + 0.8 \pm 0.3$ K.

B) Estimate from the Intensities of Orthorhombic and Axial Spectra of the Ti^{3+} Center

The Ti^{3+} center is best created by irradiating "pure" undoped $SrTiO_3$ with fast neutrons. Some of the Ti^{4+} ions are thereby "knocked-off" their B-site and become substitutional on Sr^{2+} A-sites. The Ti^{4+} captures one electron becoming three-valent and paramagnetic with configuration 3d (1). The EPR spectrum can thus be described by an effective spin Hamiltonian with $S' = 1/2$,

$$\mathcal{H} = \mathcal{H}_z + \mathcal{H}_{Hf} = (G_x H_x + G_y H_y + G_z H_z) + \mathcal{H}_{Hf} \quad . \tag{15}$$

$\mathcal{H}_z$ stands for the Zeeman interaction and $\mathcal{H}_{Hf}$ is the hyperfine term $G_i = g_i \beta$. β is the Bohr magneton. $i = x,y,z$ marks a right-handed magnetic local coordinate system. H_i are the magnetic field components. The Ti^{3+} center has been recently analyzed in detail by Schirmer and Müller (21) in the tetragonal and cubic phase of $SrTiO_3$. It couples in an entirely different way to the fluctuations of the order parameter $\varphi_\ell^\alpha(t)$, like the paramagnetic B-site ions do. The latter is a direct one as seen from equation (1), whereas the former couples quite indirectly.

1. Static properties of Ti^{3+} : orthorhombic spectrum

The radius of the Ti^{3+} of 0.75 Å is appreciably smaller than that of the Sr^{2+} of 1.10 Å it substitutes. Therefore the Ti^{3+} is not found at the ideal Sr^{2+} position but is moved "off-center" by $d = 0.03$ Å. Thereby it induces an electric dipole moment $3ed \sim 0.1$ Å. This was shown by investigating quantitatively the EPR spectra under application of external electric fields $\vec{E}$. Depending on the orientation of $\vec{E}$ some lines grow in intensity and others diminish. In the low-temperature phase of $SrTiO_3$ the Ti^{3+} moves off-center nearly perpendicular to the domain axis c (22) towards <u>one</u> of the twelve surrounding oxygens. The one oxygen the Ti^{3+} "selects" or adheres to, is one which due to the tetragonal phase moves <u>towards</u> the Ti^{3+}. This is depicted in figure 5, where a (100) lattice plane of TiO ions is shown, and projected onto it is a Sr^{2+} (100) plane lying 1.95 Å above it. In the middle the substitutional Ti^{3+} ion appears which is moved off-center 0.03 Å towards one of the next four oxygens U, L, I and II. The tetragonal domain axis is along the [001] direction.

Let us assume that the rotation φ_ℓ of the oxygen octahedra with center at titanium 1 and 2 is such that oxygen U moves out of the (100) plane towards the Ti^{3+}. Oxygen I and II remain at their ideal positions. If this motion exceeds a critical angle φ_c, i.e., $\varphi_\ell^{[001]} > \varphi_c$, the Ti^{3+} "sticks" to oxygen U. The main paramagnetic z-axis is not parallel to the [100] axis but tilted by $\tau = 2.5^o$ (not shown) towards the [001] domain axis; τ is independent of temperature. The Ti^{3+}-O^2 interaction is mutual with fixed tilting angle τ. The magnetic axes x,y orthogonal to z are

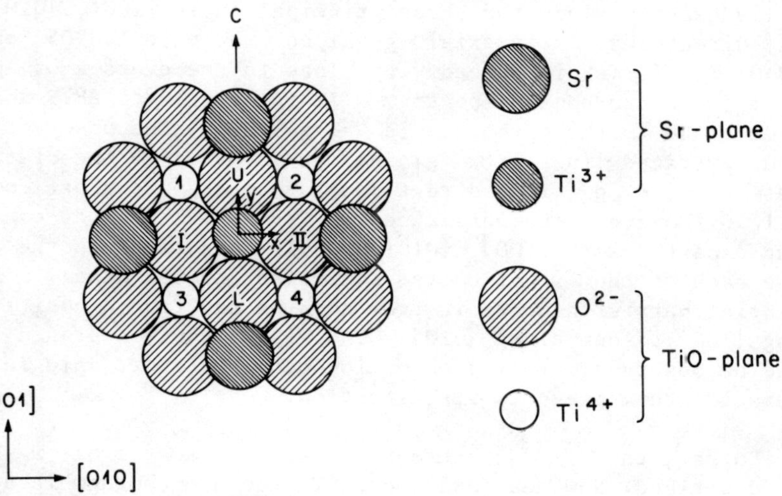

Figure 5. Projection of Sr (001) plane including a Ti^{3+} defect onto a TiO plane 1.95 Å below it in $SrTiO_3$.

shown $x \parallel [010]$ and $y \parallel [001] - \tau$ towards the [100]direction. The measured orthorhombic g values in the tetragonal phase below T_c are $g_x = 1.9228$ and $g_y = 1.8530$. $g_z = 1.9945$ close to the values of the free electron.

Of importance for the dynamics is the following: Assume that the c-axis were directed along [010]. Then oxygens U and L would be at rest, and I and II would move out of the (100) plane as the octahedral rotation is now around [010]. Then for $\varphi_\ell^{[010]} > \varphi_c$ the Ti^{3+} would stick to either oxygen I or II and the magnetic x and y axes be interchanged, for oxygen II: $x \parallel [010] - \tau$ and $y \parallel [00\bar{1}]$.

2. Dynamic properties: the tetragonal Ti^{3+} spectrum

Above T_c in the cubic-phase oxygens, I, U, II and L become equivalent on the average. The Ti^{3+} starts hopping between all four of them. This leads to an averaged axial spectrum with axis along the [100] direction under the following condition for the hopping time τ_h. The hopping frequency $\nu_h = 1/\tau_h$ must be high compared to the difference of resonance frequencies $\Delta\nu = (g_x - g_y)\beta H/h$ which the Ti^{3+} shows if it is locked to say either oxygen U or I. The difference in $g_x - g_y$ comes from the interchange of x and y axes mentioned before (for fixed magnetic field). In fact, on raising the temperature well above T_c one observes three axial spectra with $S' = 1/2$ instead of the 12 distinguishable orthorhombic spectra found for $T < T_c$. The main axes of these spectra now lie exactly along the three principal cubic [100], [010] and [001] directions. Their axial g values are $g_\perp = 1.8905$ and $g_\| = 1.9940$. The former is indeed very close to the average of g_x and g_y of the orthorhombic spectrum $1/2\,(g_x + g_y) = 1.8879$ and the latter close to the orthorhombic $g_z = 1.9945$. The observation of narrow averaged lines with $g_\perp$, narrow compared to the distance of lines g_x and g_y, is a direct proof that τ_h is short compared to $1/\Delta\nu$. However, the paramagnetic Ti^{3+} is still off-center along a particular [100], [010] or [001] direction. In the cubic phase each of them is of course equivalent to the other. A high potential barrier hinders it from hopping say from a position along [100] to one along [010]. Otherwise a cubic-averaged spectrum would be observed, or if such motion occurs a short spin lattice relaxation term prevents its detection.

So far, the Ti^{3+} spectra observed well away from T_c have been described. Now near and above T_c both orthorhombic and tetragonal spectra are seen simultaneously. On raising T the intensity of the latter increases at the expense of the former. In figure 6 the normalized ratio I_o/I_t is shown. A natural explanation is the existence of fluctuations in the order parameter which is nearly static close to T_c. If $\varphi_\ell(t) > \varphi_c$ and if it pre-

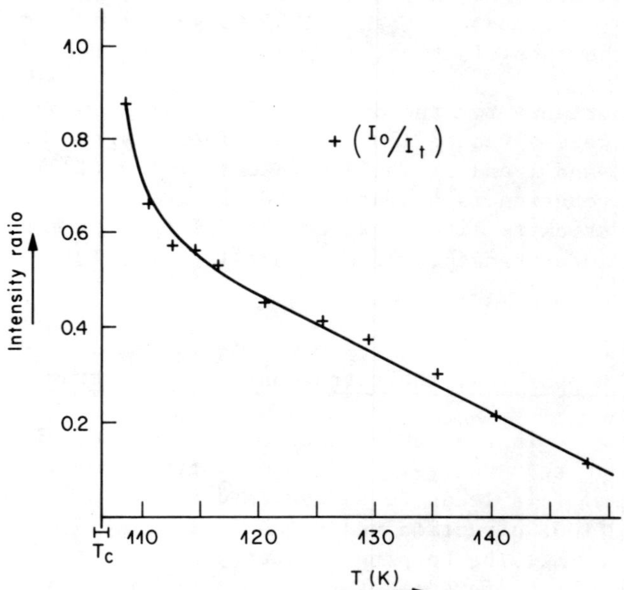

Figure 6. Temperature dependence of the intensity ratio of orthorhombic and tetragonal lines I_o/I_t above T_c for the Ti^{3+} center in $SrTiO_3$.

vails longer than $1/\Delta\nu$, the reciprocal of the frequency difference between lines g_x and g_y, an orthorhombic spectrum is observed; otherwise the lines are averaged to an axial spectrum. In the neighborhood of T_c the amplitude of the local fluctuations does not vary appreciably. We can thus assume $\varphi_\ell > \varphi_c$ and concentrate on their time dependence described by $J(\omega)$. The axial lines due to fast motion are approximately proportional to the integral $\int_{2\pi\Delta\nu}^{\infty} J(\omega)\,d\omega$, the orthorhombic spectra to $\int_0^{2\pi\Delta\nu} J(\omega)\,d\omega$. The latter integral contains the slow critical part in $J(\omega)$ which is strongly temperature dependent near T_c. While a quantitative explanation of the curve in figure 6 is still lacking, we ascribe the marked change in slope of I_o/I_t near $T_c + 4$ K to the fact that the width of the critical part of $J(\omega)$, $\Delta\omega_c$ has become of the order of $2\pi\Delta\nu$. In the experiments from figure 6 which were carried out in X band, we get $\Delta\nu = (g_x - g_y)\beta H/h = 3.5 \times 10^8$ Hz.

3. <u>Temperature dependence of $\Delta\omega$ and the width of the central peak</u>

It has been shown theoretically that the width of the central peak at $\vec{q} = \vec{q}_R$ varies with temperature as (23)

$$\Gamma_c = \Gamma_0 \; [(T - T_c)/T_c]^{\gamma} \quad , \tag{16}$$

where $\gamma = 1.28$ (18). In q-space the central peak extends over 0.03 $Å^{-1}$ (13). Due to the small region we can assume from equations (3) and (16) that

$$\frac{\Delta\omega}{2\pi} = \Delta\nu_0 \;\; [(T - T_c)/T_c]^{\gamma} \tag{17}$$

holds.

From section IV.A) we have, at $T_c + 0.8$ K, $\Delta\omega/2\pi = 70$ MHz, and calculate from equation (17) $\Delta\omega/2\pi = 4.2 \times 10^8$ Hz for $T = T_c +$ 4 K. This correlates with the size of $\Delta\nu = 3.5 \times 10^8$ Hz obtained from the Ti^{3+}-center EPR results. Thus we have shown here that the critical width in $J(\omega)$ varies with temperature approximately according to equation (17) with $\Delta\nu_0 = 10^{10}$ Hz. One estimates from equation (3) that $\Delta\nu_0$ should be of the order of Γ_0 to a factor 3. Furthermore, the linewidth broadening of the Fe^{3+}-V_0 center and the intensity ratio of the Ti^{3+} center I_o/I_t (Figure 6) is seen to vary up to 150 K. This shows that a sensible quasi-static order-parameter variation prevails up to this temperature as does the detection of the central peak (12, (13).

From figures 3 and 4 it is quite clear that the changeover from fast to slow motion for the Fe^{3+}-V_0 center has been observed. However, whether the quasi-static regime has been reached close to T_c is not entirely certain. In the next section it is shown that this regime has indeed been reached.

IV. ANISOTROPY OF FLUCTUATIONS IN A MONODOMAIN SAMPLE AT T_c

A) Transverse and Longitudinal Fluctuations from EPR

Consider a monodomain crystal in the low-temperature phase with the c-axis aligned along the [001] axis. The local fluctuations of BO_6 octahedra $\delta\varphi_c$ around the domain axis c ∥ [001] and those around the two equivalent a-axes $\delta\varphi_a$ parallel to [010] or [100] will be different due to the broken symmetry. These two types of fluctuations could be studied separately by the use of EPR of the Fe^{3+}-V_0 pair center. In section III.A) experiments to probe the fluctuations around a [001] axis have been summarized. These fluctuations $\delta\varphi_c$ are probed if the external magnetic field $\vec{H}$ is aligned in a (001) plane along a [110] direction near g = 4.5. Therefore with the magnetic field $\vec{H}$ pointing along a [101] direction one probes the fluctuations $\delta\varphi_a$ around a [010] axis. In the former case the EPR of Fe^{3+}-V_0 pair centers aligned along a [010] axis was used and for the latter Fe^{3+}-V_0 centers aligned along a [001] axis. Figure 7 shows the temperature dependence of the linewidth for fluctuations around an a-axis and those around the c-axis. The latter data are here reproduced from figure 2(a). for comparison. Well below T_c due to the tetragonality, $\Delta H(\varphi_c) \neq \Delta H(\varphi_a)$. This has been related to the dynamics of the lattice, i.e., to the singly and doubly degenerate soft modes ω_A and ω_E. The details are beyond the present context (10). On approaching T_c from below, one would expect the two widths $\Delta H(\varphi_c)$ and $\Delta H(\varphi_a)$ to become equal at and above T_c, because there is no distinction between a and c directions possible, the crystal being cubic. Unexpectedly at T_c the two widths are different. Actually one infers from the figure that less than a degree below T_c the $\Delta H(\varphi_c)$ curve crosses the one of $\Delta H(\varphi_a)$. No data points are shown there due to overlapping lines and asymmetrical shapes.

Also well above T_c where the crystal is cubic the linewidth remains anisotropic. This anisotropy is qualitatively not explainable by static symmetry properties. Approaching T_c from above the anisotropy increases. It must be connected with the growing regions of correlated rotational fluctuations. Only for temperatures close enough to T_c are these regions significantly larger than a few unit cells. Their size is given by the correlation length $\xi(\varepsilon) \equiv 1/\kappa(\varepsilon)$ which at 115 K must be of the order of four unit cells.

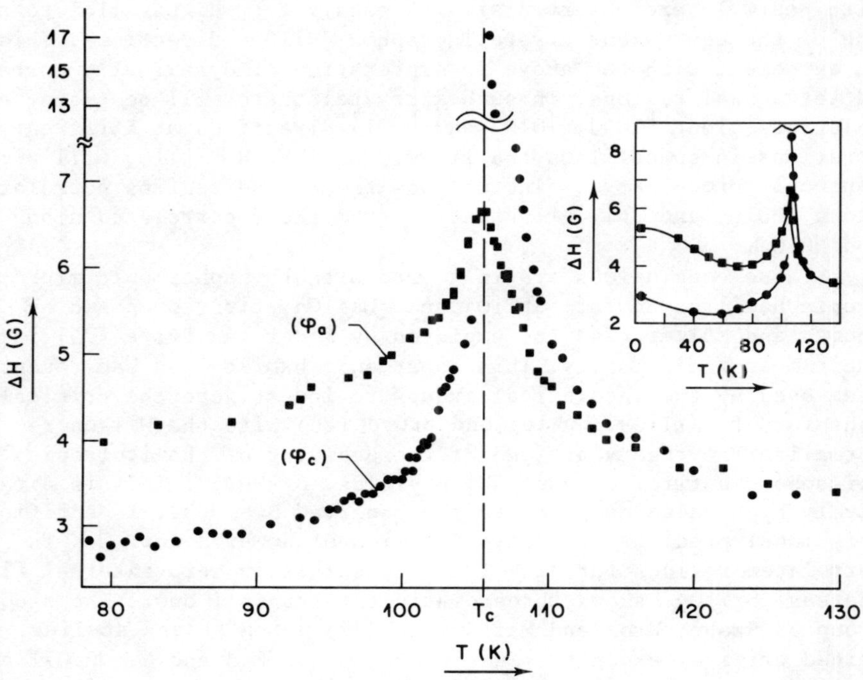

Figure 7. Anisotropy of the temperature-dependent linewidth in a monodomain sample with $\vec{c} \parallel [001]$. Circles represent data reproduced from figure 2(a), denoting broadening caused by fluctuations around the c-axis. Squares are measured for the corresponding line with $\vec{H}$ along 0.5^o from [101] and represent broadening by fluctuations around an a-axis.

This is consistent with $\kappa_0 = 0.24$ Å^{-1} used in III.A). Within a certain region, the local fluctuations pertaining to one particular component of the three equivalent $\varphi^{<100>}$ rotations are correlated in the sense of the low-temperature tetragonal phase. The dynamic symmetry in this region is therefore no longer cubic since one of the three <100> axes is distinguishable from the other two. This is the axis of correlated rotations. Hence, owing to the dominance of one particular c-type correlation, the crystal "feels" the low-temperature phase. The correlation with respect to that local fluctuation component α means that within the correlation region the static susceptibility $\chi^{\alpha\alpha}(\vec{q},\varepsilon) \sim 1/(\vec{q}^2 + \kappa^2)$ is no longer equal for the other two components α, but is larger for the correlated component, since its κ is smallest.

In a well-annealed crystal, which transforms into a polydomain sample below T_c, a narrow line with width $\Delta H(\varphi_a)$ and a broad one with $\Delta H(\varphi_c)$ are observed simultaneously for $\vec{H}$ parallel to any one of the equivalent crystallographic <110> directions. This is in agreement with the above interpretation of dynamically correlated tetragonal regions. In such a crystal there will be c-type correlations along [100], [010] and [001] directions at different locations in space. Thus the linewidth with $\vec{H} \parallel [110]$ will simultaneously probe $\delta\varphi_c$ fluctuations from those regions correlated along [001], and $\delta\varphi_a$ fluctuation from those correlated along [100] and [010].

These conclusions are in accord with the anharmonic microscopic hamiltonian for $SrTiO_3$ and $LaAlO_3$, first proposed by Thomas and Müller (24) and employed by Feder and Pytte (25) in their mean-field theory. This anharmonic hamiltonian has recently been used by two theoretical groups to investigate the critical behavior of $SrTiO_3$. Cowley and Bruce (26) with the Wilson renormalization theory arrived at the behavior of the isotropic Heisenberg antiferromagnet which yields $\beta = 0.37$. This is a relatively high value compared to the observed $\beta = 0.33$. In addition, this model predicts that any direction of domains for $T < T_c$ or correlated regions for $T > T_c$ is possible in zero external field, whereas $SrTiO_3$ shows three distinct tetragonal ones! The second group of Szabo, Droz and Malaspinas (27) using strong scaling, obtained critical exponents $\beta = 0.322$, $\nu = 0.643$ and $\alpha = 0.071$ as for the three-dimensional Ising model. The latter values are close to the ones observed as are those of the XY model mentioned earlier. The three authors also predict that these values are independent of the anharmonic anisotropy strength close to T_c and the correlations become three-dimensional. This is another limitation of the XY model apart from those already mentioned in III.A), which due to its anisotropy may apply only in the regions away from T_c where two-dimensional correlations prevail.

In the sample transforming below T_c into a monodomain with c-axis along [001], the correlations above T_c will occur predominantly with respect to the $\varphi_{[001]}$ fluctuations. It is on account of these fluctuations that the correlation length will diverge, since this type of rotation is the order parameter of the sample below T_c. For the fluctuations around [100] and [010] the correlation length will remain finite. These motions are of quasi-order-parameter type. If close to T_c the slow-motion regime is reached, then according to equations (1) and (12)

$$\Delta H_i^2 = 4\,A^2 \langle \delta\varphi_i^2 \rangle \qquad i\ :\ a\ \text{or}\ c \tag{18}$$

must hold. In the next paragraph it is shown that this is indeed the case near T_c. From equation (18) and figure 4 one gets for the mean spread at T_c

$$(\langle\delta\varphi_c^2\rangle)^{1/2} = 0.4^o \quad \text{and} \quad (\langle\delta\varphi_a^2\rangle)^{1/2} = 0.15^o \quad .$$

$(\langle\delta\varphi_c^2\rangle)^{1/2}$ is in remarkably close agreement with the value of the generalized order parameter $\langle\varphi_\ell^{[001]}\rangle$ near 100 K where the deviation from Landau behavior in the temperature dependence of the order parameter is observed (3).

B) Difference of Transverse and Longitudinal Fluctuations from Birefringence

Very recent optical birefringence measurements in $SrTiO_3$ by Courtens (28) are in agreement with the near static observation of anisotropic fluctuations $\langle\varphi_a^2\rangle$ and $\langle\varphi_c^2\rangle$ by EPR. They were carried out on one of the "monodomain" samples used for the EPR studies. They show, besides a systematic decrease of birefringence Δn on approaching T_c^-, a tail within a comparable temperature region above T_c and a small but distinct cusp at the transition temperature (see figure 8).

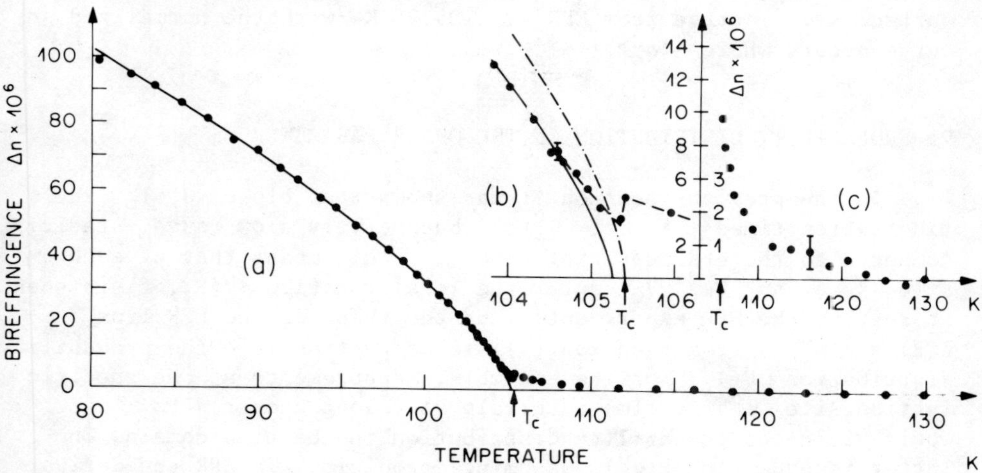

Figure 8. The birefringence data around T_c. The solid line is an exponential fit. (a) The entire temperature range. (b) The exact vicinity of T_c on a greatly expanded scale. The cusp position is shown by the arrow. The dash-dot line is a properly scaled replica of $\langle\varphi_c\rangle^2$ from reference (3). The dashed line passes through the experimental points. Taken from reference (28).

The birefringence is $\Delta n = S(<\varphi_c>^2 + <\delta\varphi_c^2> - <\varphi_a^2>)$, S being a proportionality constant. Owing to their much higher characteristic frequency ω_m these experiments "see" the fluctuations as static for all temperatures. Well below T_c, $<\delta\varphi_c^2> < <\varphi_a^2>$, and the birefringence deviates systematically to smaller values from the static value of $<\varphi_c>^2$ determined by EPR (3). For $T \to 0$, both $<\delta\varphi_c^2>$ and $<\varphi_a^2>$ are small, which allows the proportionality constant $S = 55 \times 10^{-6}$ to be obtained using the EPR $<\varphi_c>$ result. With this value of S, the peak of Δn at T_c was computed (28), inserting in the expression $\Delta n = S(<\delta\varphi_c^2> - <\varphi_a^2>)$ the fluctuation values from figure 7. This is permissible if at T_c the EPR linewidth is given by quasi-static fluctuation values. The value $\Delta n = 5.0 \times 10^{-6}$ which was obtained in this way agrees quantitatively with the peak in Δn found experimentally. This confirms by an independent experiment that the EPR linewidths yield the instantaneous spread in local fluctuations at T_c, i.e., one is well in the slow-motion regime. The observation of a birefringence tail ($\Delta n \neq 0$) for $T > T_c$ also yields an anisotropy in fluctuations $<\varphi_c^2> > <\varphi_a^2>$, in agreement with the conclusions from EPR in IV.A). The critical exponents β and ν obtained from birefringence experiments are also comparable with those obtained in III.A). Note from figure 7 that less than a degree below T_c, $(<\delta\varphi_c^2> - <\varphi_a^2>)$ changes sign. This accounts for the left side of the cusp in figure 8(b). The crossing of the dash-dotted $<\varphi_c>^2$ curve from EPR at 105.25 K with the normalized Δn curve occurs where $<\delta\varphi_c^2> = <\varphi_a^2>$.

V. PROBABILITY DISTRIBUTION OF THE ORDER PARAMETER

In the previous section it was shown that close to T_c the fluctuation times τ_s of $\varphi_\ell^\alpha(t)$ become very slow (quasi-static) compared to the EPR measuring time τ_m. This means that at each site ℓ of an Fe^{3+}-V_0 center the local rotation $\varphi^\alpha(\vec{R}_\ell, t)$ is seen at rest by the EPR experiment. Thus the shape of the EPR line $L(H) = L(\varphi_\ell^\alpha)$ (from equation (1)) is proportional to the probability distribution $P(\varphi_\ell^\alpha)$ of the ensemble. Measurements at one specific lattice site φ_ℓ^α at time intervals very long compared to τ_s would yield the probability distribution in the time domain. The latter is equal to $P(\varphi_\ell^\alpha)$ assuming ergodicity. In EPR the derivatives $dL(H)/dH$ of the absorbtion lines $L(H)$ are recorded using Zeeman modulation. For those not familiar with the technique, dL/dH and $L(H)$ are shown in figure 9 for $<\varphi> = 0$ above T_c. For T very near T_c the curves are proportional to $P(\varphi_\ell)$ and $dP/d\varphi_\ell$, respectively.

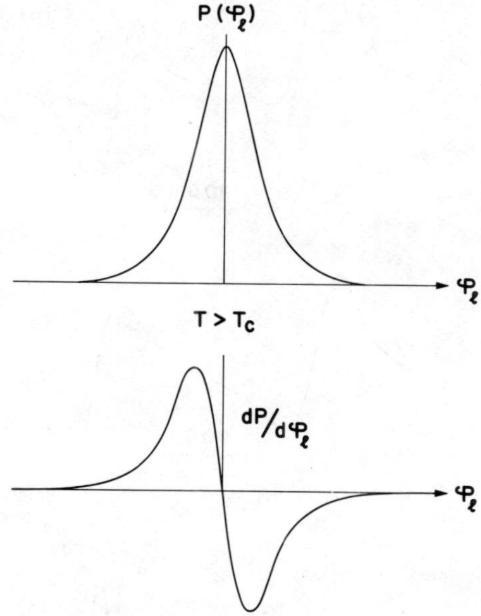

Figure 9. Probability distribution $P(\varphi_\ell)$ and the derivative $dP/d\varphi_\ell$ which in the slow-motion regime are proportional to the line shape $L(H)$ and its derivative dL/dH, respectively.

A) Critical Asymmetry for $T \to T_c^-$

EPR experiments are now summarized showing that $P(\varphi_\ell)$ is asymmetric for $\langle\varphi\rangle \neq 0$, i.e., $T < T_c$ (29). With $\vec{H}$ pointing along [110], rotations around the [001] axis are observed as mentioned in II. Because of the nonvanishing order parameter, $\langle \varphi^{[001]} \rangle \equiv \langle\varphi_c\rangle \neq 0$ and the alternate rotation of octahedra in the antiferrodistortive phase (1), two lines corresponding to $\langle\varphi_+\rangle$ and $\langle\varphi_-\rangle$ are observed below T_c. These are shown in figure 10. Let the maximum and minimum slopes be designated by $A = |dL/dH|_{max}$ and $B = |dL/dH|_{min}$. Well below T_c in the fast-motion regime (Figure 10(c)) each of these lines is symmetric and $A = B$. On approaching T_c, $A/B \neq 1$, which is not due to overlap (Figures 10(b) and 10(a)), and one sees that the lines become asymmetric, but show mirror symmetry with respect to H_o, corresponding to $\langle\varphi^{[001]}\rangle = 0$, equation (1). In order to have an asymmetry parameter a_s which vanishes for a symmetric line,

$$a_s = A/B - 1 \tag{19}$$

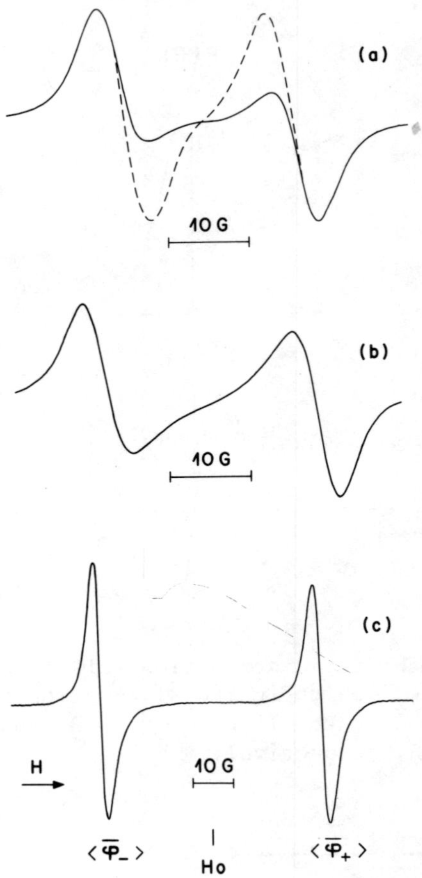

Figure 10. EPR lines of the Fe^{3+} -V_O center at K band for $\vec{H}$ || [110] in a monodomain $SrTiO_3$ sample below T_c, c-axis || [001] at (a) $T = T_c - 0.75$ K; (b) $T = T_c - 1.00$ K; (c) $T = T_c - 10.25$ K. Note scale change in magnet field sweep. $H_o = 3090$ G.

was defined (29). Figure 11 shows the variation of a_s, where for $T \rightarrow T_c^-$ the critical increase is clearly seen. Assuming a divergence in a_s with temperature $a_s \propto (T_c - T)^{-n}$, $n = 1.05 \pm 0.15$ for $|\epsilon| < 2 \times 10^{-2}$ was found. In the inset of figure 11, $a_s^{-1.05}$ is plotted against $|\epsilon|$ in the critical region.

The asymmetry in the EPR line shape for $T < T_c$ (Figure 11) becomes appreciable in the region of $|\epsilon| = |T - T_c|/T_c$ near 10^{-2}, where from the $T > T_c$ linewidth data analysis in III.A) one knows that the regime changes from fast to slow. Thus close to T_c, L(H)

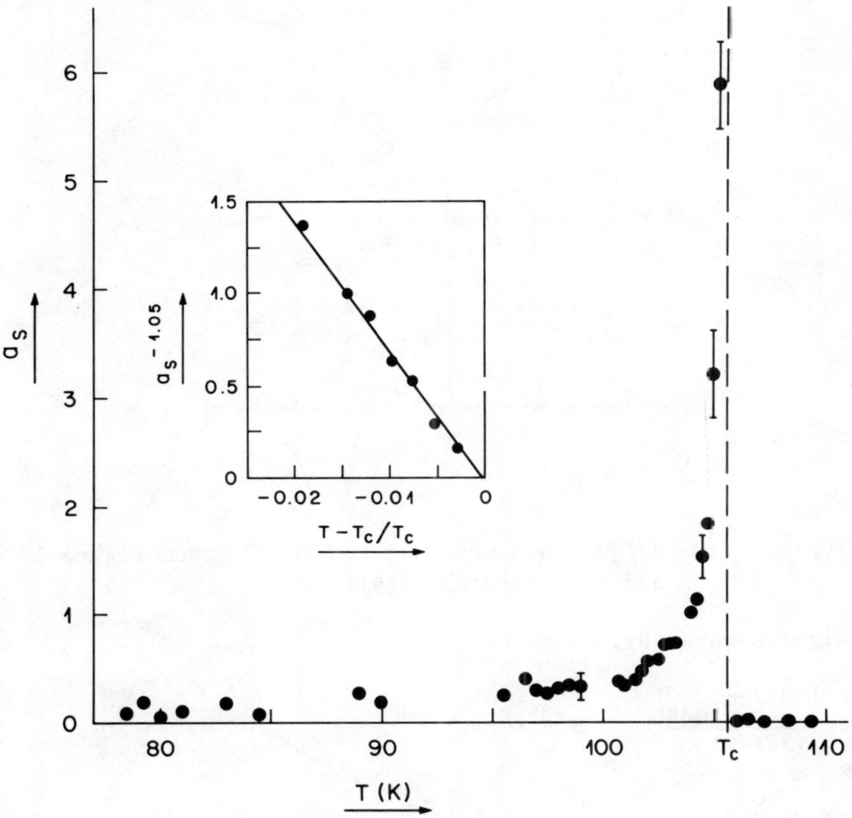

Figure 11. Asymmetry parameter a_s in $SrTiO_3$ for $T \to T_c$.

must be proportional to the probability distribution $P(\varphi_\ell)$ of φ_ℓ. We note that φ_ℓ^m, at which $P(\varphi_\ell)$ is a maximum, is sited at $dL(H)/dH = 0$ and differs from $\langle\varphi\rangle$. This means that if $\langle\varphi\rangle \neq 0$, then odd moments in $\varphi_\ell - \varphi_\ell^m$ also differ from zero. From figure 10(a) one also sees that $P(\varphi_\ell)$ is larger in the region between φ_ℓ^m and $\varphi_\ell = 0$. The octahedra fluctuate more easily towards $\langle\varphi\rangle = 0$ than away from it. This is borne out by the integrated derivative of figure 10(a) as shown in figure 12.

The asymmetry of $P(\varphi)$ as such can easily be obtained by writing down the classical probability distribution of φ,

$$P(\varphi,T) = C''(T) \exp\left[- \Delta F(\varphi, T)\right] , \qquad (20)$$

where $\Delta F(\varphi, T) = F(\varphi, T) - F_o(T) = a(T)\varphi^2 + b\varphi^4$ is the Landau Ansatz for the free energy (29). With it one calculates in a

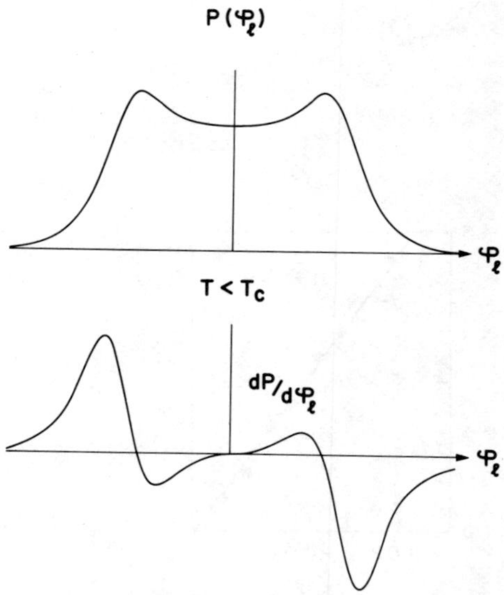

Figure 12. $dP/d\varphi_\ell$ proportional to dL/dH from figure 10(a) and its integral $P(\varphi)$.

straightforward way, close to T_c ,

$$\left|\frac{(dP/d\varphi)_{max}}{(\partial P/\partial\varphi)_{min}}\right| \sim \langle\varphi\rangle^{-3} \sim (-\varepsilon)^{-3\beta} , \qquad (21)$$

where $\beta = 1/2$, i.e., $n = 1.5$. In the critical region this analysis is not valid; there $\beta = 0.33 \pm 0.01$ as was previously obtained (3). The experimental result, $n = 1.05 \pm 0.15$, suggests that in the critical region $n = 3\beta$ may still hold. In the slow-motion regime the linewidth is dominated by the central mode of $S(\vec{q},\omega)$ for $T \lessgtr T_c$. Therefore the asymmetry a_s for $T < T_c$ must be reflected in the form of the central mode (12).

From the results one foresees that a breaking in symmetry of local fluctuations is present in all distortive-type (1) phase transitions whenever the overall symmetry is broken, i.e., $T < T_c$ and the order parameter $\langle\eta\rangle$ differs from zero. This means that the symmetry breaking of local fluctuations, in principle, also occurs in weak first-order phase transitions (1) as the only requirement is $\langle\eta\rangle \neq 0$. Now the latter requirement is of course the one which is common to the majority of phase transitions like the gas-fluid, the ferromagnetic, etc. Therefore, one may postulate this broken symmetry of local fluctuations for nonzero order parameter occurs quite generally.

B) Deviations from Gaussian Distribtuion for $T \geq T_c$

To terminate, an account is given of the most recent experiments on EPR line-shape analysis for $T \to T_c^+$. From these a deviation from Gaussian distribution on $P(\varphi_\ell^c)$ was found. To prove this, considerable care had to be taken in the experiments to ensure that: a) the temperature was above the transition $T \geq T_c$, i.e., $\langle\varphi\rangle = 0$, and b) to still be at a temperature T well below the changeover temperature T_{ch} to the fast-motion regime. T_{ch} was obtained by the procedure discussed in section III.A). It is necessary to have $T \ll T_{ch}$, in order that the line shape $L(H) = L(\varphi_\ell)$ is proportional to the probability distribution $P(\varphi_\ell)$. Conditions a) and b) together require a temperature about 0.2 K above T_c. This was quite delicate to realize for the following reason: Suppose $T \lessapprox T_c$, then one has two asymmetrical lines $L(H)$ or probability distributions $P(\varphi_\ell)$ on either side of $\pm\langle\varphi\rangle$. They are superposed on one another for near-vanishing order parameter $\langle\varphi\rangle$. This certainly yields an overall probability $P(\varphi_\ell)$ centered on $\langle\varphi\rangle = 0$ which deviates from a Gaussian distribution. No theories for the asymmetry of $P(\varphi_\ell)$ in the critical region are known (30). Therefore reaching of T_c cannot be obtained analyzing the lines by superposing two known forms of $P(\varphi_\ell)$.

Two independent empirical methods were used to determine T_c. In the first, the known temperature dependence $\varphi(\varepsilon) = \varphi_0 \{(T_c-T)/T_c\}^\beta$ of the static order parameter was employed. The quantity $(\varphi(\varepsilon)/\varphi_0)^{1/\beta}$ was extrapolated to zero value, yielding T_c, from the region where the EPR lines are still sufficiently symmetric to assume reliable values of $\langle\varphi\rangle \equiv \varphi(T)$. (The asymmetry of the lines close to T_c was the reason why at the time no closer values of $\varphi(\varepsilon)$ than $\varepsilon = 0.001$ were reported (3).) The second procedure is still more empirical. It was found that in the steep slope of the line derivatives dL/dH a less steep portion occurred exactly in the middle near H_0 (i.e., $\varphi = 0$) when T_c was not yet reached. This is shown in figure 13(a) for $T = T_c - 0.10$ K. It corresponds to the existence of two minima in the second derivative d^2L/d^2H of the lines. When just one minimum in d^2L/d^2H was observed corresponding to the disappearance of the less steep portion in dL/dH in figure 13(a) T_c was reached. T_c determined in these two ways was the same within the accuracy of temperature measurements of 0.03 K. Measurements of dL/dH just above T_c in the slow-motion regime show that the shoulders of the lines fall more rapidly to zero than a Gaussian. In figure 13(b) a record at $T = T_c + 0.25$ K is shown together with the derivative of a Gaussian having the same width at maximum slope and nearly the same amplitude. It is clear that for the measured curve the outer wings fall down steeper. Analyses undertaken on such curves for T close to T_c show that

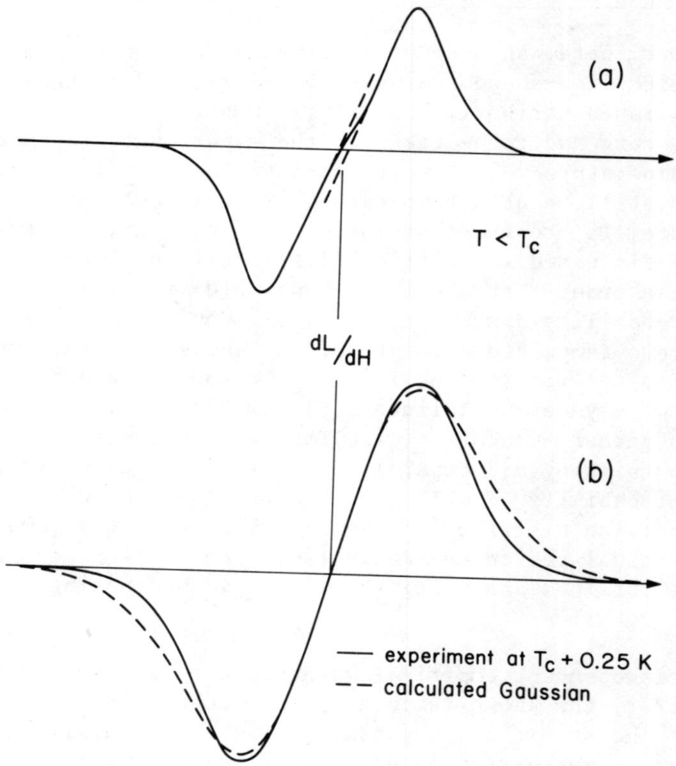

Figure 13. Experimental line-shape derivative of the $Fe^{3+}-V_0$ center for $H \parallel [110]$ at K-band $g = 4.5$ a) for $T = T_c - 0.1$ K, and b) for $T = T_c + 0.25$ K and computed Gaussian with the same width at maximum slope.

$P(\varphi_\ell)$ is better described by adding a term $e^{-\varphi_\ell^4}$. Close to T_c, up to 20 % of a normalized function of this form had to be used. Further analysis is underway right now including computation of even moments of $L(H)$. Such observations on local fluctuations have not been made in any other system to the best of our knowledge. The reason for the deviation appears to be the same as the occurrence of an asymmetry just below T_c. The latter result from cooperativity and thus the long-range correlations present. However another possibility must also be considered here. This is the finite local rotation φ_ℓ. The maximum average $\langle\varphi\rangle$ at $T = 0$ has been measured to be 2.1^o (2). Thus near T_c, φ_ℓ can fluctuate at least by this amount, but probably up to 4°. In IV.A) at T_c the mean fluctuation $(\langle\delta\varphi_c^2\rangle)^2 = 0.4$ was obtained. Therefore this quantity is an order of magnitude smaller than the extrema in φ_ℓ and the marked deviation from a Gaussian in figure 13(b), which is of the order of 20 %, is unlikely to result from the finite range in φ_ℓ, but rather due to the cooperative anharmonic potential.

VI. CONCLUSIONS

The width $\Delta\omega$ of the spectral density $J(\omega)$ of the local order parameter $\varphi_\ell(t)$ is strongly temperature dependent near T_c. This is concluded from studies of the temperature dependence of the EPR linewidth of the Fe^{3+}-V_O center and the intensity ratio of orthorhombic and tetragonal Ti^{3+} EPR spectra. The $\Delta\omega(T)$ dependence is compatible with an equation $\Delta\omega/2\pi = \Delta\nu_0\ [(T - T_c)/T_c]^\gamma$ with $\gamma = 1.28$, $\Delta\nu_0 = 1 \times 10^{10}$ Hz. $\Delta\omega$ should, within a factor of three, correspond to the width Γ_c of the central peak in $S(\vec{q},\omega)$ at $\vec{q}_R$. Thus the latter must be temperature dependent. Therefore the central peak whose frequency width Γ_c has not been resolved in neutron scattering is an intrinsic property. Were the central peak created by impurities (31) then it would not be a critical quantity near T_c. At T_c the mean spread $(\langle\delta\varphi_c^2\rangle)^{1/2}$ of the local rotational order parameter is 0.4°. In a sample transforming to a monodomain the mean spread of rotation perpendicular to the c-axis $(\langle\delta\varphi_a^2\rangle)^{1/2}$ is 0.15°. The values agree with recent birefringence measurements. They prove that with the EPR of the Fe^{3+}-V_O center near T_c a regime is reached, where the motions of $\varphi_\ell(t)$ are slow compared to the measuring time τ_m (instantaneous snapshot).

From the analysis of the line shapes of the Fe^{3+}-V_O center in the slow-motion regime above and below T_c it is found that the probability $P(\varphi_\ell)$ to observe φ_ℓ deviates from a Gaussian above and below T_c. This is ascribed to the existence of long-range correlations in the critical regime. For $T > T_c$, $P(\varphi_\ell)$ is better described by adding an $e^{-\varphi_\ell^4/\langle\varphi_\ell^4\rangle}$-dependent term. For $T < T_c$ where $\langle\varphi\rangle \neq 0$, and the symmetry is broken, $P(\varphi_\ell)$ is asymmetric. The asymmetry a_s is a critical quantity diverging for $T \to T_c^-$ like $a_s \propto \varepsilon^{-1}$.

To conclude, one can state the following: From the determination of the mean width in local spectral density $J(\omega)$ one gets an estimate of the width of the central peak. Despite the usefulness of this result it would appear that if $S(\vec{q},\omega)$ could be determined with sufficient resolutions, this would be superior to the investigations of local fluctuations. The main reason is that the correlation length ξ is becoming very long near T_c and is best probed by measuring wave-vector dependencies. Such an argument is however only partially true because due to these long-range correlations the statistics of the local fluctuations are changed. Well away from T_c the statistics follow Gaussian distributions in the time and therefore in the ensemble. Near T_c they are fundamentally altered. Thus the observation of local order-parameter fluctuations is complementary in nature to high-resolution information on $S(\vec{q},\omega)$ if the latter become available.

ACKNOWLEDGEMENTS

The author has benefited highly from discussions with P. Helle: and Th. von Waldkirch as well as from J. Axe, M. Blume, E. Courtens J. Feder, A. Rigamonti, O. Schirmer, T. Schneider, G. Shirane, H. Schwabl, M. Szabo, H. Thomas and many others interested in this work. He is grateful to Th. von Waldkirch for allowing him to reproduce an unpublished figure from his thesis. W. Berlinger recorded the new data of figure 13 and integrated the line shape derivatives. C. West performed preliminary APL computer fits to the line shapes close to and above T_c. He is indebted to R.A. Cowley, J. Feder, and M. Szabo for sending him preprints of their recent work.

Note Added After the Conference

Heller (32) has calculated Γ_c from the amplitude ΔH_f of the extrapolated fast-motion linewidth (straight line in figure 4). That this is possible, in principle, is easily recognized from equations (3), (7), and (8) and has also been suggested by Schwabl at this conference. However using equation (8) requires knowledge of the absolute magnitude of the susceptibility $\chi(\vec{q},\omega) = \chi_o/(q_x^2 + q_y^2 + \Delta q_z^2 + \kappa^2) = \langle\varphi_q^2 = q_R\rangle/\ kT$. Noting that in the magnetic system FeF_2 an analogous procedure gives correct results using (φ or magnetic order parameter)

$$\langle\varphi^2_{\vec{q}\,=\,\vec{q}_R}\rangle = C^+\ \varepsilon^{-\gamma}\ \langle\varphi_\ell(T = 0)\rangle^2 \quad \text{with} \quad C^+ \cong 1 \qquad (22)$$

and employing values γ, ν and κ_o given in these lectures, he estimates $\Gamma_c \simeq \Delta\omega/s$ with s between 2.6 and 4. One can also obtain $\langle\varphi^2_{\vec{q}\,=\,\vec{q}_R}\rangle$ from $(\langle\delta\varphi_c^2\rangle)^{1/2} = 0.4^o$ at T as given in section IV.A) and the q-width of the central peak as measured by Shapiro et al. (13). With this procedure one arrives at the same result for $\langle\varphi^2_{\vec{q}\,=\,\vec{q}_R}\rangle$ (33) as equation (22). These findings support well the conclusions made in the two lectures.

REFERENCES

1. For a discussion on the nomenclature used see: H. Gränicher and K.A. Müller, Mat. Res. Bull., 6, 977 (1971).

2. K.A. Müller in Structural Phase Transitions and Soft Modes edited by E.J. Samuelsen, E. Andersen and J. Feder (Oslo, Universitetsforlaget, 1971) p. 85.

3. K.A. Müller and W. Berlinger, Phys. Rev. Letters, 26, 13 (1971).

4. E.F. Steigmeier and H. Auderset, Solid State Commun., 12, 565 (1973).

5. E.S. Kirkpatrick, K.A. Müller and R.S. Rubins, Phys. Rev., A86, 135 (1964).

6. Th. von Waldkirch, K.A. Müller and W. Berlinger, Phys. Rev. B 5, 4324 (1972).

7. A. Abragam, The Principles of Nuclear Magnetism (Oxford, Clarendon, 1961) Chapter 10.

8. A.M. Gottlieb and P. Heller, Phys. Rev. B 3, 3615 (1971).

9. Th. von Waldkirch, K.A. Müller, W. Berlinger and H. Thomas, Phys. Rev. Letters, 28, 503 (1972).

10. Th. von Waldkirch, K.A. Müller and W. Berlinger, Phys. Rev. B 7, 1052 (1973).

11. H. Schwabl, Phys. Rev. Letters, 28, 500 (1972); Z. Physik, 254, 57 (1972).

12. T. Riste, E.J. Samuelsen, K. Otnes and J. Feder, Solid State Commun., 9, 1455 (1971).

13. S.M. Shapiro, J.D. Axe, G. Shirane and T. Riste, Phys. Rev. B 6, 4332 (1972); W.G. Stirling, J. Phys. C 5, 2711 (1972).

14. A. Rigamonti (private communication).

15. K.A. Müller, W. Berlinger, M. Capizzi and H. Gränicher, Solid State Commun., 8, 549 (1970).

16. J.K. Kjems, G. Shirane, K.A. Müller and H.J. Scheel, Phys. Rev., (to be published).

17. H.E. Stanley in Ref. 2, p. 271.

18. T. Schneider and E. Stoll in Ref. 2, p. 383; T. Schneider, R. Shrinivasan and E. Stoll, Helv. Phys. Acta, 45, 629 (1972).

19. K. Fossheim and B. Berre, Phys. Rev. B 5, 3292 (1972).

20. W. Rehwald, Solid State Commun., 8, 607 (1970).

21. O.F. Schirmer and K.A. Müller, Phys. Rev. B 7, 2986 (1973).

22. The motion along the domain axis is hindered by an activation energy of 11 meV. This was determined from electric-field induced intensity changes (see Ref. 21).

23. T. Schneider, Phys. Rev. B 7, 201 (1972).

24. H. Thomas and K.A. Müller, Phys. Rev. Letters, 21, 1256 (1968)

25. E. Pytte and J. Feder, Phys. Rev., 187, 1077 (1969); J. Feder and E. Pytte, Phys. Rev. B 1, 4803 (1970).

26. R.A. Cowley and A.D. Bruce (to be published).

27. M. Szabo, M. Droz and A. Malaspinas (to be published).

28. E. Courtens, Phys. Rev. Letters, 29, 1380 (1972).

29. K.A. Müller and W. Berlinger, Phys. Rev. Letters, 29, 715 (1972).

30. A starting point is a recent paper by J. Feder (to be published). There in a local microscopic theory, coupling of order-parameter and entropy fluctuations is considered.

31. This possibility has been mentioned recently, J. Axe (this School).

32. P. Heller (private communication).

33. P. Heller and K.A. Müller (to be published).

HYDRODYNAMICS NEAR PHASE TRANSITIONS OF SECOND ORDER+

F. Schwabl

Institut für Festkörperforschung, KFA Jülich GmbH, D. 517 Jülich, Germany and Institut für Physik, Hochschule Linz, A-4045 Linz, Austria

I. INTRODUCTION

On approaching T_c, the correlation length ξ of the order parameter fluctuations becomes infinite. The increase of the range of the correlated regions implies that the characteristic time for the evolution of the order parameter also becomes infinite. Near the transition, microscopic details are irrelevant, and the structure of the dynamics depends only on the symmetry of the system.

There are two basic characteristics which follow solely from the symmetry of the Hamiltonian. Firstly, the order parameter may or may not be a conserved quantity. Secondly, the symmetry broken in the low temperature phase may be either discrete or continuous. Combining these two characteristics it is convenient to classify the various systems into the following groups:

I) The broken symmetry is continuous
   a) Order parameter is conserved
      (Heisenberg ferromagnet)
   b) Order parameter is not conserved
      (Heisenberg antiferromagnet, Lambda transition of He^4)

II) The broken symmetry is discrete
   a) Order parameter is conserved
      (Uniaxial ferromagnet, liquid-gas transition)
   b) Order parameter is not conserved
      (Uniaxial antiferromagnet, $SrTiO_3$, ferroelectric transition)

+ Work supported by the Fonds zur Förderung der wissenschaftlichen Forschung, Austria

Table I: Broken symmetry and critical mode

	order parameter	broken symmetry	critical mode
uniaxial ferromagnet	$\langle M^z \rangle$	rotation by π	spin diffusion
isotropic ferromagnet	$\langle \vec{M} \rangle$	rotation	spin wave
superfluid Helium	$\langle \Psi \rangle$	gauge invariance	second sound
uniaxial antiferromagnet	$\langle N^z \rangle$	rotation by π	relaxation of $N^z_{\vec{q}}$
isotropic antiferromagnet	$\langle \vec{N} \rangle$	rotation	spin wave

In most of the lectures of this meeting, the case IIb will be considered. In this lecture we shall review the other three, where the order parameter is conserved or a continuous symmetry broken. The relevant dynamics is then hydrodynamics, which is an exact description of the long time behavior of the densities of the conserved variables.

In section II I will discuss the critical dynamics of the Heisenberg ferromagnet and introduce the concept of dynamical scaling. In section III I will discuss the critical dynamics of the antiferromagnet and in section IV I will consider briefly the anisotropic ferromagnet. Since spin models bear more relation to the main topics of this school, we shall emphasize them and mention the lambda transition of Helium only very briefly in section III. In this lecture I shall not attempt to derive the hydrodynamic equations for any of these systems. The derivation of hydrodynamic and other collective excitations by the Mori theory will be discussed in the next lecture. For the static quantities the reader is referred to Ref.(1) and (2).

II. ISOTROPIC FERROMAGNET

Let us first consider the isotropic Heisenberg ferromagnet, which in order to be specific we describe by the Hamiltonian

$$H = -\frac{1}{2}\sum_{\vec{\ell},\vec{\ell}'} J(\vec{\ell}-\vec{\ell}')\, \vec{S}_{\vec{\ell}}\vec{S}_{\vec{\ell}'} \tag{2.1}$$

where $J(\vec{\ell}-\vec{\ell}')$ are the exchange constants and $\vec{\ell}$ denotes the lattice sites. As a consequence of the rotational invariance of the Hamiltonian, the total magnetization

$$\vec{M} = \sum_{\vec{\ell}} \vec{S}_{\vec{\ell}} \tag{2.2}$$

is a constant of motion.

The hydrodynamic equations for the local magnetization $\vec{M}(\vec{x})$ are diffusion equations for $T > T_c$

$$\vec{M}(\vec{x}) = D_M \nabla^2 \vec{M}(\vec{x}). \tag{2.3}$$

These describe the dynamics for long wave-length and small frequency. It is convenient to introduce Fourier transformed variables

$$\vec{M}_{\vec{q}} = \int d^3x\, e^{-i\vec{q}\vec{x}}\, \vec{M}(\vec{x})$$

in terms of which (2.3) is written as

$$\vec{M}_{\vec{q}} = -q^2 D_M\, \vec{M}_{\vec{q}}\ . \tag{2.4}$$

The hydrodynamic excitations above T_c are three diffusion modes characterized by

$$\omega = -iD_M q^2\ . \tag{2.5}$$

In the low temperature phase the continuous rotation symmetry is broken. Suppose that the direction of magnetization is in the positive z-direction. (For the proper definition of the Bogoliubov quasi averages see Ref.(3).) Then the hydrodynamic equations have the form

$$\frac{d}{dt} M^{\genfrac{}{}{0pt}{}{x}{y}}_{\vec{q}} = \pm\, \omega_o(q)\, M^{\genfrac{}{}{0pt}{}{y}{x}}_{\vec{q}} - q^4 \Lambda M^{\genfrac{}{}{0pt}{}{x}{y}}_{\vec{q}} \tag{2.6a}$$

$$\frac{d}{dt} M^z_{\vec{q}} = -Dq^2 M^z_{\vec{q}} \quad . \tag{2.6b}$$

Whereas the z component obeys still a diffusion equation, the broken continuous symmetry implies an oscillatory mode for the x and y components. The excitation spectra of these three equations are

$$\omega = \pm\, \omega_o(q) - i\Lambda q^4 \tag{2.7a}$$

$$\omega = -\, iDq^2 \quad . \tag{2.7b}$$

The frequency $\omega_o(q)$ is determined by static quantities

$$\omega_o(q) = \frac{M}{\chi^T_q} \quad . \tag{2.8}$$

where $M = \langle M^z \rangle$ is the average magnetization and χ^T_q the transverse susceptibility:

$$\chi^T_q = \frac{bM^2\xi}{q^2} \quad . \tag{2.9}$$

Here ξ is the correlation length and b a coefficient of order one. The hydrodynamic equations (2.6) can either be derived from the Bloch equations (4) or by use of the Mori theory (5)(6)(7). In going through T_c the oscillatory spin wave mode (2.7a) has to transform into the diffusive mode (2.5). There will be a region close to T_c, where both the diffusion and spin wave pictures fail. A fundamental requirement for hydrodynamics to hold is that the system is in <u>local equilibrium</u> (8). This immediately implies that the hydrodynamic equations apply only for disturbances of wave lengths which are much larger than the correlation length, $q^{-1} \gg \xi$

$$\xi = \xi_o^< \varepsilon^{-\nu'} \;(T < T_c) \;, \quad \xi = \xi_o^> \varepsilon^{-\nu} \;(T > T_c) \tag{2.10}$$

with $\varepsilon = |1-T/T_c|$. (We restrict ourselves to the case of zero external field.) This replaces the standard condition $q \ll a^{-1}$, where a is a microscopic distance.

The different dynamical regimes are most easily discussed by referring to the diagram in Fig. 1.(9) The inverse correlation length ξ^{-1} is plotted on the abcissa and the wave number q on the ordinate. The high temperature phase ($T > T_c$) is plotted on the right and the low temperature phase ($T < T_c$) is plotted on the left. The origin $\xi^{-1} = 0$ is the critical point $T = T_c$. In the shaded region I the condition $q\xi \ll 1$ is fulfilled and the hydro-

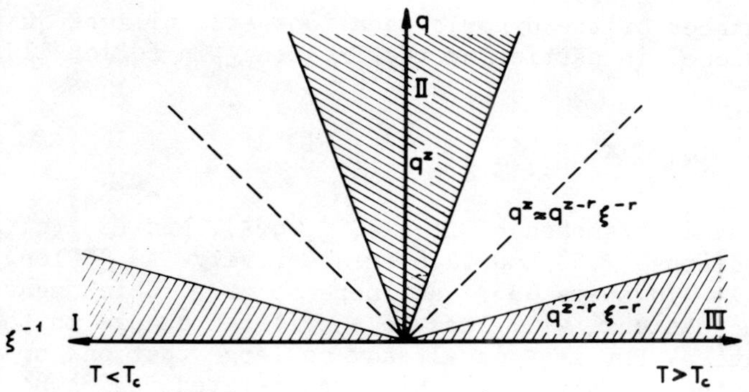

Fig. 1: The $q-\xi^{-1}$ plane showing the three characteristic regions. (I and III hydrodynamic and II non hydrodynamic)

dynamic equations (2.6) apply. The same is true in the high temperature side for the shaded region III, where the diffusion equations apply.

In the so-called non-hydrodynamic critical region II, in which $q\xi >> 1$, hydrodynamics is no longer valid. Although the wave lengths we consider are large on the usual microscopic scale, they are small compared to the critical correlation length in II.

The problem is now to find the structure of the dynamics in region II and in the regimes joining I and II, and II and III. The second problem is the behavior of the coefficients in the hydrodynamic equations, D_M, Λ etc., in the hydrodynamic regimes. The phenomenological theory which gives an answer to these questions is the dynamical scaling hypothesis, which I would like to make plausible with the following remarks.

First of all let me note that the spin wave frequency $\omega_0(q)$ has the following dependence on q and ξ:

$$\omega_0(q) = D q^2 \propto q^2 \xi^{-\frac{1}{2}(1-\eta)} = q^{\frac{5-\eta}{2}} (q\xi)^{-\frac{1}{2}(1-\eta)} . \quad (2.11)$$

This follows from inserting (2.9) into (2.8) and use of the static scaling law (1)(2) $2\beta - \nu' = \eta\nu'$. The stiffness constant of the spin wave frequency decreases like the inverse square root of the correlation length on approaching T_c.

Let me denote the complex resonance frequency which describes the excitations in the whole (macroscopic) $q-\xi^{-1}$plane by $\Omega(q,\xi)$. In region II the correlation length is practically ∞, which is of course exactly true only for $T = T_c$. Hence static correlation functions, resonances of the dynamic equations etc. will be just functions of q alone, in particular the frequency in region II is of the form

$$\Omega_{II}(q,\xi) \approx \Omega(q,\infty) \equiv \Omega_{II}(q).$$

The hydrodynamic branches of $\Omega(q,\xi)$, $\Omega_I(q,\xi)$, and $\Omega_{III}(q,\xi)$ are given by equations (2.5) and (2.6) respectively. As long as the wave length $\lambda = 1/q$ obeys $\lambda \gg \xi$ an increase of ξ, which measures the range of fluctuations, will lead to a critical change in the dynamical quantities. But if ξ is already so large that one or several wave lengths lie already within a correlated region no drastic change can be expected by further increase of ξ. This makes plausible that the critical temperature variation of $\Omega_I(q,\xi)$ and $\Omega_{III}(q,\xi)$ implied by the dependence on ξ, terminates when ξ^{-1} reaches the value $\xi^{-1} \approx q$. This suggests the <u>matching conditions</u>

$$\Omega_{II}(q) = d_I \Omega_I(q, \xi = \frac{1}{q}) \qquad (2.12a)$$

$$\Omega_{II}(q) = d_{III} \Omega_{III}(q, \xi = \frac{1}{q}) \qquad (2.12b)$$

between the asymptotic forms of $\Omega(q,\xi)$. d_I and d_{III} are constants of order one. Equations (2.12) are one possible formulation of the <u>dynamical scaling hypothesis</u>. This formulation was introduced in the context of the lambda transition of Helium (10).
Using (2.12a) we find (9)

$$\mathrm{Re}\,\Omega_{II}(q) = a\, q^z \qquad (2.13)$$

with

$$z = \frac{5-\eta}{2} \qquad (2.14)$$

Since the response changes from a propagating to a diffusive type in going through T_c, the resonances must be overdamped at T_c. This suggests that $\mathrm{Im}\,\Omega_{II}(q)$ is a homogenous function of the same degree as $\mathrm{Re}\,\Omega_{II}(q)$. Using the matching condition for the imaginary part we find

$$\Lambda \propto \xi^{\frac{3+\eta}{2}} \qquad (2.15)$$

and

$$D_M \propto \xi^{-\frac{1-\eta}{2}} \qquad (2.16)$$

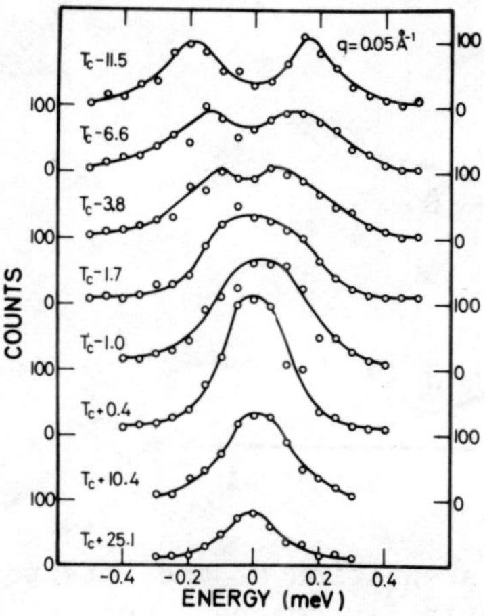

Fig. 2: Neutron scattering cross section for iron, showing the transition from spin wave to diffusive response. After Collins et al (13).

Table II: Dynamical critical exponents

		theory	iron	nickel
$T < T_c$	$\mathcal{D} \propto \varepsilon^{\sigma'}$	$\sigma' = \frac{\nu'}{2}(1-\eta) \approx 1/3$	0,37 ± 0,03	0,39 ± 0,04
	$\Lambda(q) \propto q^4 \varepsilon^{-b}$	$b = \nu'(\frac{3+\eta}{2}) \approx 1$	?	?
$T = T_c$	$\Omega_{II}(q) \propto q^z$	$z = \frac{5-\eta}{2}$	2,7 ± 0,3	2,46 ± 0,25
$T > T_c$	$D_M \propto \varepsilon^{\sigma}$	$\sigma = \frac{\nu}{2}(1-\eta) \approx 1/3$	0,36 ± 0,06	0,51 ± 0,05

The first critical neutron scattering experiments were carried out by Riste et al (11) and Jacrot et al (12). In table II the theory is compared with the experimental results on iron and nickel of Collins et al, Minkiewicz et al, and the Saclay group (13)(14)(15). See Fig. 2.

If there are no logarithmic singularities the matching condition implies that $\Omega(q,\xi)$ is a homogenous function

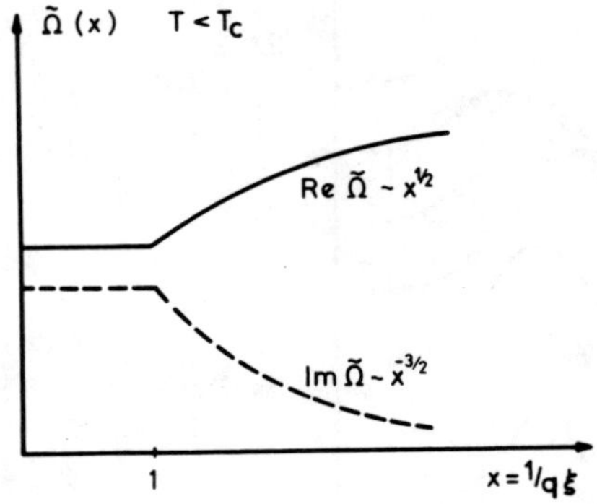

Fig. 3: Schematic representation of the interpolation function $\tilde{\Omega}(x)$ for the isotropic ferromagnet, neglecting the exponent η. In this plot the asymptotic laws are extended right to the point $x=(q\xi)^{-1}= 1$.

$$\Omega(q,\xi) = q^z \tilde{\Omega}(1/q\xi) \quad . \tag{2.17}$$

For this case the dynamical scaling hypothesis can be generalized by replacing the matching condition (2.12) by equation (2.17) and by allowing for general interpolation functions $\tilde{\Omega}(x)$.(9)
($\lim_{x\to\infty} \tilde{\Omega}(x)$ = const; in the limit $x\to 0$ $\tilde{\Omega}(x)$ behaves such that the q dependence of $\Omega(q,\xi)$ coincides with the hydrodynamical solutions).
The realistic interpolation function $\tilde{\Omega}(x)$ will be smoother and can have more structure than the function shown in Fig. 3, which is obtained from matching precisely at x = 1.

Having introduced the scaling hypothesis for the eigenmodes, we can easily obtain a scaling structure for the correlation functions. The <u>dynamic</u> <u>form</u> <u>factor</u> of the transverse spin components has for $T < T_c$ the structure

$$S^{xx}(q,\omega,\xi) = T\chi^T_{\vec{q}} \left\{ \frac{\mathrm{Re}\Lambda(q,\xi,\omega)}{[\omega-\omega_o(q,\xi)-\mathrm{Im}\Lambda(q,\xi,\omega)]^2+[\mathrm{Re}\Lambda(q,\xi,\omega)]^2} + \right.$$

$$\left. + (- \leftrightarrow +) \right\} \tag{2.18}$$

The resonances of the system are given by the zeros of the denominator of equation (2.18). Inserting for ω $\Omega(q,\xi) = q^z \tilde{\Omega}(1/q\xi)$ one sees that $\Lambda(q,\xi,\omega)$ must be of the form

$$\Lambda(q,\xi,\omega) = q^z \Lambda'(q\xi,\omega/q^z).$$

Hence the homogeneity property of the resonances implies that

$$S^{xx}(q,\omega,\xi) = q^{-2+\eta-z} \bar{S}^{xx}(q\xi,\omega/q^z) . \qquad (2.19)$$

Consequently $S^{xx}(q,\omega,\xi)$ changes under the scale transformation $q\to Lq$, $\omega\to L^z\omega$ and $\varepsilon\to L^{1/\nu}\varepsilon$ by the scale factor $L^{-2+\eta-z}$:

$$S(Lq, L^z\omega, L^{-1}\xi) = L^{-2+\eta-z} S(q,\omega,\xi) \qquad (2.20)$$

III. ISOTROPIC ANTIFERROMAGNET

As a second example, let us consider the isotropic antiferromagnet, which is also described by the Hamiltonian (2.1). Now the exchange constants in Eq.(2.1) are negative such that they favour alternating alignment of the spins on the two sublattices. The order parameter in this case is the staggered magnetization

$$\vec{N} = \sum_{\vec{\ell}} e^{i\vec{q}_o\vec{\ell}} \vec{S}_{\vec{\ell}} \qquad (3.1)$$

where $\vec{q}_o$ is the super-lattice vector. As in the ferromagnet, the total magnetization is a hydrodynamic variable. Below T_N, a component of N, say N^z, assumes a finite value. This breaks the rotational symmetry of the Hamiltonian and consequently the transverse susceptibility (the susceptibility of the components $N^x_{\vec{q}}$ and $N^y_{\vec{q}}$) becomes infinite as $\vec{q}\to 0$. (3.2)

This or an equivalent property implies that $N^x_{\vec{q}}$ and $N^y_{\vec{q}}$ obey hydrodynamic equations below T_N, although they do not obey a microscopic conservation law. The hydrodynamic equations give spin wave excitations

$$\omega = \pm\,\omega(q) - \frac{i}{2} Dq^2 \qquad (3.3)$$

with

$$\omega(q) = \frac{\langle N^z\rangle}{\sqrt{\chi^{TM}(q)\chi^{TN}(q)}} \propto \xi^{-\frac{1}{2}} q \qquad (3.4)$$

The staggered transverse susceptibility is given by an expression like (2.9) with M replaced by $\langle N^z\rangle$. The uniform transverse sus-

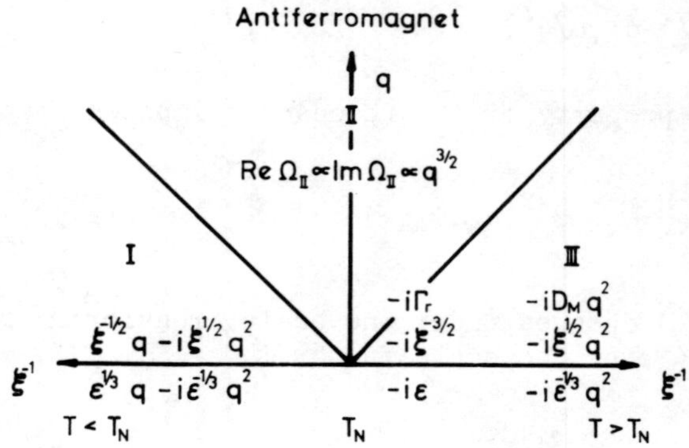

Fig. 4: The asymptotic q and $\xi(\varepsilon)$ dependence of the critical modes in the isotropic antiferromagnet. In the line below the abscissa ν and ν' are replaced by 2/3. Only the dependence on q and ξ or ε is shown and constant factors have been omitted.

ceptibility χ^{TM} is finite at T_N for external field H = 0. The dynamical scaling hypothesis gives in the non hydrodynamic region

$$\mathrm{Re}\Omega_{II}(q) \propto \mathrm{Im}\Omega_{II}(q) \propto q^{3/2} \qquad \text{(i.e.: } z = 3/2\text{)}.$$

Consequently the damping coefficient D is divergent

$$D \sim \xi^{1/2} \sim \varepsilon^{-\nu'/2} .$$

On the high temperature side scaling predicts for the diffusion coefficient of the magnetization $D \sim \xi^{1/2}$ and for the relaxation rate of the staggered magnetization $\Gamma_r \sim \xi^{-3/2} \sim \varepsilon^{(3/2)\nu}$. Fig. 4 shows the asymptotic laws with ν replaced by 2/3. Lau et al (16) find for $T > T_N$, $\Gamma_r \sim \varepsilon^{-1.4 \pm 0.2}$ and for $T = T_N$ $z = 1.4$. These authors have also tested the homogeneity property of $S(q,\omega)$ for $T = T_N$ ($\xi = \infty$). Then Eq. (2.19) reads

$$S(q,\omega) = q^{-2+\eta-z}\,\tilde{S}(\omega/q^z). \qquad (3.5)$$

Taking into account the corrections coming from the resolution function, the experiments confirm the scaling relation for $S(q,\omega)$. (See Fig. 5).

The <u>lambda transition</u> of He^4 shows similar critical behavior as the antiferromagnet. The new hydrodynamic variable below T_λ

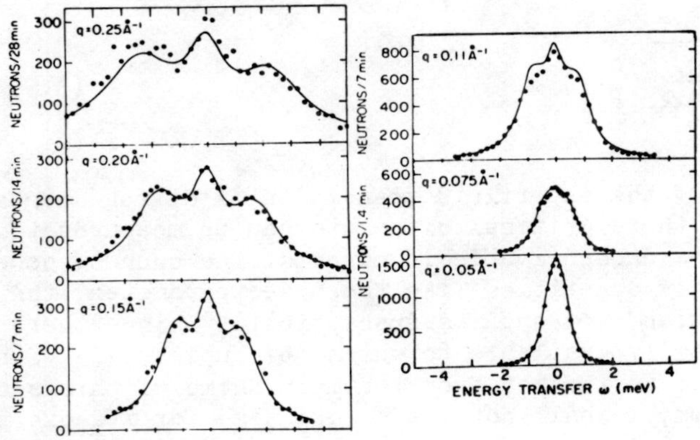

Fig. 5: Neutron scattering cross section for $RbMnF_3$ (Lau et al(16)) at T_N as a function of frequency for various values of q. The finite resolution function masks the three peak structure for small q. The same theoretical function $\tilde{S}(\omega/q^z)$ folded with the resolution function (solid line) is used for all values of q. The parameters of the solid line are fitted to the experimental points in the top diagram.

resulting from the broken gauge invariance is the phase of the order parameter. The critical mode corresponding to the spin wave in the antiferromagnet is <u>second sound</u>, the velocity of which is

$$c_2 = \sqrt{\frac{TS^2\rho_s}{C_P^< \rho_n}} \tag{3.6}$$

where ρ_s, ρ_n and $C_P^{\gtrless}$ are the superfluid and normal density and the specific heat at constant pressure for $T \gtrless T_\lambda$. In the critical region II the characteristic frequency is

$$\omega_{II}(q) \sim q^{3/2}/\sqrt{C_P^<(1/q)} \ . \tag{3.7}$$

By matching the regions II and III

$$D_T(\xi)q^2 = \omega_{II}(q)\Big|_{q = \xi^{-1}}$$

one finds

$$D_T \sim \sqrt{\frac{\xi}{C_P^<(\xi)}} \ . \tag{3.8}$$

This implies the divergence of the thermal conductivity (10)

$$\Lambda_T \sim \sqrt{\frac{\xi}{C_P^<(\xi)}}\; C_P^>(\xi)\ . \tag{3.9}$$

The study of the superfluid transition is of fundamental interest, since there critical exponents can be measured to an accuracy which is unconceivable in magnets. One cannot, however, determine directly quantities like the order parameter, the superfluid wave function $\langle\psi\rangle$, and its susceptibility since there is no field experimentally available to which it couples (17). Since superfluid Helium is further from the main theme of this school than spin systems, I shall not spend more time for its discussion.

IV. ANISOTROPIC MAGNETS

In the isotropic systems, one could determine the dynamical critical exponents by means of the dynamical scaling hypothesis. This was possible since below T_c the critical mode was a propagating hydrodynamic mode, the velocity or stiffness constant of which was expressed by static (thermodynamic) quantities which we assume to be known.

In anisotropic systems like a fluid near T_c or the anisotropic antiferromagnet the structure of the hydrodynamic equations is the same above and below T_c. In the <u>uniaxial ferromagnet</u> the z-component of the magnetization obeys

$$M^z_{\vec q} = -q^2 D\, M^z_{\vec q} \tag{4.1}$$

above and below T_c for $q\xi \ll 1$. The diffusion coefficient is expressed in terms of the magnetic conductivity Λ_M and the susceptibility χ as

$$D = \frac{\Lambda_M}{\chi}\ . \tag{4.2}$$

The dynamical scaling hypothesis gives <u>relations</u> between the critical exponents of D for $T > T_c$ and $T < T_c$ and the exponent z in region II but can determine none of them. The mode coupling theory shows that Λ_M is uncritical hence

$$D \sim \xi^{+2-\eta}$$

and

$$z = 4 - \eta \quad . \tag{4.3}$$

In this case, as is readily understood by the mode coupling theory, the van Hove hypothesis (19) that the Onsager coefficients are uncritical is fulfilled. In the isotropic cases, as we have seen, and in some anisotropic systems the Onsager coefficients are critical. For instance the thermal conductivity near the liquid gas transition is proportional to ξ.

V. CONCLUDING REMARKS

Given the static thermodynamic quantities and the structure of the hydrodynamic equations, we can determine the dynamic response in the whole critical region for those cases where a continuous symmetry is broken. In the anisotropic case additional assumptions about the temperature dependence of the transport-coefficients must be made.

The results of the dynamical scaling hypothesis are confirmed by the mode-coupling theory (20)(21). In the mode coupling theory the transport coefficients are calculated by considering the decay of a hydrodynamic mode into other hydrodynamic modes. The contribution of these processes becomes large on approaching T_c, because either the vertex for the decay process or the energy denominators become large. The mode coupling theory not only allows the calculation of the asymptotic forms of $\Omega(q,\xi)$ but also gives the interpolation function $\tilde{\Omega}(x)$.

The limited space does not allow us to describe the mode coupling theory in any detail. We should like, however, to explain in this section, why the Onsager coefficients are singular for the isotropic ferromagnet and not for the uniaxial ferromagnet. For simplicity, we consider the high temperature phase and start with the Kubo formula for the generalized diffusion coefficient of $M^z_{\vec{q}}$:

$$D^z(q,\omega)q^2 = \frac{1}{\chi_q} \int_0^\infty dt \; e^{i\omega t} \langle \dot{M}^z_{\vec{q}}(t), \; \dot{M}^z_{\vec{q}}(0) \rangle . \tag{5.1}$$

The Green's function is expressed by $D^z(q,\omega)$ through

$$G(\vec{q},\omega) = -i \int_0^\infty dt \; e^{i\omega t} \langle M^z_{\vec{q}}(t), \; M^z_{\vec{q}}(0) \rangle$$

$$= \chi_q \frac{1}{\omega + iD^z(q,\omega)q^2} \tag{5.2}$$

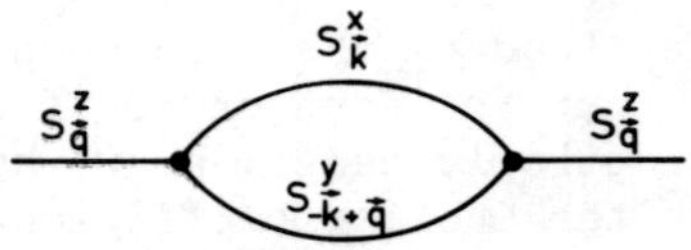

Fig. 6: Decay process of the $M^z_{\vec{q}}$ mode into a $M^x_{\vec{q}}$ and a $M^y_{\vec{q}}$ mode.

The microscopic equation of motion for $M^z_{\vec{q}}$ reads

$$\dot{M}^z_{\vec{q}} = \frac{1}{2N} \sum_{\vec{q}'} (J(\vec{q}') - J(\vec{q}-\vec{q}'))\, M^x_{\vec{q}'}\, M^y_{\vec{q}-\vec{q}'} , \qquad (5.3)$$

where $J(\vec{q})$ is the Fourier transformed exchange interaction for the x and y components which is identical to the exchange interaction of the z component in the isotropic case.

In the uniaxial ferromagnet $M^x_{\vec{q}}$ and $M^y_{\vec{q}}$ are uncritical and hence the Onsager coefficient, which is represented by the time integral in Eq. (5.1), is regular. In the isotropic ferromagnet all components are <u>equivalent</u> and the slowing down manifests itself also in the Onsager coefficient. The decay of the $M^z_{\vec{q}}$ mode into a $M^x_{\vec{k}}$ and $M^y_{\vec{k}-\vec{q}}$ mode (Fig. 6) contributes

$$D^z(\vec{q},\omega)q^2 = \frac{J^2T}{\chi_{\vec{q}}} \sum_{\vec{q}'} \int_{-\infty}^{\infty} \frac{d\omega'}{2\pi} (\vec{q}.\vec{q}')^2\, G(\vec{q}' + \frac{\vec{q}}{2}, \omega' + \frac{\omega}{2})$$

$$G(-\vec{q}' + \frac{\vec{q}}{2}, -\omega' + \frac{\omega}{2}) . \qquad (5.4)$$

If the ω dependence of $D^z(q,\omega)$ is neglected the simpler integral equation ($\Gamma(q) \equiv D^z(q,0)q^2$)

$$\Gamma(\vec{q}) = \frac{J^2T}{\chi_{\vec{q}}} \int \frac{d^3q'}{(2\pi)^3} (\vec{q}'.\vec{q})^2 \frac{\chi_{\vec{q}'+\frac{\vec{q}}{2}}\ \chi_{\vec{q}'-\frac{\vec{q}}{2}}}{\Gamma(\vec{q}'+\frac{\vec{q}}{2}) + \Gamma(\vec{q}'-\frac{\vec{q}}{2})} \qquad (5.5)$$

is obtained, which has been solved numerically by Resibois and Piette (22). In order that the correct dependence of the exponent z on η is found $J^2(\vec{q}.\vec{q}')^2$ is to be replaced by

$$\frac{1}{4}\left[\chi(\vec{q}' + \frac{\vec{q}}{2})^{-1} - \chi(\vec{q}' - \frac{\vec{q}}{2})^{-1}\right]^2 .$$

With this modification the asymptotic results in the hydrodynamic and critical region are in agreement with the dynamical scaling hypothesis as can be seen directly from (5.5).

The mode coupling theory has been applied to the high temperature phase of the ferromagnet and antiferromagnet (23)(24)(28), the low temperature phase of the ferromagnet (26)(27), the liquid gas transition (21), binary mixtures (20)(29), and the lambda transition of liquid Helium (30).

Acknowledgement: I should like to thank Dr. P. Kortman for reading the manuscript.

REFERENCES

1. L.P. Kadanoff, W. Götze, D. Hamblen, R. Hecht, E.A.S. Lewis, V.V. Palciauskas, M. Rayl, J. Swift, D. Aspnes, J. Kane, Rev. Mod. Phys. 39, 315 (1967)

2. M.E. Fisher, Reports on Progress in Physics, Vol. XXX, II, p. 615 (1967)

3. N.N. Bogoliubov, Physica 26, 51 (1960); Phys. Abh. SU 6, 1, 113, 229 (1962)

4. B.I. Halperin, P.C. Hohenberg, Phys. Rev. 188, 898 (1969)

5. F. Schwabl and K.H. Michel, Phys. Rev. B2, 189 (1970)

6. H. Mori, Progr. Theor. Phys. 33, 423 (1965)

7. K. Kawasaki and H. Mori, Progr. Theor. Phys. (Kyoto) 25, 1043 (1961)

8. L.D. Landau, E.M. Lifshits, Fluid Mechanics (Pergamon, New York (1959))

9. B.I. Halperin and P.C. Hohenberg, Phys. Rev. Lett. 19, (1967) 700; Phys. Rev. 177, 952 (1969)

10. R.A. Ferrell, N. Menyhárd, H. Schmidt, F. Schwabl, P. Szépfalusy, Phys. Rev. Lett. 18, 891 (1967), Phys. Lett. 24A, (1967) 493, Ann. Phys. (N.Y.) 47, 565 (1968)

11. T. Riste and A. Wanic, J. Phys. Chem. Solids 17, 318 (1961); T. Riste, J. Phys. Chem. Solids 17, 308 (1960)

12. D. Cribier, B. Jacrot, G. Parette, J. Phys. Soc. Jap. 17, Suppl. B III (1962)

13. M.F. Collins, V.J. Minkiewicz, R. Nathans, L. Passel and G. Shirane, Phys. Rev. 179, 417 (1969)

14. V.J. Minkiewicz, M.F. Collins, R. Nathans and F. Shirane, Phys. Rev. 182, 624 (1969)

15. Le Groupe de Diffusion Inélastique des Neutrons, Saclay, Phys. Lett. 31A, 561 (1970)

16. H.Y. Lau, L.M. Corliss, A. Delapalme, J.M. Hastings, R. Nathans and A. Tucciarone, Phys. Rev. Lett. 23, 1225 (1969)

17. Neutron scattering with large momentum transfer allows the determination of the condensate wave function at low temperature where it is comparatively large. O.K. Harling, Phys. Rev. Lett. 24, 1064 (1970)

18. P.C. Hohenberg, Varenna Lectures on Critical Phenomena (1970) to be published

19. L. van Hove, Phys. Rev. 93, 1374 (1954)

20. K. Kawasaki, Phys. Rev. 150, 291 (1966), Ann. Phys. (N.Y.) 61, 1 (1970)

21. L.P. Kadanoff and J. Swift, Phys. Rev. 166 89 (1968)

22. P. Resibois and C. Piétte, Phys. Rev. Lett. 24, 514 (1970)

23. K. Kawasaki, J. Phys. Chem. Solids 28, 1277 (1967); Progr. Theor. Phys. (Kyoto) 39, 285, 1133 (1968)

24. F. Wegner, Z. Physik 216, 433 (1968), 218, 260 (1969)

25. P. Resibois, and M. DeLeener, Phys. Rev. 178, 806 (1969)

26. J. Villain, Phys. Stat. Sol. 26, 501 (1968)

27. F. Schwabl, Z. Physik 246, 13 (1971)

28. J. Hubbard, J. Phys. C, 4, 53 (1971)

29. R.A. Ferrell, Phys. Rev. Lett. 24, 1169 (1970)

30. L.P. Kadanoff and J. Swift, Ann. Phys. (N.Y.) 50, 312 (1968)

REVIEW OF METHODS FOR OBTAINING A CENTRAL MODE+

F. Schwabl

Institut für Festkörperforschung, KFA Jülich GmbH, D 517 Jülich, Germany and Institut für Physik, Hochschule Linz, A-4045 Linz, Austria

I. INTRODUCTION

The study of structural phase transitions has attracted increasing experimental and theoretical interest in the past few years (1)-(7). Among the most extensively studied are the transitions of ABO_3 pervoskites from the cubic high temperature phase to a tetragonal or trigonal low temperature phase. The most prominent example for this transition is $SrTiO_3$, which undergoes a second order transition at $T_c = 105^{o}K$. Below T_c the TiO_6 octahedra are rotated around one of the cubic axes in an alternating way. In the aluminates, e.g.: $LaAlO_3$, the rotation is around a cube diagonal (see table I), in $KMnF_3$ the low temperature structure is the same as in $SrTiO_3$.

An interesting feature of these transitions is that they exhibit non classical critical fluctuations as demonstrated by EPR (4) and neutron scattering (3) experiments. It is remarkable that the theory of structural transitions was for a couple of years not affected by developments like static scaling (10,11), dynamical scaling (12,13,14) and the mode coupling theory (15,16). It was recognized only recently that such concepts have to be applied also at structural transitions. There are of course structural transitions where the molecular field theory or Landau theory is adequate. For long ranged (dipole) interactions the Landau theory may be valid very close to T_c (17). One may, however, anticipate that non classical critical fluctuations will be found in many structural transitions.

+ Work supported by the Fonds zur Förderung der wissenschaftlichen Forschung, Austria

A second reason which makes transitions like the 105°K transition of $SrTiO_3$ interesting is the existence of a central peak in addition to the propagating mode above T_c. This effect was first revealed by neutron scattering experiments of Riste et al (3) in $SrTiO_3$ and was found later in $KMnF_3$ (18) and $LaAlO_3$ (19).

In the next section some definitions and a Hamiltonian will be introduced. In section 3 we shall briefly review Mori's theory of Brownian motion, which is then used in the following section to derive the dynamic response function. The central peak will be introduced in a semi-phenomenological manner by assuming that there exists a slowly relaxing quantity. In section 5 the dynamic form factor and its resonances will be discussed. Having discussed the general properties of the three pole propagator for $SrTiO_3$, we list in section 6 other situations where a central peak is found. In view of the possibility that part of the central peak in $SrTiO_3$ is an impurity effect we shall discuss in section 7 those experimental consequences of the theory which might help to discriminate between dynamical and static effects. In the final section we give a brief outline of the lattice dynamical theories for the central peak.

II. BASIC DEFINITIONS AND HAMILTONIAN

In the theory of displacive phase transitions, it is usually possible to represent the important degrees of freedom by a small number of normal coordinates. We shall denote the Fourier transforms of these normal coordinates by $\phi^{\alpha}_{\vec{q}}$. In addition to these variables, there are secondary variables, e.g.: the acoustic modes and other non critical modes at a ferroelectric transition or at the 105°K transition of $SrTiO_3$.
Let us denote the operator conjugate to $\phi^{\alpha}_{\vec{q}}$ by $P^{\alpha}_{\vec{q}}$

$$P^{\alpha}_{\vec{q}} = I^{\alpha}(\vec{q})\dot{\phi}^{\alpha}_{\vec{q}} \tag{2.1}$$

where the generalized mass density $I^{\alpha}(\vec{q})$ is in general non local, i.e.: dependent on $\vec{q}$.

In order to illustrate these definitions let me note that for the <u>ferroelectric transition</u> in $BaTiO_3$ the collective coordinate is approximately given by

$$\phi^{\alpha}_{\vec{q}} = \frac{1}{\sqrt{N}}\sum_{\vec{\ell}} e^{-i\vec{q}\vec{\ell}}\, x^{\alpha Ti}_{\vec{\ell}} \tag{2.2a}$$

where $x^{\alpha Ti}_{\vec{\ell}}$ is the displacement of the Ti-ions. The mass density $I^{\alpha}(\vec{q})$ is equal to the titanium mass in this case.

At the <u>105°K transition of $SrTiO_3$</u> and similar systems the coordinate $\phi^{\alpha}_{\vec{q}}$ describes the <u>staggered rotation</u> of the oxygen octahedra. Let us denote the microscopic angle of rotation of the octahedron at lattice site $\vec{\ell}$ around the cube axis $\hat{e}^{\alpha}$ by $\phi^{\alpha}_{\vec{\ell}}$. Then $\phi^{\alpha}_{\vec{q}}$ is given by

$$\phi^{\alpha}_{\vec{q}} = \frac{1}{\sqrt{N}}\sum_{\vec{\ell}} e^{-i(\vec{q}+\vec{q}_R)\vec{\ell}}\,\phi^{\alpha}_{\vec{\ell}}. \qquad (2.2b)$$

with $\vec{q}_R = (\pi/a,\pi/a,\pi/a)$. The mass density is in this case the generalized moment of inertia of Pytte and Feder (6):

$$I^{\alpha}(\vec{q}) = I\,\frac{1}{2}\Big(1 - \frac{1}{2}\cos q_{\alpha}a + \frac{1}{2}\sum_{\lambda=1}^{3}\cos q_{\lambda}a\Big) \qquad (2.3)$$

with $I = \frac{1}{2}M_o a^2$ given in terms of the oxygen mass M_o and the lattice constant a.

The total Hamiltonian can be decomposed into a part containing the primary variables and a part containing the secondary variables plus an interaction term:

$$H = \sum_{\vec{q}\alpha} \frac{1}{2I^{\alpha}(\vec{q})} P^{\alpha}_{\vec{q}} P^{\alpha}_{-\vec{q}} + \sum_{n}\frac{1}{n!}\sum_{\alpha_1 \ldots \alpha_n} V^{(n)}\begin{pmatrix}\alpha_1 \ldots \alpha_n\\ \vec{q}_1 \ldots \vec{q}_n\end{pmatrix} \phi^{\alpha_1}_{\vec{q}_1} \ldots \phi^{\alpha_n}_{\vec{q}_n} +$$

$$H(\text{secondary}) + H_{int} \qquad (2.4)$$

We shall consider harmonic interactions of the type

$$V^{(2)}\begin{pmatrix}\alpha_1 & \alpha_2\\ \vec{q}_1 & \vec{q}_2\end{pmatrix} = \Big(c\vec{q}_1^{\,2} - v^{\alpha_1}(\vec{q}_1)\Big)\Delta(\vec{q}_1+\vec{q}_2)\,\delta^{\alpha_1\alpha_2} \qquad (2.5)$$

where c is a constant and $v^{\alpha}(o)>0$. The first term in Eq.(2.5) comes from the interaction of the displacements in different cells. The second term leads together with a positive definite four phonon interaction

$$\sum_{\vec{q}_i} V^{(4)}\begin{pmatrix}\alpha_1 \ldots \alpha_4\\ q_1 \ldots q_4\end{pmatrix} \phi^{\alpha_1}_{\vec{q}_1}\phi^{\alpha_2}_{\vec{q}_2}\phi^{\alpha_3}_{\vec{q}_3}\phi^{\alpha_4}_{\vec{q}_4} > 0 \qquad (2.6)$$

to a local double well potential.

In some cases it is convenient to use instead of the oscillator variables ϕ^{α} and P^{α} quasi spin variables. In particular if the double well potential is very deep one takes into account only the two lowest (symmetric and antisymmetric) levels. The effective spin $\frac{1}{2}$ Hamiltonian derived from Eq.(2.4) has the form (20)-(23)

$$H = -\frac{1}{2}\sum_{\vec{\ell}\,\vec{\ell}'} J(\vec{\ell}-\vec{\ell}')S^z_{\vec{\ell}}\,S^z_{\vec{\ell}'} - h^x\sum_{\vec{\ell}} S^x_{\vec{\ell}} \qquad (2.7)$$

The interaction with other phonons can also be considered (24). The elimination of the uncritical degrees of freedom leads to renormalized coupling coefficients in (2.4) or (2.7).

One of the standard theoretical approaches used in the theory of structural transitions is the random phase approximation (RPA) or time dependent Hartree approximation. The RPA gives a valid description as long as one is not too close to T_c. The critical exponents are the molecular field theory exponents, e.g.: $\gamma = \gamma' = 1$, $\beta = \frac{1}{2}$. In the RPA the short distance correlations are not treated self-consistently. In the self-consistent Hartree approximation the results get worse. Then one finds spherical model exponents, $\gamma = 2$, and more over the transition becomes of 1st order. This fact is known from spin systems (25) and has been discussed recently in context with structural transitions (26).

In the following sections we shall describe a theory which relates the dynamical properties to exact static properties. We imagine that the static correlation functions and critical exponents are known from numerical calculations or from the Wilson expansion (27).

III. MORI'S IDENTITY OR THEORY OF BROWNIAN MOTION

In the Mori theory (28)-(30) one decomposes the variables (operators) characterizing the system into two groups. The first group contains a finite number of collective variables $X^c_{\vec{q}}$ and the second group the "fluctuating" non collective variables. It is usually convenient to redefine these variables such that they are orthonormal

$$<X^c_{\vec{q}}, X^{c'}_{\vec{q}}> = \chi^{cc'}(\vec{q}) = \delta^{cc'}, \qquad (3.1)$$

where the scalar product is given by the static susceptibility $\chi^{cc'}(\vec{q})$.

Let me first write down the equations of motion for the collective variables:

$$\dot{X}^c_{\vec{q}} = -iC^{cc'}(\vec{q})\, X^{c'}_{\vec{q}} + f^c_{\vec{q}} . \qquad (3.2)$$

Here the force acting on the right hand side has been decomposed into a collective part and a fluctuating part $f^c_{\vec{q}}$. The requirement that $f^c_{\vec{q}}$ is no more proportional to $X^{c'}_{\vec{q}}$ gives

$$C^{cc'}(\vec{q}) = i<\dot{X}^c_{\vec{q}}, X^{c'}_{\vec{q}}> = <[X^c_{\vec{q}}, X^{c'\dagger}_{\vec{q}}]>. \qquad (3.3)$$

Then one can show that the dynamical susceptibility $\chi^{cc'}(\vec{q},\omega)$ of the collective variables $X^c_{\vec{q}}$ and $X^{c'}_{\vec{q}}$ is given by

$$\overleftrightarrow{\chi}(\vec{q},\omega) = \left[-\overleftrightarrow{C}(\vec{q}) + i\overleftrightarrow{\Gamma}(\vec{q},\omega)\right]\left[\omega\overleftrightarrow{1} - \overleftrightarrow{C}(\vec{q}) + i\overleftrightarrow{\Gamma}(q,\omega)\right]^{-1} \qquad (3.4)$$

While the coefficients $\overleftrightarrow{C}(\vec{q})$ are given by static expectation values (3.3), the damping coefficients can formally be expressed by Kubo formulas involving the forces $f^c_{\vec{q}}$:

$$\Gamma^{cc'}(\vec{q},\omega) = \int_0^\infty dt\, e^{i\omega t} \langle f^c_{\vec{q}}(t),\, f^{c'}_{\vec{q}}(o)\rangle \qquad (3.5)$$

Here $\langle A(t),B(o)\rangle$ denotes the Kubo-relaxation function. Eq. (3.4) is an exact relation, but the problem is to find a reliable approximation for $\overleftrightarrow{\Gamma}(\vec{q},\omega)$. It is important to note that the evolution of $f^c_{\vec{q}}(t)$ is not governed by the usual equations of motions, but by

$$\dot{f}^c_{\vec{q}} = i\left[H,\, f^c_{\vec{q}}\right] - i\langle\left[H,\, f^c_{\vec{q}}\right],\, X^{c'}_{\vec{q}}\rangle\, X^{c'}_{\vec{q}} \qquad (3.6)$$

In the right hand side of Eq.(3.6) the collective part of $\dot{f}^c_{\vec{q}}$ is projected off. Therefore in some cases a relaxation time approximation for $f^c_{\vec{q}}(t)$ will be possible.

There is one limit where the choice of the "collective" variables is unique. This is the long wave length-long time limit. Then the densities of the conserved variables have much slower characteristic time than all the other quantities and one can obtain from (3.3) and (3.4) the hydrodynamic response functions. Eq.(3.2) and (3.5) show that the damping terms have a wave number dependence which is proportional to $[C(\vec{q})]^2$.

Eqs.(3.4) and (3.5) have also been used as a starting point for the mode coupling theory (15). Considering as collective variables the one and two mode operators Kawasaki has derived from the Mori identity a kinetic equation, the lowest order perturbation theoretical solution of which gives again the mode coupling theory (15).

IV. DYNAMIC RESPONSE FUNCTION FOR THE SOFT MODE

In the case of a displacive structural transition we want to know the correlation function of $\phi^\alpha_{\vec{q}}$. Thus the obvious collective variables are $\phi^\alpha_{\vec{q}}$ and $P^\alpha_{\vec{q}}$, which obey the following equations of motion

$$(d/dt)\phi'^{\alpha}_{\vec{q}} = \omega^\alpha_o(\vec{q})P'^{\alpha}_{\vec{q}} \qquad (4.1a)$$

$$(d/dt)P'^{\alpha}_{\vec{q}} = -\omega^\alpha_o(\vec{q})\phi'^{\alpha}_{\vec{q}} + R^\alpha_{\vec{q}}. \qquad (4.1b)$$

Here

$$\phi'^{\alpha}_{\vec{q}} = \phi^\alpha_{\vec{q}} / \sqrt{\chi^{\alpha\alpha}(\vec{q})} \qquad (4.2a)$$

and

$$P'^{\alpha}_{\vec q} = P^{\alpha}_{\vec q} / \sqrt{I^{\alpha}(\vec q)} \qquad (4.2b)$$

are the normalized variables, where the property $\langle P^{\alpha}_{\vec q}, P^{\alpha}_{\vec q}\rangle = I^{\alpha}(\vec q)$ was used. The frequency ω^{α}_{o} is in terms of the static order parameter susceptibility $\chi^{\alpha\alpha}(\vec q)$ and $I^{\alpha}(\vec q)$ given by

$$\omega^{\alpha}_{o}(\vec q) = \left\langle (d/dt)\phi'^{\alpha}_{\vec q}, P'^{\alpha}_{\vec q} \right\rangle = \left(I^{\alpha}(\vec q)\chi^{\alpha\alpha}(\vec q)\right)^{-1/2} .$$

The dynamic susceptibility for $\phi^{\alpha}_{\vec q}$ following from (4.1) is

$$\chi^{\alpha\alpha}(\vec q,\omega) = \chi^{\alpha\alpha}(\vec q)\, \frac{-\left[\omega^{\alpha}_{o}(\vec q)\right]^2}{\omega^2 - \left[\omega^{\alpha}_{o}(\vec q)\right]^2 + i\Gamma^{\alpha}(\vec q,\omega)\omega} . \qquad (4.3)$$

The internal force acting on the octahedra

$$R^{\alpha}_{\vec q} = i\left[H, P'^{\alpha}_{\vec q}\right] + \omega^{\alpha}_{o}(\vec q)\phi'^{\alpha}_{\vec q} \qquad (4.4)$$

will always contain portions which vary much faster than $\omega^{\alpha}_{o}(\vec q)$. If this were all, we could represent $\Gamma^{\alpha}(\vec q,\omega)$ by a constant. However, if $R^{\alpha}_{\vec q}$ contains also contributions which have a slow characteristic time, the function $\Gamma^{\alpha}(\vec q,\omega)$ will show some structure. Suppose $R^{\alpha}_{\vec q}$ can be decomposed into two parts which are proportional to operators $R'^{\alpha}_{1\vec q}$ and $R'^{\alpha}_{2\vec q}$ having different time scales

$$R^{\alpha}_{\vec q} = b^{\alpha}_{1}(\vec q)\, R'^{\alpha}_{1\vec q} + b^{\alpha}_{2}(\vec q)\, R'^{\alpha}_{2\vec q}. \qquad (4.5)$$

The coefficients $b^{\alpha}_{i}(\vec q)$ are given by $b^{\alpha}_{i}(\vec q) = \langle \dot P^{\alpha}_{\vec q}, R'^{\alpha}_{i\vec q}\rangle$.
The equations of motions for $R'^{\alpha}_{i\vec q}$ have the structure

$$\dot R'^{\alpha}_{i\vec q} = - b^{\alpha}_{i}(\vec q)\, P'^{\alpha}_{\vec q} - S^{\alpha}_{i\vec q}.$$

Let us assume now that the characteristic time of $R'^{\alpha}_{1\vec q}$ is small. Then $R'^{\alpha}_{1\vec q}$ has to be treated on the same level as $\phi^{\alpha}_{\vec q}$ and $P^{\alpha}_{\vec q}$ and has to be included in the set of collective variables. Then one obtains for $\Gamma^{\alpha}(\vec q,\omega)$

$$\Gamma^{\alpha}(\vec q,\omega) = \frac{i\left[b^{\alpha}(\vec q)\right]^2}{\omega + i\gamma^{\alpha}(\vec q,\omega)} + \sigma^{\alpha}(\vec q,\omega) \qquad (4.6)$$

($\gamma^{\alpha}(\vec q,\omega)$ and $\sigma^{\alpha}(\vec q,\omega)$ can formally be expressed by Kubo formulas. Cross relaxation terms between R_2 and S_1 have been neglected.)

If $R_{1\vec q}$ is the only slow part and hence $R'^{\alpha}_{2\vec q}$ and $S^{\alpha}_{1\vec q}$ are "fast" variables, the frequency dependence of $\gamma^{\alpha}(\vec q,\omega)$ and $\sigma^{\alpha}(\vec q,\omega)$ may be neglected and we obtain ($b^{\alpha} \equiv b^{\alpha}_{1}$)

$$\Gamma^{\alpha}(\vec q,\omega) = \frac{i\left[b^{\alpha}(\vec q)\right]^2}{\omega + i\gamma^{\alpha}(\vec q)} + \sigma^{\alpha}(\vec q) \qquad (4.6').$$

In $SrTiO_3$ the quantity $R^{\alpha}_{\vec{q}}$ does not contain a conserved contribution, therefore $b^{\alpha}(o)$ and $\gamma^{\alpha}(o)$ are finite. In analogy to anisotropic spin systems one may argue that $\gamma^{\alpha}(\vec{q})$ and $b^{\alpha}(\vec{q})$ are uncritical; i.e.: they remain finite at T_c (9). This is born out also by our microscopic calculation (33).

Eqs.(4.6) and (4.3) were obtained under the assumption that $R^{\alpha}_{\vec{q}}$ contained a slow contribution $R'^{\alpha}_{1\vec{q}}$, which was treated together with $\phi^{\alpha}_{\vec{q}}$ and $P^{\alpha}_{\vec{q}}$ as a collective variable. $S^{\alpha}_{1\vec{q}}$ and $R'^{\alpha}_{2\vec{q}}$ were assumed to decay rapidly.

A special case of the approximation considered here is the third long time approximation of the continued fraction representation (third L.T.A.). There one considers as collective variables $\phi^{\alpha}_{\vec{q}}$, $P^{\alpha}_{\vec{q}}$ and $R^{\alpha}_{\vec{q}}$ and does not allow for the possibility that $R^{\alpha}_{\vec{q}}$ contains parts which are governed by different time scales. In this approximation the second term in Eq.(4.6) is missing and the whole damping function $\Gamma^{\alpha}(\vec{q},\omega)$ is approximated by a single relaxation function. Although the importance of allowing for different time scales in this problem was explained in Ref. 9, the more restrictive third L.T.A. was subsequently considered by T. Schneider (36). In order to judge the implications let me consider the sum rule

$$\int_{\infty}^{\infty}\frac{d\omega}{2\pi}\,\mathrm{Re}\,\Gamma^{\alpha}(\vec{q},\omega) = \sum_i \left[b_i^{\alpha}(\vec{q})\right]^2 \qquad (4.7)$$

$$= \langle \dot{P}^{\alpha}_{\vec{q}}, \dot{P}^{\alpha}_{\vec{q}}\rangle / TI^{\alpha}(\vec{q}) - \left[\omega_o^{\alpha}(\vec{q})\right]^2 ,$$

which follows from $b_i^{\alpha}(\vec{q}) = -\,i\langle[P'^{\alpha}_{\vec{q}}, R'^{\alpha}_{i\vec{q}}]\rangle$. Obviously if $\Gamma^{\alpha}(\vec{q},\omega)$ is replaced by a single relaxation function the strength of this relaxation function $[b^{\alpha}(\vec{q})]^2 \equiv [b_1^{\alpha}(\vec{q})]^2$ is overestimated, if it is determined from (4.7). Hence Schneider's "physical identification" of $[b^{\alpha}(\vec{q})]^2$ is an overestimation. It has also been noted by Cowley and Coombs that the high frequency portions of $\Gamma^{\alpha}(\vec{q},\omega)$ will contribute much more to the sum rule than the central peak (37). Schneider's criticism that the presence of $\sigma^{\alpha}(\vec{q})$ in $\Gamma^{\alpha}(\vec{q},\omega)$ violates the sum rule (4.7) is therefore irrelevant. It is generally implied and has even been mentioned in Ref. 9, that σ^{α} is not constant for all frequencies but only in the range of interest.

The assumption that $R^{\alpha}_{1\vec{q}}$ is just a purely relaxing mode can be too restrictive in some cases. Instead $R^{\alpha}_{1\vec{q}}$ may be a sum of two terms both of which are oscillating. For instance, if the anharmonic terms in the Hamiltonian allow a nonlinear coupling of the soft mode to two almost degenerate excitations, $R^{\alpha}_{1\vec{q}}$ is of the form

$$R^{\alpha}_{1\vec{q}} = b^{\alpha}(\vec{q})\sum_{\vec{k}}(A^{+}_{\vec{k}}\,B^{-}_{-\vec{k}+\vec{q}} + B^{+}_{-\vec{k}+\vec{q}}\,A^{-}_{\vec{k}})\,c(\vec{k},\vec{q}). \qquad (4.8)$$

Here $A^{\pm}_{\vec{k}}$, $B^{\pm}_{\vec{k}}$ are creation and annihilation operators of these modes and $c(\vec{k},\vec{q})$ is a matrix element. Rewriting (4.8) in the form

$$R^{\alpha}_{1\vec{q}} = b^{\alpha}(\vec{q})X^{\alpha}_{\vec{q}} \qquad (4.8')$$

and introducing $\Pi^{\alpha}_{\vec{q}}$, the momentum conjugate to $X^{\alpha}_{\vec{q}}$, we find equations of motion for the nonequilibrium average which have the structure

$$-i\omega\delta\phi' = \omega_o\delta P' \quad , \qquad -i\omega\delta X = -\omega_1\delta\Pi - \gamma_1\delta X - b\delta\Pi$$

$$-i\omega\delta P' = -\omega_o\delta\phi' + b\delta X \ , \qquad -i\omega\delta\Pi = \omega_1\delta X - \gamma_2\delta\Pi. \qquad (4.9)$$

These equations are equivalent to the equations of motion for the Green's functions of the dynamical susceptibility since $\delta\phi'$ etc. are related linearly to the susceptibility. From Eq.(4.9) we obtain

$$\Gamma^{\alpha}(\vec{q},\omega) = \frac{(i\omega-\gamma^{\alpha}_2)\ \left[b^{\alpha}(\vec{q})\right]^2}{\omega^2-\left[\omega^{\alpha}_1(\vec{q})\right]^2+i\omega\left[\gamma^{\alpha}_2(\vec{q})+\gamma^{\alpha}_1(\vec{q})\right]-\gamma^{\alpha}_1(\vec{q})\gamma^{\alpha}_2(\vec{q})} + \sigma^{\alpha}(\vec{q}) \qquad (4.10)$$

In the limit $\omega^{\alpha}_1(\vec{q})\to 0$, Eq.(4.10) goes over into the relaxation function (4.6).

In our microscopic calculation (33) it turns out that $R^{\alpha}_{1\vec{q}}$ is a sum of three terms and hence $\Gamma^{\alpha}(\vec{q},\omega)$ is also a sum of three terms of type (4.6) or (4.10) depending on the magnitude of $\omega^{\alpha}_1(\vec{q})$. The characteristic quantities $b^{\alpha}(\vec{q}),\gamma^{\alpha}(\vec{q})$ are then replaced by $b^{\alpha}_{\sigma}(\vec{q})\ b^{\alpha}_{\sigma}(\vec{q})$ $(\sigma=1,2,3)$. These coefficients become equal for small $\vec{q}$

V. DYNAMIC FORM FACTOR AND RESONANCES

The dynamic form factor at $\vec{q}_R+\vec{q}$ is given by

$$S^{\alpha\alpha}(\vec{q},\omega) = \frac{\omega(1-e^{-\omega/T})^{-1}\chi^{\alpha\alpha}(\vec{q})(\omega^{\alpha}_o(\vec{q}))^2\,\mathrm{Re}\Gamma^{\alpha}(\vec{q},\omega)}{(\omega^2-(\omega^{\alpha}_o(\vec{q}))^2-\omega\mathrm{Im}\Gamma^{\alpha}(\vec{q},\omega))^2+(\omega\mathrm{Re}\Gamma^{\alpha}(\vec{q},\omega))^2} \ . \qquad (5.1)$$

Its resonances are determined by the zeros of the denominator. Firstly there is a central resonance of width (9)

$$\Gamma^{\alpha}_c(\vec{q}) = \gamma^{\alpha}(\vec{q})\ (\omega^{\alpha}_o(\vec{q})/\omega^{\alpha}_s(\vec{q}))^2. \qquad (5.2)$$

Secondly, there is a propagating contribution to the soft mode, which we call for brevity <u>rotaton</u> , of frequency

$$\omega^{\alpha}_{\pm}(\vec{q}) = \pm\omega^{\alpha}_s(\vec{q}) - \frac{i}{2}\Gamma^{\alpha}_s(\vec{q}) \qquad (5.3)$$

with

$$\omega_s^\alpha(\vec{q}) = (\omega_o^\alpha(\vec{q})^2+b^\alpha(\vec{q})^2)^{1/2}, \quad \Gamma_s^\alpha(\vec{q}) = \gamma^\alpha(\vec{q})(b^\alpha(\vec{q})/\omega_s^\alpha(\vec{q}))^2+\sigma^\alpha. \quad (5.4)$$

The condition for a central resonance is $\gamma^\alpha(\vec{q})<<\omega_s^\alpha(\vec{q})$. In order that the rotaton is propagating and not overdamped the condition $\sigma^\alpha<<\omega_s^\alpha$ has to be fulfilled. A typical consequence of the central peak is that the square of the rotaton frequency is not given by the inverse of the susceptibility but shifted by $(b^\alpha(\vec{q}))^2$. The integrated intensity of the central peak is (9)

$$I_c^\alpha(\vec{q}) = \chi^{\alpha\alpha}(\vec{q})(b^\alpha(\vec{q})/\omega_s^\alpha(\vec{q}))^2 \quad (5.5a)$$

and of both wings of the propagating soft mode (rotaton)

$$I_s^\alpha(\vec{q}) = \chi^{\alpha\alpha}(\vec{q})(\omega_o^\alpha(\vec{q})/\omega_s^\alpha(\vec{q}))^2. \quad (5.5b)$$

The three-peak structure is exhibited by the following expression for the dynamical form factor, which is valid under the condition $\gamma^\alpha<<\omega_s^\alpha$ and $\gamma^\alpha\sigma^\alpha<<(\omega_s^\alpha)^2$:

$$S^{\alpha\alpha}(\vec{q},\omega) = \omega(1-e^{-\omega/T})^{-1}\chi^{\alpha\alpha}(\vec{q})\left\{\left(\frac{b^\alpha(\vec{q})}{\omega_s^\alpha(\vec{q})}\right)^2 \frac{\Gamma_c^\alpha(\vec{q})}{\omega^2+(\Gamma_c^\alpha(\vec{q}))^2} + \right.$$

$$\left. + \left(\frac{\omega_o^\alpha(\vec{q})}{\omega_s^\alpha(\vec{q})}\right)^2 \frac{(\omega_s^\alpha(\vec{q}))^2\Gamma_s^\alpha(\vec{q})-\Gamma_c^\alpha(\vec{q})\ (\omega^2-\omega_s^\alpha(\vec{q})^2)\ (b^\alpha(\vec{q})/\omega_o^\alpha(\vec{q}))^2}{(\omega^2-\omega_s^\alpha(\vec{q})^2)^2+(\omega\Gamma_s^\alpha(\vec{q}))^2}\right\} \quad (5.6)$$

Eqs.(5.1) to (5.6) are completely general and independent of any assumption about the temperature dependence of γ_1^α, b^α and σ^α. We shall discuss them now under the supposition that γ_1^α and b^α are uncritical and that far away from T_c the coupling of the central resonance to $R_{\vec{q}}^\alpha$ is weak. We may distinguish two regimes.

A) Far from T_c we have $\omega_o^\alpha(\vec{q})>>b^\alpha(\vec{q})$ and the frequency of the propagating soft mode is __________ given by

$$\omega_s^\alpha(\vec{q}) = \omega_o^\alpha(\vec{q}) + \frac{1}{2}\,(b^\alpha(\vec{q})/\omega_o^\alpha(\vec{q}))^2\omega_o^\alpha(\vec{q}). \quad (5.7)$$

The width of the central resonance is γ^α and its strength rises proportional to $(\chi^{\alpha\alpha}(\vec{q}))^2$ on approaching T_c. The propagating soft mode moves to $b^\alpha(\vec{q})$ and its strength is transferred to the central resonance.

We expect that in most cases the damping Γ_s^α of the propagating soft mode is dominated by the second term in (5.4), but it should be noted that the first term rises like $\chi(\vec{q})$ as long as $\omega_o^\alpha>>b^\alpha$ on approaching T_c.

B) Close to T_c we have $\omega_o^\alpha(\vec{q}) << b^\alpha(\vec{q})$ and the frequency of the soft mode is

$$\omega_s^\alpha(\vec{q}) = b^\alpha(\vec{q}) + \frac{1}{2} b^\alpha(\vec{q}) \; (\omega_o^\alpha(\vec{q})/b^\alpha(\vec{q}))^2 .$$

The width of the central resonance is then given by

$$\Gamma_c^\alpha(\vec{q}) = \gamma^\alpha(\vec{q}) \; (\omega_o^\alpha(\vec{q})/b^\alpha(\vec{q}))^2 \propto (\chi^{\alpha\alpha}(\vec{q}))^{-1} \; . \quad (5.8)$$

In this region the dynamic form factor is almost saturated by the central resonance. It is important to realize that for this reason dynamical scaling (13,14) holds in the regime close to T_c. However, it does not hold in the regime described first, since Γ_c^α, ω_s^α and Γ_s^α scale differently. Deviations from dynamical scaling occur in other situations where slow uncritical degrees of freedom couple to the order parameter as in antiferromagnets below T_N(38).

In Fig.1 we plot $S(0,\omega)\;(1-e^{-\omega/T})/(\omega\chi(0))$ for ε = 0.01, 0.03, 0.05, 0.13, 0.2, taking $\omega_o = 6\varepsilon^{2/3}$ and $\gamma^\alpha = 0.01$, $b = 0.5$, $\sigma = 0.5$. ($\varepsilon = |1-T/T_c|$) (44). The total area under each of these curves is the same. One sees from this graph that by approaching T_c the propagating soft mode moves to b and its strength is transferred to the central resonance.

In Table 2 we have compiled the frequencies and amplitudes of the resonances at the R point ($\vec{q} = 0$). The critical exponent of the susceptibility γ should not be confused with γ^α. We have introduced $b = b^\alpha(0)$, $\gamma^\alpha = \gamma^\alpha(0)$, $\chi_1 = \chi_o\kappa_o^{-2+\eta}$ and the characteristic temperature ε_b defined through

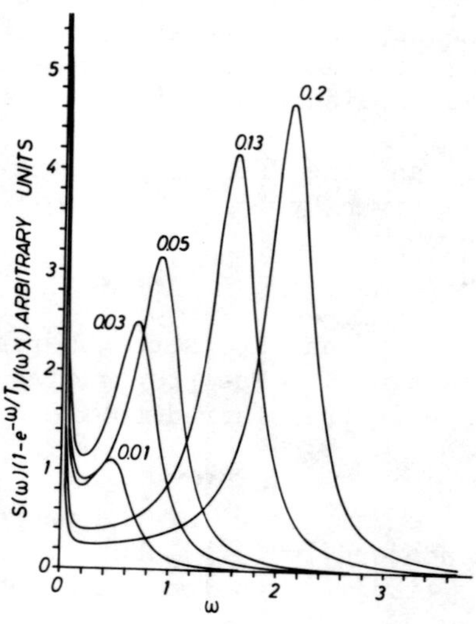

Fig. 1: Dynamic form factor divided by the static susceptibility for several values of ε.

$$\omega_o(\vec{q} = 0, \ \varepsilon = \varepsilon_b) = b(0, \varepsilon = \varepsilon_b) \ . \qquad (5.9)$$

Table 2: Frequencies and Amplitudes of the resonances

	$\Gamma_c(0)$	$I_c(0)/\chi_1$	$\omega_s(0)$	$I_s(0)/\chi_1$
A) $\varepsilon \gg \varepsilon_b$	γ^α	$\varepsilon^{-2\gamma}\ \varepsilon_b^{\ \gamma}$	$(\varepsilon/\varepsilon_b)^{\gamma/2} b$	$\varepsilon^{-\gamma}(1-(\frac{\varepsilon_b}{\varepsilon})^\gamma)$
B) $\varepsilon \ll \varepsilon_b$	$\gamma^\alpha(\frac{\varepsilon}{\varepsilon_b})^\gamma$	$\varepsilon^{-\gamma} - \varepsilon_b^{-\gamma}$	b	$\varepsilon_b^{-\gamma}$

It should be stressed that regime A ($\varepsilon \gg \varepsilon_b$) may be outside the critical region and therefore may not be appropriate for the determination of the asymptotic critical exponents.

The behaviour described after Eq.(5.6) is a typical hybridisation phenomenon. (See discussion of Ref. 9).

There will be cases, where the rotaton is overdamped (i.e. $\Gamma_s^\alpha \gg \omega_s^\alpha$). Then the dynamic structure factor (5.6) is a sum of a narrow and of a broad central component:

$$S^{\alpha\alpha}(\vec{q},\omega) = \chi^{\alpha\alpha}(\vec{q})\left\{\left(\frac{b^\alpha(\vec{q})}{\omega_s^\alpha(\vec{q})}\right)^2 \frac{\Gamma_c^\alpha(\vec{q})}{\omega^2+(\Gamma_c^\alpha(\vec{q}))^2} + \left(\frac{\omega_o^\alpha(\vec{q})}{\omega_s^\alpha(\vec{q})}\right)^2 \frac{\Gamma_v^\alpha(\vec{q})}{\omega^2+(\Gamma_v^\alpha(\vec{q}))^2}\right\} T$$

(5.10)

with $\Gamma_v^\alpha(\vec{q}) = (\omega_s^\alpha(\vec{q}))^2/\sigma$. This is still under the condition $\gamma^\alpha\sigma^\alpha \ll (\omega_s^\alpha)^2$. If σ is so much larger than ω_s^α that $\gamma^\alpha\sigma \gg (\omega_s^\alpha)^2$, there is only one central peak of width $(\omega_o^\alpha(\vec{q}))^2/\sigma$.

In addition to the crossover at ε_b, which is a consequence of the central peak, there may be a second crossover. Since neighboring octahedra share an oxygen atom, within the plane perpendicular to the axis of rotation, these octahedra will be strongly coupled though only weakly coupled in a direction parallel to the rotation axis. Hence these systems can be visualized as consisting of two-dimensional layers which are only weakly coupled. If there were no coupling the system would behave two-dimensionally. On the other hand any finite coupling assures that the system behaves three-dimensionally at and close to T_c. Hence in general (for small coupling) there will be a crossover temperature ε_Δ. For $\varepsilon \gg \varepsilon_\Delta$ the system will behave three-dimensionally, whereas it will behave two-dimensionally for $\varepsilon \ll \varepsilon_\Delta$. (9,49) This change in critical behaviour is accompanied by a change of the critical exponents. Because of the existence of three rotation axes and the resulting inter-

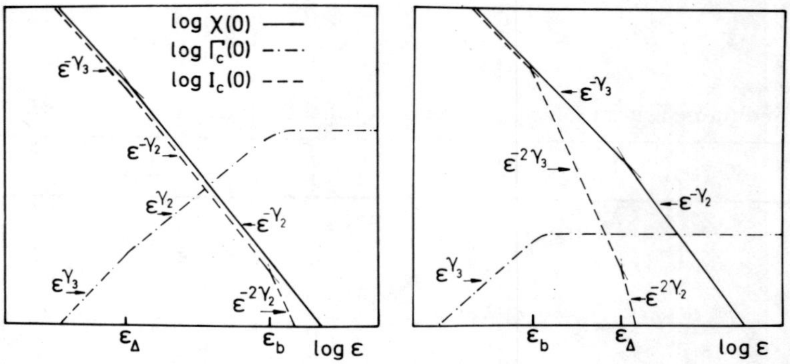

Fig. 2: Crossover effects: At ε_Δ the two-dimensional critical behaviour, valid for $\varepsilon >> \varepsilon_\Delta$, changes to three-dimensional behaviour for $\varepsilon << \varepsilon_\Delta$. The crossover at ε_b is a consequence of the central peak.

actions, the critical exponents in the two-dimensional region may be quite different from Ising exponents. In Fig.2 we show the most general crossover behaviour for $\varepsilon_\Delta < \varepsilon_b$ and $\varepsilon_\Delta > \varepsilon_b$. As examples we plot the static susceptibility, the width and the strength of the central peak.

VI. OTHER CASES OF A CENTRAL PEAK

Having discussed the structure of the dynamic form factor in such detail, it may be worth while to point out analogies to other situations where a relaxational mode couples to a sound wave. Let me first emphasize that the relations of the last section were obtained under the condition

$$\gamma^\alpha(\vec{q}) << \omega_s^\alpha(\vec{q}).$$

One may easily see that if the relaxation rate is faster than the frequency of the sound wave i.e.:

$$\gamma^\alpha(\vec{q}) >> \omega_s^\alpha(\vec{q}) ,$$

the response function assumes the form

$$\chi(\vec{q},\omega) = \chi(\vec{q}) \frac{-\left(\omega_o(\vec{q})\right)^2}{\omega^2 - \left(\omega_o(\vec{q})\right)^2 + i\omega(\sigma + b^2/\gamma)} \qquad (6.1)$$

Under these circumstances there is no central resonance and the relaxation mode produces just an additional damping of the oscillating wave, the frequency of which is given for small damping by

$$\omega = \pm\, \omega_o(\vec{q}) - \frac{1}{2}\, i(\sigma + b^2/\gamma). \qquad (6.2)$$

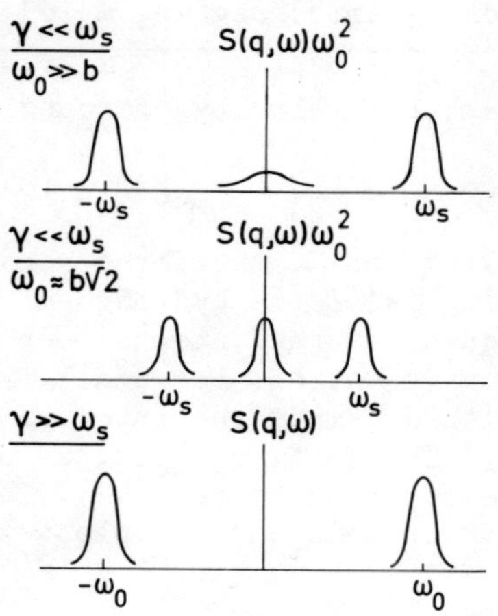

Fig.3: Schematic representation of the dynamic form factor
a) $\gamma<<\omega_s$, $\omega_o>>b$ $(\omega_s\approx\omega_o)$; b) $\gamma<<\omega_s$, $\omega_o\approx b/\sqrt{2}$; c) $\gamma>>\omega_s$.

The possible situations are shown schematically in Fig.3. Let me now list some of the more familiar situations, where a relaxation damping is important.

A) Optical soft mode coupled to a relaxation mode

(b(0) = const., b >> γ). This has been considered in the last section.

B) Hydrodynamic sound wave, weakly coupled to internal degrees of freedom ξ.

The internal degrees of freedom can be the energy of internal vibrational levels, or orientational degrees of freedom (39)-(41). Another example is a gas in which a chemical reaction takes place. (35). Since ω_o and b are proportional to the wave number : $\omega_o = c_o q$, $\sigma = \sigma' q^2$, $b = b'q$ and γ = const., there are two different regimes:

a) High frequency regime, $\omega_s(q) >> \gamma$

In this regime there is a central peak in the dynamic form factor (density-density correlation function) and the velocity of sound

$$c_s \equiv c_\infty = \sqrt{\left.\frac{\partial P}{\partial \rho}\right)_\xi} \tag{6.3}$$

is given by the compressibility at fixed ξ. The damping in the high frequency region is $\gamma(b'/c_s)^2 + \sigma' q^2$.

b) Small frequency (hydrodynamic) regime, $\omega_s(q) \ll \gamma$

There is no central peak; the velocity of sound

$$c_o = \sqrt{\frac{\partial P}{\partial \rho}\Big)_{\Xi=0}} \qquad (6.4)$$

is given by the compressibility for Ξ, the force conjugate to ξ, equal to zero. Now the damping is of the hydrodynamic form $(\sigma'+(b')^2/\gamma)q^2$. For high frequencies the internal degrees of freedom cannot follow the sound wave motion. For small frequency local equilibrium is established, since the internal degrees of freedom follow adiabatically. The fact, $c_\infty > c_o$, is an immediate consequence of the principle of Le Chatelier (42). The density-density correlation function for this case is plotted in Fig. 4.

C) Soft Acoustic Mode (Nb_3Sn)

This is a special case of B. Since $c_o \propto \chi^{-1/2} \to 0$ there are intermediate wave numbers which are of the type b) far away from T_c and of the type a) close to T_c. The softening of c_o also implies that strength is transferred from the propagating mode to the central peak in wave number region b).

D) Coupling of heat diffusion to a sound wave

In the hydrodynamic regime the condition

$$D_T q^2 \ll cq \qquad (6.5)$$

is fulfilled, where D_T is the heat diffusivity. Hence the sound velocity is given by the adiabatic compressibility.

$$c_s^2 = \frac{\partial P}{\partial \rho}\Big)_s \qquad (6.6)$$

Heat diffusion is too slow to produce an isothermal situation. The density-density correlation function consists of the Brillouin doublet and a central Rayleigh component. The intensity ratio is governed by the famous Landau-Placzek formula

$$\frac{I_{Rayleigh}}{I_{Brillouin}} = \frac{c_s^2 - c_T^2}{c_T^2} = (C_p - C_v)/C_v, \qquad (6.7)$$

where

$$c_T^2 = \frac{\partial P}{\partial \rho}\Big)_T \qquad (6.8)$$

and C_v and C_p are the specific heats at constant volume and pressure.

In the low temperature phase of magnetic and structural transitions the heat fluctuations are coupled to the order-parameter. Below T_c the order parameter correlation function contains a diffusive contribution

$$(\chi_T - \chi_s)\,\frac{D_T q^2}{\omega^2+(D_T q^2)^2} \quad ,$$

where D_T is the thermal diffusivity and χ_T and χ_s the isothermal and adiabatic susceptibilities. This coupling has been observed by neutron scattering experiments in MnF_2 and K_2NiF_4 (38). It has been suggested that this coupling could be used to measure the "second magnon", which is a propagating thermal wave in antiferromagnets, by neutron scattering (43).

VII. EXPERIMENTAL CONSEQUENCES

In this section we shall discuss those experimental aspects which can help to understand the nature of the central peak and allow to distinguish between a static and a dynamical effect. The main part of this section will be devoted to ultrasonic attenuation.

A. Neutron scattering

Despite the agreement with the dynamical theory, it is obvious that part of the central peak seen by neutron scattering can be due to impurities (45). Defects which produce locally the low temperatur structure lead to a staticcentral peak at the R point. The detailed examination of this type of Huang scattering shows that the qualitative critical behaviour is similar to the dynamical central peak considered in the previous sections. As long as the separation of the "R-point active" defects is much larger than the critical correlation length, one can show that the intensity of the Huang scattering (scattering due to the static deformation of the lattice by the defects) is given by

$$J^\alpha(\vec{q}) \;\propto\; c^\alpha\; \chi^{\alpha\alpha}(\vec{q})^{\;2}. \qquad (7.1)$$

c^α is the concentration of defects which induce an alternating rotation around the axis $\hat{e}^\alpha$. When the correlation length becomes comparable to the separation of the defects, a departure from the quadratic dependence on $\chi^{\alpha\alpha}$ results. (Note the similar qualitative behaviour in the dynamical theory.)

In order to discriminate between the dynamical and static effects, it would be very important to have precise experimental results on the temperature and impurity dependence of the coefficient $b^\alpha(\vec{q})$, over a wide temperature region.

B. Electron Spin Resonance

The anomalies of the EPR line width have been considered theoretically in Ref. 9. Similar to NMR it has been shown that in a certain frequency and temperature regime the line is Lorentzian. The critical contribution to the width of this Lorentzian, $\delta\nu_{crit}$, is proportional to the quantity

$$\Gamma^{\alpha}(0,0) = \left[b^{\alpha}(0)\right]^{2}/\gamma^{\alpha}(0) + \sigma^{\alpha}(0). \qquad (7.2)$$

Since the other factors in Eq. (4.4) of the second of Ref.9 are known, one may determine $\Gamma^{\alpha}(0,0)$ from an absolute measurement of $\delta\nu_{crit}$.(47) Without central peak $\Gamma^{\alpha}(0,0)$ is of the order of a phonon damping. If there is a dynamical central peak with $b^{\alpha}>>\gamma^{\alpha}$ the quantity $\Gamma^{\alpha}(0,0)$(and $\delta\nu_{crit}$) will be much larger. Hence one can decide between both effects from EPR line width data (46).One has, of course, to make sure that the shape is really Lorentzian.

C. Ultrasonic Attenuation

Sound waves are scattered by the order parameter fluctuations. The interaction between a sound wave and the order parameter fluctuations in $SrTiO_3$ is linear in the strain field $E_{ij}(\vec{q})$ and bilinear in $\phi_{\vec{k}}$.(48):

$$H_{ph,s} = (N)^{-1/2}\sum_{\vec{k},\vec{q}}\left[A\phi_{\vec{k}}^{(1)}\phi_{-\vec{k}-\vec{q}}^{(1)} + B(\phi_{\vec{k}}^{(2)}\phi_{-\vec{k}-\vec{q}}^{(2)} + \phi_{\vec{k}}^{(3)}\phi_{-\vec{k}-\vec{q}}^{(3)})\tilde{E}_{11}(\vec{q}) + \right.$$

$$\left. + 2C\phi_{\vec{k}}^{(1)}\phi_{-\vec{k}-\vec{q}}^{(2)}\tilde{E}_{12}(\vec{q})\right] + \text{cyclic permutations.} \qquad (7.3)$$

The terms which are obtained by cyclic permutation of the suffixes (1,2,3) must be added.

$$\tilde{E}_{ij}(\vec{q}) = N^{-1/2}\sum_{\vec{\ell}} e^{-i\vec{q}.\vec{\ell}}E_{ij}(\vec{\ell}) \qquad (7.4)$$

is the Fourier transform of the strain field. A, B and C are coupling coefficients.

For simplicity we have omitted all $\vec{q}$ and $\vec{k}$ dependent factors in (7.3) since they do not modify the critical behaviour of the attenuation.(50) We shall evaluate the coefficient of attenuation by the mode coupling theory in a formulation, which is similar to Kawasaki's theory (49) for ferromagnets.

The starting point of our analysis is the general hydrodynamic formula for the coefficient of sound attenuation of a sound wave with wave vector $\vec{q}$ and polarisation λ:

Fig. 4: Density-density correlation function $S(q,\omega)$ multiplied by q for a liquid with internal degrees of freedom, neglecting the coupling to the heat diffusion. ($c_o=1$, $\gamma=0.2$, $\sigma=0.4q^2$, $b=0.5q$). a) The high wave number regime, $0.01 \leq q \leq 1$. b) The small wave number regime, $0.01 \leq q \leq 0.2$.

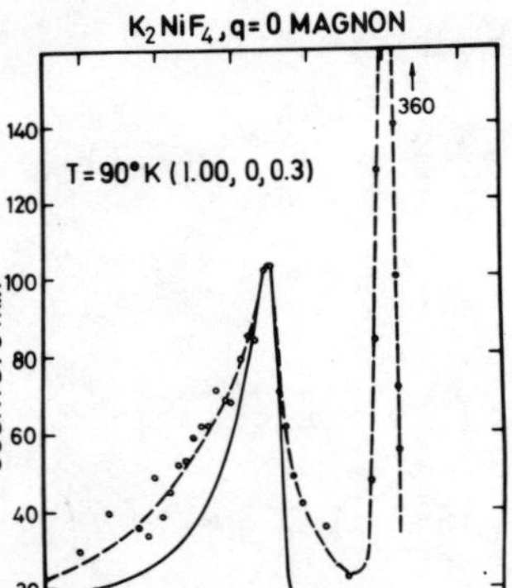

Fig. 5: Neutron scattering cross section of K_2NiF_4 for $T<T_N$ showing the magnon and a central (diffusive) peak. R.J. Birgenau, J.Skalyo and G.Shirane (38).

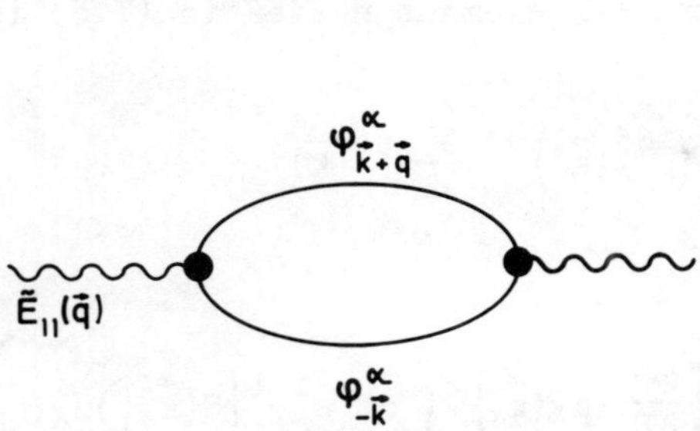

Fig. 6: Decay of a longitudinal sound wave in the [100] direction into two order parameter fluctuations.

$$\alpha^{\lambda}_{\vec{q}}(\omega) + i\beta^{\lambda}_{\vec{q}}(\omega) = \sum_{i,j} \frac{q_i q_j}{2c_\lambda(\vec{q})} \int_o^\infty dt\, e^{i\omega t} < f^{\lambda}_i(\vec{q},t) f^{\lambda}_j(\vec{q},o)^{\dagger} >. \quad (7.5)$$

Here $f^{\lambda}_i(\vec{q},t)$ is the internal force acting on the sound wave normal coordinate $Q_{\vec{q},\lambda}$. In terms of these quantities the interaction Hamiltonian can be written in the form

$$H_{ph,s} = \sum_{\vec{q},\lambda,i} iq_i f^{\lambda}_i(\vec{q})\, Q_{\vec{q},\lambda}. \quad (7.3')$$

For a longitudinal wave in the [100] direction one finds

$$f^{\ell}_i((q,o,o)) = \frac{\delta_{io}}{\sqrt{NM}} \sum_{\vec{k}} \left\{ A\phi^{(1)}_{\vec{k}} \phi^{(1)}_{-\vec{k}-\vec{q}} + B(\phi^{(2)}_{\vec{k}} \phi^{(2)}_{-\vec{k}-\vec{q}} + \phi^{(3)}_{\vec{k}} \phi^{(3)}_{-\vec{k}-\vec{q}}) \right\}$$

$$= \frac{\delta_{io}}{\sqrt{NM}} \sum_{\vec{\ell}} e^{-i\vec{q}.\vec{\ell}} \left\{ A\phi^{(1)}_{\vec{\ell}} \phi^{(1)}_{\vec{\ell}} + B(\phi^{(2)}_{\vec{\ell}} \phi^{(2)}_{\vec{\ell}} + \phi^{(3)}_{\vec{\ell}} \phi^{(3)}_{\vec{\ell}}) \right\}. \quad (7.6)$$

Let me introduce the orthonormal two mode operator

$$\Phi^{\alpha\alpha}_{\vec{k},\vec{q}-\vec{k}} = (T\chi^{\alpha\alpha}(\vec{k})\chi^{\alpha\alpha}(\vec{k}-\vec{q}))^{-1/2}\, \phi^{\alpha}_{\vec{k}} \phi^{\alpha}_{\vec{q}-\vec{k}} \quad (7.7)$$

The decay process of Fig. 6 contributes to

$$\alpha^{\ell}_{[100]}(\omega) = \text{Re}\; q^2 \sum_{\delta} \int dt\, e^{i\omega t} | < f^{\ell}(q,o,o),\, \Phi^{\delta\delta}(\vec{k},\vec{q}-\vec{k}) > |^2$$
$$(T\chi^{\delta\delta}(\vec{k})\chi^{\delta\delta}(\vec{k}-\vec{q}))^{-1} S^{\delta\delta}(\vec{k},t) S^{\delta\delta}(-\vec{k}+\vec{q},t) \quad (7.8)$$

We consider now the experimentally relevant limit that the wave number of the ultrasonic wave q is smaller than the inverse of the correlation length. Then the matrix element needed in (7.8) is given by

$$\lim_{q\to o} < f^{\ell}_i(q,o,o),\, \Phi^{\delta\delta}_{\vec{k},\vec{q}-\vec{k}} > = d_\delta \left(\chi^{\delta\delta}(\vec{k})\right)^{-1} \frac{\partial}{\partial T} \chi^{\delta\delta}(\vec{k}) \quad (7.9)$$

This finally gives:

$$\alpha^{\ell}_{[q,oo]}(\omega) = \text{Re}\, \frac{\omega^2}{\rho}\, d^2 \int \frac{d^3k}{(2\pi)^3} \left[\frac{\partial}{\partial T} \ln\chi(\vec{k})\right]^2 \int_o^\infty dt\, e^{i\omega t} \left| S(\vec{k},t)/\chi(\vec{k}) \right|^2. \quad (7.10)$$

Expression (7.7) can be evaluated by inserting $S(\vec{k},t)$ of section 5.

Let me denote the maximal wave number for which a central peak is present by q_c and assume that $\xi^{-1} << q_c$. Under the condition $\omega << \Gamma_c(k)$ one finds

$$\alpha(\omega) \propto \omega^2 \varepsilon^{-\rho}, \qquad (7.11)$$

In region B where the central peak dominates one obtains for the critical exponent

$$\rho = \gamma_3 - 3\nu_3 + 2 = \gamma_3 + \alpha_3 \qquad (7.12)$$

for three-dimensional correlations, and

$$\rho = \gamma_2 - 2\nu_2 + 2 = 2 - \nu_2 \eta_2 \qquad (7.13)$$

for two-dimensional correlations. Going away from T_c one enters region A in which the exponents are raised by γ. The crossover of the exponent ρ at $\varepsilon = \varepsilon_b$ would be a direct indication of a dynamical central peak. A complication arises from the condition $\kappa < q_c$. Its violation leads also to a stronger temperature variation of the attenuation. One will find a rounding of the ultrasonic anomaly and deviation from the ω^2 law if

$$\Gamma_c(\kappa) \approx \omega \ . \qquad (7.14)$$

In pure samples this rounding effect would allow determination of the width $\Gamma_c(\kappa)$. (50).

(The dynamical exponents are related to the static two- or three-dimensional exponents, for which this type of theory can give no prediction.)

D. Velocity of Sound

We should like to close this section with some remarks about the velocity of sound. In perturbation theory one finds for the critical shift of the velocity of sound

$$\frac{\Delta c_\ell}{c_\ell} = - < f^\ell(o) f^\ell(o) > / \, 2c_\ell MT \ , \qquad (7.15)$$

where c_ℓ is the longitudinal sound velocity in the [100] direction. Since $f^\ell(o)$ scales like the energy density one obtains that Δc_ℓ is proportional to the negative of the specific heat.

We compare now this perturbation theoretical result, the validity of which may be questioned in the hydrodynamic region, with the thermodynamic relation. The velocity of sound is given by

$$c(T) = \frac{\partial\sigma}{\partial E}\Big)_s , \qquad (7.16)$$

the adiabatic derivative, where σ and E are the stress and strain fields. Following the analysis of Ahlers (52) for the lambda transition of Helium one finds

$$c(T) - c(T_c) = B\left(\frac{1}{C_\sigma(T)} - \frac{1}{C_\sigma(T_c)}\right), \qquad (7.17)$$

where B is a coefficient and C_σ is the specific heat at constant stress.

If the specific heat is characterized by a <u>positive critical exponent</u> ($\alpha > 0$)

$$C_\sigma(T) = C_\sigma^{\text{uncrit.}} + A\varepsilon^{-\alpha},$$

one finds for temperatures for which the anomalous part is much smaller than $C_\sigma^{\text{uncrit.}}$

$$c(T) - c(T_c) \approx B\,\frac{1}{C_\sigma^{\text{uncrit.}}} - \frac{B}{C_\sigma^{\text{uncrit.}}}\,\frac{A\varepsilon^{-\alpha}}{C_\sigma^{\text{uncrit.}}} \qquad (7.18)$$

For a <u>negative critical exponent</u> α:

$$C_\sigma(T) = C_\sigma(T_c) - A'\varepsilon^{+|\alpha|}$$

one could use the same expansion, or an expansion which is valid very close to T_c:

$$c(T) - c(T_c) = B\left(\frac{1}{C_\sigma(T_c)}\right)^2 A'\varepsilon^{|\alpha|} .$$

Hence in both cases the perturbation theoretical result is recovered <u>approximately</u>. In general the thermodynamic relation (7.17) should be used.

VIII. MICROSCOPIC LATTICE DYNAMICAL THEORY

The general relation for the damping $\Gamma(\vec{q},\omega)$ was based on the assumption that there exists a slowly relaxing variable which couples to the force acting on the rotation angles. We shall now review some of the microscopic suggestions for this slow motion. (Part of this section was discussed in the summary).

As before we disregard here the more trivial cases where a conserved variable couples to the order parameter and consider solely the Γ_{25} transitions.

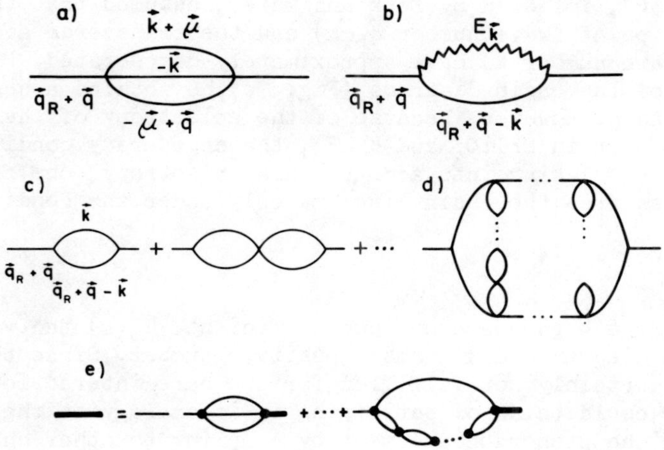

Fig. 7: Decay processes of the zone boundary phonon.

The first suggestion by Riste et al (3) and Feder (8) was that there is a linear coupling to the temperature fluctuations, resulting in a difference between the adiabatic and isothermal susceptibilities. For symmetry such a linear coupling is not present above T_c.

Silberglitt (53) has suggested that the central peak results self-consistently from decay into two acoustic modes and into a central peak. (See Fig. 7a). He has to restrict the integration over the intermediate wave number $\vec{\mu}$ to very small values in order that the energy difference of the acoustic phonons $\omega(\vec{k})-\omega(\vec{k}-\vec{\mu})$ is small. It is hard to see a physical reason for such a cutoff. One may suspect that a small cutoff would reduce the strength $(b^{\alpha})^2$ of the central peak and that the integration over all $\vec{\mu}$ would wash out the central peak. It is also obvious that in the regime $\varepsilon >> \varepsilon_b$, where $\Gamma_c = \gamma$, no selfconsistent solution for γ can be found from this lowest order process.

Another possibility one may consider is that the slow operator is the product of energy and ϕ fluctuations, or equivalently that the central peak is bootstrapped from the decay into an energy fluctuation and a central peak (9). Energy-fluctuations are slow, but only for small wave vector. Hence the integration over all $\vec{k}$ gives a very broad central peak. The mode coupling calculation shows that the γ_α of this process is an average in k-space of $Dk^2+\lambda$, where D is the energy diffusivity of the soft mode system and λ the rotator-lattice relaxation rate. Only if these quantities are extremely small, this process can produce a central peak.

Enz (54), followed by Beck and Meier, assumed that the phonons near the R point (wave number $\vec{q}_R - \vec{k}$) and the transverse acoustic phonons (wave number $\vec{k}$) have approximately degenerated. From the summation of the chain diagrams (Fig. 7c) he obtains a negative $[b(\vec{q})]^2$ and γ. However, because of the anisotropy of the zone boundary phonon in $SrTiO_3$ and $KMnF_3$, the degeneracy condition is not fulfilled. Taking into account this anisotropy, one finds a central peak from the chain diagrams only under the condition

$$g_4 G_2^o(0) = -1 + \varepsilon.$$

Here g_4 is the 4 phonon coupling coefficient, $G_2^o(\omega)$ the value of the bubble diagram and ε a small positive number. Since this condition is certainly not fulfilled for the bare interaction coefficient, one could think of satisfying it in analogy to the Kanamori theory for the Stoner enhancement by coupling to other channels. But it seems hard to find a formal or physical reason why the resulting parquet diagrams (Fig. 7d) should give a central peak.

The author's recent microscopic theory (33) starts from the observation that the zone boundary phonons form a two phonon resonance in the region where the dispersion is linear.The central peak results from the decay into a phonon and a two phonon resonance. (Fig. 7e).

Acknowledgement: We are grateful to Dr. P. Kortman for careful reading of the manuscript and to Mrs. v. Studnitz for supplying the plot in Fig. 4.

REFERENCES

1. H. Unoki and T. Sakudo, J.Phys.Soc.Jap. 23, 546(1967)

2. G. Shirane and Y. Yamada, Phys.Rev. 177, 858(1969); J.D. Axe, G. Shirane and G. Müller, Phys.Rev. 183, 820(1969); V.J. Minkiewicz and G. Shirane, J.Phys.Soc.Jap. 26, 674 (1969)

3. T. Riste, E.J. Samuelsen, K. Otnes and J. Feder, Solid State Comm. 9 , 1455(1971)

4. K.A. Müller, Nato Advanced Study Institute, Structural Phase Transitions and Soft Modes, Geilo, Norway; K.A. Müller and W. Berlinger, Phys.Rev.Lett. 26, 13 (1970)

5. H. Thomas, K.A. Müller, Phys.Rev. Letters 21, 1256 (1968)

6. E. Pytte and J. Feder, Phys.Rev. 187, 1077 (1969); J. Feder and E. Pytte, Phys. Rev. B1, 4803 (1970)

7. H. Thomas, in Structural Phase Transitions and Soft Modes, Ed. E.J. Samuelsen, E. Andersen and J. Feder, Universitetsforlaget, Oslo, p. 15 (1971)

8. J. Feder, Solid State Comm. 9, 2021 (1971)

9. F. Schwabl, Phys. Rev. Letters 28, 500 (1972), Z. Physik 254, 57 (1972)

10. B. Widom, J. Chem. Phys. 43, 3892 (1965)

11. L.P. Kadanoff, Physics 2, 263 (1966)

12. R.A. Ferrell, N. Menyhárd, H. Schmidt, F. Schwabl, P. Szépfalusy, Phys. Lett. 24A, 493 (1967)

13. R.A. Ferrell, N. Menyhárd, H. Schmidt, F. Schwabl and P. Szépfalusy, Ann. Phys. (N.Y.), 47, 565 (1968), Phys. Rev. Lett. 18, 891 (1967)

14. B. Halperin and P.C. Hohenberg, Phys. Rev. Lett. 19, 700 (1967); Phys. Rev.

15. K. Kawasaki, Phys. Rev. 150, 291 (1966); Annals of Physics (N.Y.) 61, 1 (1970); Progress Theor. Phys. (Kyoto) 40, 11(1968)

16. L.P. Kadanoff and J. Swift, Phys. Rev. 166, 89 (1968)

17. V.L. Ginzburg, Soviet Physics Solid State 2, 1824 (1960) It should be noted that the Ginzburg criterion is derived for a dipolar fluid. The result is different for a lattice (A. Hüller, private communication).

18. J.D. Axe, G. Shirane, D. Shapiro, T. Riste, Phys. Rev. B6, 4332 (1972)

19. J.K. Kjems, G. Shirane, K.A. Müller and H.J. Scheel to be published

20. P.G. De Gennes, Solid State Comm. 1, 132 (1963)

21. V.G. Vaks and A.I. Larkin, Soviet Physics J.E.T.P. 22, 678 (1966)

22. R. Blinc and B. Zeks, Report on Progress in Physics, to be published

23. R.J. Elliott, in Structural Phase Transitions and Soft Modes, Ed. E.J. Samuelsen, E. Andersen, J. Feder, Univer itetsforlaget, p. 235 (1971)

24. K.K. Kobayashi, J. Phys. Soc. Japan 24 , 497 (1968)

25. J. Coopersmith 1961, quoted in G. Horwitz and H. Callen, Phys. Rev. 124, 1757 (1961); F. Schwabl, Annals of Physics 54, 1 (1969)

26. E. Pytte, Phys. Rev. Letters 28, 895 (1972) and references cited therein

27. R.A. Cowley and A.D. Bruce, to be published

28. H. Mori, Progr. Theor. Phys. (Kyoto) 33, 423 (1965)

29. H. Mori, in Many Body Theory, Ed. R. Kubo, Syokabo, Tokyo (1966) p. 17

30. F. Schwabl and K.H. Michel, Phys. Rev. B2, 189 (1970) and references cited therein

31. F. Schwabl, Conference on "Magnetische Umwandlungen und kritische Erscheinungen", Weinheim, 4.-6.Oct. 1971

32. Equation (2.10') has first been proposed in Ref.31 and in the first of Ref. 9 and later in G. Shirane, J.D. Axe, Phys. Rev. Lett. 27, 1803 (1971) for the central peak in Nb_3Sn. It came to our attention that this structure for the phonon self energy was also suggested by P.C. Hohenberg at a Gordon meeting 1971.

33. F. Schwabl, Solid State Comm., to be published

34. The Mori theory was applied to ferroelectric transitions already by K. Tani, J. Phys. Soc. Jap. 26, 93 (1969); For the static correlation function the Ornstein Zernike form was assumed to be valid.

35. L.D. Landau and E.M. Lifshits, Fluid Mechanics, p. 305, Pergamon Press 1959

36. T. Schneider, Phys. Rev. B7, 201 (1973)

37. R.A. Cowley and F.J. Coombs, J. Phys. C6, 143 (1973)

38. M.P. Schulhof, P. Heller, R. Nathans and A. Linz, Phys. Rev. Lett. 24, 1184 (1970); R.J. Birgeneau, J. Skalyo, Jr. and G. Shirane, Phys. Rev. B3, 1736 (1971)

39. K.F. Herzfeld and T.A. Litovitz, Absorbtion and Dispersion of Ultrasonic Waves, Academic Press New York 1959

40. R.D. Mountain, J. Res. Nat. Bur. Stand. Sect. A 70, 207 (1966)

41. R.D. Mountain, Critical Review in Solid State Sciences 1, 5 (1970)

42. L.D. Landau and E.M. Lifshits, Statistical Physics, Pergamon Press 1959

43. K.H.Michel and F. Schwabl, Phys. Rev. Lett. 26, 1568 (1971)

44. Review for static critical properties are M.E.Fisher: Reports on Progress in Physics, Vol. XXX, II p. 615 (1967), L.P. Kadanoff, W. Götze, D. Hamblen, R. Hecht, E.A.S. Lewis, V.V. Palciauskas, M. Rayl, J. Swift, D. Aspnes, J. Kane, Rev. Mod. Phys. 39, 315 (1967)

45. It has been suggested by W. Schilling (T. Riste, Private communication) that the C.P. results from defects and imperfections which bring about locally a static staggered rotation.

46. Th. v. Waldkirch, K.A. Müller, W. Berlinger and H. Thomas, Phys. Rev. Lett. 28, 503 (1972) and to be published

47. In the right hand side of Eq. (4.4) in Ref. 9 a factor T is missing.

48. E. Pytte, Phys. Rev. B1, 924 (1970)

49. K. Kawasaki, International J. Magnetism 1, 171 (1971)

50. F. Schwabl, Phys. Rev. B 7, 2038 (1973)

51. K. Fossheim and B. Berre, Phys. Rev. B5, 3292 (1971)
W. Rehwald, Phys. Kondens. Materie 14, 21 (1971) and References cited therein

52. G. Ahlers, Phys. Rev. 182, 352 (1969)

53. R. Silberglitt, Solid State Comm. 11, 247 (1972)

54. C.P. Enz, Phys. Rev. B6, 4695 (1972); H. Beck and P.F. Meier, to be published

NONLINEAR FLUCTUATIONS AND THE CENTRAL MODE

Jens Feder*

Institute of Physics, University of Oslo, Norway

ABSTRACT

The fluctuations of a simple model Hamiltonian for a structural phase transition is discussed. Particular attention is paid to the nonlinear interaction and relaxation of the orderparameter fluctuations, and it is shown how a central mode in the spectrum of the orderparameter correlation function may be obtained using nonlinear fluctuation theory.

INTRODUCTION

At an increasing number of structural phase transitions one finds a central mode in the spectrum of the orderparameter correlation function (1).

Cowley (2) originally showed that a central mode was to be expected at ferroelectric transitions on the basis of weakly anharmonic lattice theory. Recently Cowley and Coombs (3) have extended this work and found that for $T < T_c$ a quasielastic central mode in addition to the soft modes is expected for ferro electric-, improper ferroelectric and antiferrodistortive transitions. For $T > T_c$, however, their approach yields no central mode for the antiferrodistortive case such as $SrTiO_3$. They suggest, however,

---

* Part of this work was done at IBM T.J. Watson Research Center, Yorktown Heights, N.Y. 10598.

that critical effects may give rise to a central mode at all structural transitions even above T_c.

It is the purpose of this paper to contribute to the understanding of how a central mode may arise in materials like $SrTiO_3$, in particular above T_c where the simple approach predicts no central peak. Since we will discuss the effect of nonlinear relaxation and interaction of fluctuations - a little developed and highly controversial field - we shall have to concentrate on rather fundamental aspects of fluctuation theory, and our discussion of the central peak problem will be incomplete. Also, our approach is related to the Green's function method (4) used by for instance Silberglitt (5). However, a discussion of similarities and differences has yet to be done.

We originally suggested (6) that a central peak is to be expected when fluctuations of the orderparameter are taken into account. In the high temperature phase the orderparameter vanishes on the average, but due to fluctuations one may find regions that have the low temperature structure, and in such regions the orderparameter couples to the entropy fluctuations just as in the low temperature phase - and a central mode results.

Fluctuations are generally important only in small systems - their relative importance decreasing with increasing size of the system under observation. Consider for instance an EPR experiment (7), where the orderparameter is measured at the site of a paramagnetic impurity. The system under observation is thus essentially of the order of one lattice unit cell, i.e., only 1/N of the whole crystal. Thus one would expect that the fluctuations in the orderparameter as observed either by considering one impurity as function of time, or observing the orderparameter in an ensemble of impurities at one instant of time, would give fluctuations in the orderparameter of the same order of magnitude as the orderparameter itself. This is indeed the case in EPR experiments on $SrTiO_3$ (7).

Similarly in an inelastic neutron scattering experiment, one ideally observes the behaviour of a single mode out of the N-modes that describe the state of the crystal. Again only 1/N of the crystal is the system under observation, the rest is acting as a heat bath.

In the following sections we shall concentrate on the simplest case where the orderparameter is observed locally in real space, - which means that the model Hamiltonian will be discussed in the molecular field approximation (MFA). The situation where the orderparameter is observed locally in q-space, - and where the model Hamiltonian has to be discussed in the random phase approximation (RPA), is somewhat more complicated, and will be discussed in more detail later.

Let us at this point stress that the center of interest in this paper is the nonlinear fluctuations - and we shall not try to improve on the MFA or RPA approximations, in spite of the fact that these approximations treat fluctuations only approximately. We are therefore only trying to achieve qualitatively correct results.

FLUCTUATION THEORY

In a bath at temperature T, we observe a system that may be described in terms of a density matrix ρ, (8). The expectation value of any observable A for the system is then given by

$$\langle A\rangle = \mathrm{Tr}\,\rho\, A \qquad (1)$$

The density matrix may be a function of the coordinate operators $\underline{Q} = Q_1 \cdots Q_N$ of the system, and depend on a number of parameters $\underline{x} = x_1 \cdots x_n$, that characterize the density matrix. Such a density matrix is written $(\underline{Q};\underline{x})$, and an example is given in eq.(16). For a given ρ, we may calculate the free energy of the system,

$$F = F\{\underline{x}\} = E - TS = \mathrm{Tr}\,\{\rho\, H + \frac{1}{\beta}\rho\ln\rho\} \qquad (2)$$

where $\beta = 1/kT$ and H is the system Hamiltonian. The free energy is thus formally a function of temperature and the parameters$\{\underline{x}\}$- that characterize the state of the system. The probability for finding the system with a density matrix $\rho(\underline{Q};\underline{x})$, when the system is in thermal equilibrium with a bath at temperature T, is, (9),

$$p_o\{\underline{x}\} \sim \exp^{-\beta F\{\underline{x}\}} \qquad (3)$$

In order to find the most probable set of parameters $\underline{x}$, one maximises $p_o\{\underline{x}\}$, or equivalently requires $\partial F\{\underline{x}\}/\partial\underline{x} = 0$, subject to the constraint,

$$\mathrm{Tr}\,\rho = 1 \qquad (4)$$

If the set of parameters $\underline{x}$ is so large that any possible variation in the density matrix may be obtained, the most probable density matrix is found to be the Gibb's distribution,

$$p_o = \exp^{-\beta H}/\mathrm{Tr}\,\exp^{-\beta H} \qquad (5)$$

In the Schrödinger picture the density matrix develops with time according to the equation of motion, (8),

$$i\frac{\partial}{\partial t}\rho = [H,\rho] \qquad (6)$$

where H is the Hamiltonian of the system including its interaction with the bath. We note that the Gibb's distribution, eq. (5), is time independent by eq.(6). For the sake of discussion, assume that H has the form

$$H = H_o(\underline{Q}) - \underline{Q} \cdot \underline{K}(t) \tag{7}$$

Here $\underline{Q}$ represents the system coordinates, and $\underline{K}(t)$ the conjugate bath forces. Since the bath itself fluctuates the best we can do is to give a probability distribution $P(\underline{K})$ for the forces $\underline{K}(t)$, (10). In the interaction representation, i.e., the Heisenberg representation of the system Hamiltonian $H_o(\underline{Q})$, the density matrix has only the weak time dependence caused by the interaction with the bath, we therefore continue the discussion in the interaction representation.

Lacking any detailed information as to the state of the bath, and the precise form of the interaction Hamiltonian which in general does not have the simple form given in eq.(7), we represent the dynamical effect of the bath on the system in terms of transition probabilities. Thus the effect of the bath may be represented by giving the probability

$$W(\underline{x}'|\underline{x})dt$$

that the system will go from a state described by the density matrix $\rho(\underline{x})$ to the one described by $\rho(\underline{x}')$ in the time interval dt. We let $W(\underline{x}'|\underline{x})$ satisfy the detailed balance condition

$$W(\underline{x}'|\underline{x})\ p_o(\underline{x}) = W(\underline{x}|\underline{x}')\ p_o(\underline{x}') \tag{8}$$

which ensures that the dynamics of the fluctuations is consistent with eq.(3), so that a set of parameters $\underline{x}$, occurs with a frequency proportional to $p_o(\underline{x})$ for a system at equilibrium.

Having the transition probabilities, we construct the master equation for the time development of the probability distribution $p(\underline{x},t)$ for finding the system with a density matrix $\rho(\underline{x})$ at time t.

$$\frac{\partial}{\partial t} p(\underline{x},t) = \int d\underline{x}'\ [W(\underline{x}|\underline{x}')\ p(\underline{x}',t) - W(\underline{x}'|\underline{x})\ p(\underline{x},t)] \tag{9}$$

The integral is over all possible values of the parameter $\underline{x}$. We see that the detailed balance equation, eq.(8), gives as a result that the equilibrium distribution $p_o(\underline{x})$, eq.(3), is stationary.

In order to proceed further, one must have more information as to the state and the fluctuations of the bath. Lacking such information one assumes that the transition probability $W(\underline{x}'|\underline{x})$ falls off rapidly with increasing $\Delta\underline{x} = \underline{x}' - \underline{x}$, and varies slowly with $\underline{x}$, so that eq.(9) may be approximated by a Fokker-Planck equation (11). Letting $W(\underline{x}'|\underline{x}) = W(\Delta\underline{x};\underline{x})$, eq.(9) is first rewritten

$$\frac{\partial}{\partial t} p(\underline{x},t) = \tfrac{1}{2}\int d\Delta\underline{x} \Big[W(\underline{x}|\underline{x}+\Delta\underline{x})\; p(\underline{x}+\Delta\underline{x},t) - W(\underline{x}+\Delta\underline{x}|\underline{x})\; p(\underline{x},t) - W(\underline{x}|\underline{x}-\Delta\underline{x})\; p(\underline{x}-\Delta\underline{x},t) + W(\underline{x}-\Delta\underline{x}|\underline{x})\; p(\underline{x},t)\Big]$$

Then using eq.(8) we find,

$$\frac{\partial}{\partial t} p(\underline{x},t) = \tfrac{1}{2}\int d\Delta\underline{x}\; W(\Delta\underline{x};\underline{x})\; p_o(\underline{x})\left[\frac{p(\underline{x}+\Delta\underline{x},t)}{p_o(\underline{x}+\Delta\underline{x})} - \frac{p\;(\underline{x},t)}{p_o(\underline{x})}\right]$$

$$- \tfrac{1}{2}\int d\Delta\underline{x}\; W(\Delta\underline{x};\underline{x}-\Delta\underline{x})\; p_o(\underline{x}-\Delta\underline{x}) \cdot \left[\frac{p(\underline{x}-\Delta\underline{x},t)}{p_o(\underline{x}-\Delta\underline{x})} - \frac{p(\underline{x},t)}{p_o(\underline{x})}\right]$$

Here we expand the terms in brackets and find

$$\frac{\partial}{\partial t} p(\underline{x},t) \simeq \tfrac{1}{2}\int d\Delta\underline{x}\Big[W(\Delta\underline{x};\underline{x})\; p_o(\underline{x})\; \Delta\underline{x}\cdot\frac{\partial}{\partial\underline{x}}\cdot\frac{p(\underline{x},t)}{p_o(\underline{x})} - W(\Delta\underline{x};\underline{x}-\Delta\underline{x})\; p_o(\underline{x}-\Delta\underline{x})\Delta\underline{x}\cdot\frac{\partial}{\partial\underline{x}}\cdot\frac{p(\underline{x}-\Delta\underline{x},t)}{p_o(\underline{x}-\Delta\underline{x})}\Big]$$

Expanding the last term of the integral we then finally obtain the Fokker-Planck equation.

$$\frac{\partial}{\partial t} p(\underline{x},t) = \frac{\partial}{\partial\underline{x}}\cdot \underline{\underline{D}}(\underline{x})\; p_o(\underline{x})\cdot\frac{\partial}{\partial\underline{x}}\;\frac{p(\underline{x},t)}{p_o(\underline{x})} \qquad (10)$$

where the "diffusion" tensor is given by

$$\underline{\underline{D}}(\underline{x}) = \tfrac{1}{2}\int d\Delta\underline{x}\;\Delta\underline{x}\; W(\underline{x}+\Delta\underline{x}|\underline{x})\Delta\underline{x} \qquad (11)$$

We thus have arrived at a formulation where the fluctuating effects of the bath on the system is described in terms of a diffusion constant $\underline{\underline{D}}(\underline{x})$. In the next sections we shall treat $\underline{\underline{D}}$ as a phenomenological constant, which may be calculated if more detailed information about the interaction with and fluctuation of the bath is available.

With the initial condition

$$p(\underline{x},t = 0) = \delta(\underline{x} - \underline{x}_o) \tag{12}$$

the solution $p(\underline{x},t)$ of the Fokker-Planck equation is its fundamental solution and is to be interpreted as the conditional probability distribution $p(\underline{x},t|\underline{x}_o)$ for the parameters $\underline{x}$, knowing they were $\underline{x}_o$ at $t = 0$.

The time dependent density matrix - in the Heisenberg representation - is then given by

$$\rho(t) = \int d\underline{x}\; p(\underline{x},t)\; \rho(\underline{x}) \tag{13}$$

so that the time dependent expectation value of an operator of the system is

$$\langle A(t)\rangle = \mathrm{Tr}\;\rho(t)A = \int d\underline{x}\; p(\underline{x},t)\;\mathrm{Tr}\;\rho(\underline{x})A \tag{14}$$

In the next sections we shall make use of these results in order to discuss the fluctuations of a simple model.

THE MODEL HAMILTONIAN

We shall use a simplified model Hamiltonian (12) which still contains the essential aspects of the model Hamiltonian originally used in the discussion of the $SrTiO_3$ transition (13). Let Q_ℓ be a "local normal coordinate" (12) at lattice site ℓ. In $SrTiO_3$ one would think of Q_ℓ as being proportional to the rotation angle of the oxygen octahedron. With P_ℓ being the conjugate momentum the Hamiltonian is

$$H = \tfrac{1}{2}\Sigma_\ell P_\ell^2 + \Sigma_\ell(\tfrac{1}{2}\Omega_o^2\, Q_\ell^2 + \tfrac{1}{4}\gamma\, Q_\ell^4) - \tfrac{1}{2}\Sigma_{\ell\ell'}\, Q_\ell\, v_{\ell\ell'}\, Q_{\ell'} \tag{15}$$

Here the kinetic energy term is not important in the following discussion, and dropping it corresponds to considering overdamped modes only.

We then introduce a trial density matrix of the form (12),

$$\rho = \rho\,\{\overline{Q}_\ell\;\sigma_\ell\} = \Pi_\ell\, \rho_\ell\;;\; \rho_\ell = \frac{1}{\sqrt{2\pi\sigma_\ell}}\exp\left(-\frac{1}{2\sigma_\ell}(Q_\ell-\overline{Q}_\ell)^2\right) \tag{16}$$

Here the parameters $\overline{Q}_\ell$ and σ_ℓ characterize this gaussian density matrix, and they correspond to the parameters $\underline{x}$ of the previous section.

With this simple density matrix one may actually evaluate the free energy of the total system. For our purposes classical statistical mechanics is sufficient, since we shall only be interested in frequencies ω so that $\hbar\omega/kT \ll 1$. Thus the traces may be replaced by integrals and we find

$$E\{\rho\} = \mathrm{Tr}\,\rho H = \int_{-\infty}^{\infty} dQ_1 \cdots dQ_N\, \rho\, H = \Sigma_\ell\, \epsilon_\ell \tag{17}$$

where

$$\epsilon_\ell = \tfrac{1}{2}(\Omega_o^2 + 3\gamma\,\sigma_\ell)\overline{Q}_\ell^2 + \tfrac{1}{4}\gamma\,\overline{Q}_\ell^4 + \tfrac{1}{2}\Omega_o^2\,\sigma_\ell + \tfrac{3}{4}\gamma\,\sigma_\ell^2 - \tfrac{1}{2}\,\overline{Q}_\ell\,\Sigma_{\ell'}\,\overline{Q}_{\ell'}\,v_{\ell\ell'} \tag{18}$$

The entropy of the total system is

$$S = -k\,\mathrm{Tr}\,\rho\ln\rho = \Sigma_\ell\, S_\ell + \text{const.} \tag{19}$$

where

$$S_\ell = \tfrac{1}{2}k\ln\sigma_\ell \tag{20}$$

Now we intend to consider one lattice cell - say the ℓ-th as our system and the rest of the crystal as the bath. One might of course also be interested in the properties of a somewhat larger system, say the ℓ-th cell and its nearest neighbours, as might be more appropriate for the EPR experiment. However, such a larger system requires many more coordinates for its description and since we only want qualitative results, we discuss just the case of one cell here.

We then need the energy E_ℓ of the system. But this energy can of course only be obtained in an approximate way due to the interaction term in eq.(18). The best we can do here is to replace, in eq.(18), $\overline{Q}_{\ell'}$ by its most probable value for a prescribed value of $\overline{Q}_\ell$. Above T_c we expect $\overline{Q}_{\ell'}$ to be proportional to $\overline{Q}_\ell$ so that we may write

$$\overline{Q}_\ell\,\Sigma_{\ell'}\,\overline{Q}_{\ell'}\,v_{\ell\ell'} \simeq \overline{Q}_\ell^2\,\Sigma_{\ell'}\,v_{\ell\ell'}\,\alpha_{\ell'} \tag{21}$$

Here the constant of proportionality $\alpha_{\ell'}$ is unknown, but as we shall see, we obtain the selfconsistency equations of the molecular field approximation if we let $\alpha_\ell = 1$.

Below T_c we should take into account the fact that at large distances $\overline{Q}_{\ell'} \to Q^o$ - the equilibrium value of the order parameter. Thus one expects

$$\overline{Q}_{\ell'} - Q^o = \alpha_{\ell'}(\overline{Q}_\ell - Q^o) \tag{22}$$

so that for $T < T_c$ one has for the coupling term

$$\overline{Q}_\ell \Sigma_{\ell'} \overline{Q}_{\ell'} v_{\ell\ell'} \simeq \overline{Q}_\ell^2 \Sigma_{\ell'} v_{\ell\ell'} \alpha_{\ell'} + \\ + \overline{Q}_\ell Q^o \Sigma_{\ell'} v_{\ell\ell'}(1 - \alpha_{\ell'}) \tag{23}$$

Now introducing the notation

$$v_o = \Sigma_{\ell'} v_{\ell\ell'} \tag{24}$$

and

$$\nu = 2(1 - v_o^{-1} \Sigma_{\ell'} v_{\ell\ell'} \alpha_{\ell'}) \tag{25}$$

we may write the free energy of the system cell

$$F_\ell = \tfrac{1}{2}(\Omega_o^2 - v_o + 3\gamma \sigma_\ell)\overline{Q}_\ell^2 + \tfrac{1}{4}\gamma \overline{Q}_\ell^4 + \tfrac{1}{2}\Omega_o^2 \sigma_\ell + \\ + \tfrac{3}{4}\gamma \sigma_\ell^2 - \tfrac{1}{2} kT \ln \sigma_\ell + \tfrac{1}{2}\nu v_o \left[\overline{Q}_\ell^2 - \overline{Q}_\ell Q^o\right] \tag{26}$$

Assuming $\nu = 0$; i.e. $\alpha_{\ell'} = 1$, and minimizing the free energy in eq.(26) we find the wellknown MFA self-consistency equations (12) for the most probable values of $\overline{Q}_\ell$ and σ_ℓ which are Q^o and σ^o respectively:

$$Q^o(\Omega_o^2 - v_o + 3\gamma\sigma^o + \gamma Q^{o2}) = 0 \tag{27}$$

$$- kT + \sigma^o(\Omega_o^2 + 3\gamma\sigma^o + 3\gamma Q^{o2}) = 0 \tag{28}$$

These equations give rise to a second order phase transition for certain values of the model parameters. The transition temperature is

$$T_c = v_o^2(1 - \Omega_o^2/v_o)/3\gamma \tag{29}$$

and $Q^o = 0$ for $T > T_c$, whereas below T_c one finds $Q^o \neq 0$.

If instead we had used a $\nu \neq 0$, we would obtain equations similar to eqs.(27) and (28) but v_o would be modified. Above T_c this implies only a shift of T_c whereas below T_c the form of F_ℓ is changed by the last term in eq.(26). This change will be discussed in some detail later, at present we only note that using $\nu \neq 0$ is inconsistent with the molecular field approximation and in fact one finds a first order transition unless ν is assumed to be temperature dependent in such a way that $\nu(T_c) = 0$. We are mostly interested in the results above T_c and shall use $\nu = 0$.

Introducing the reduced parameters

$$x_o = Q^o/\sqrt{\sigma^o}\ , \ x = (\overline{Q}_\ell - Q^o)/\sqrt{\sigma^o}\ , \ y = (S_\ell - S_\ell^o)/k \tag{30}$$

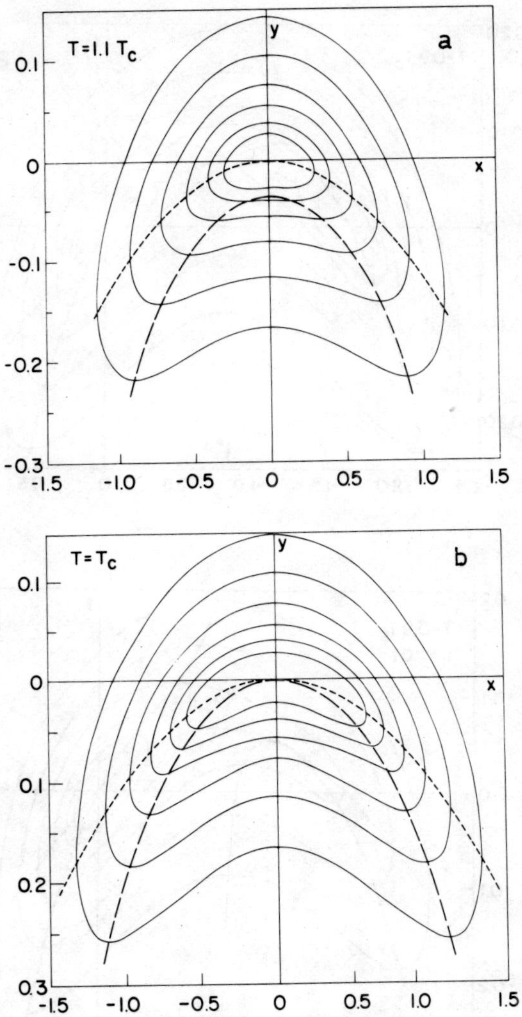

Figure 1. Contourmaps of $\Delta F(xy)/kT$. The dashed and dotted curves are the lines of vanishing driving forces $F_x = -\partial\Delta F/\partial x = 0$ and $F_y = -\partial\Delta F/\partial y = 0$ respectively. The contours shown are for $\Delta F/kT = 1,2,4,8,16,32 \times 10^{-3}$. a - Contourmap for $T = 1.1\ T_c$, b - $T = 1.0\ T_c$, c - $T = 0.9\ T_c$. The model parameters are $\Omega_o^2 = 0.7$, $v_o = 1$, $\gamma = 0.001$, $\nu = 0$.

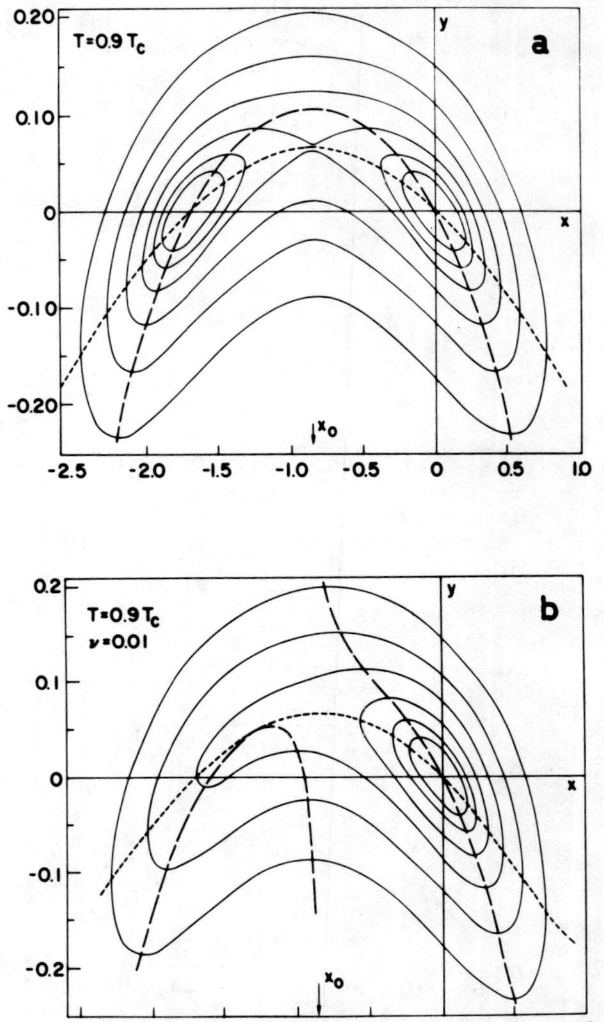

Figure 2. Contourmap of $\Delta F(xy)/kT$ at $T = 0.9\ T_c$, as in figure 1, but with different values of ν. a - $\nu = 0$, b - $\nu = 0.01$.

the change in free energy of the system by changing the system state from a density matrix given by x = 0, y = 0 to one given by x,y is using eqs.(26) to (30),

$$\beta\,\Delta F(x,y) = \beta(F_\ell(x,y) - F_\ell(0,0))$$
$$= Ax^2 + Bx^4 + (C + 6B(x + x_o)^2)(e^{2y} - 1)$$
$$+ 3B(e^{4y} - 1) + 2x_o x(A + 2B(x^2 - 2x_o^2)) - y \qquad (31)$$

where $A = (1 - 1/t)/2$, $t = T\sigma_c/T_c\sigma_o$, $B = \gamma\sigma_o^2/T$, $C = \Omega_o^2\sigma_o/2T$, and $\sigma_c = \sigma_o(T_c)$.

The probability for finding the system in a state described by the parameters x and y is then by equation (3), at equilibrium

$$p_o(xy) \sim \exp^{-\beta\Delta F(x,y)} \qquad (32)$$

Contour maps of $\beta\Delta F(x,y)$ are presented in figs. 1 and 2, with the model parameters Ω_o^2, v_o and γ chosen to have the values of 0.7, 1 and 0.001, respectively. For this set of model parameters the model has a second order phase transition. The effect of finite ν on the free energy surface may be seen in fig. 2. The main effect is that the minimum at $-2x_o$ in fig. 2 a has disappeared, otherwise the figure is qualitatively the same. Above T_c there are only small quantitative differences.

Integrating eq.(32) over y we obtain the probability distribution $p_o(x)$. The EPR lines as observed by Müller and Berlinger (14) measure the derivative of the absorption line L(H) with respect to the magnetic field. The measured quantity, dL(H)/dH, is proportional to $\frac{d}{dx}(p_o(x + x_o) + p_o(-(x + x_o))$ which is plotted in fig. 3 for the model parameters chosen. Müller and Berlinger (14) have used an asymmetry parameter

$$a_s = (A/B) - 1$$

where $A = (dL/dH)_{max}$, $B = -(dL/dH)_{min}$. For the parameter chosen we find $a_s \simeq 16$ at $T/T_c = 0.9$, which is much larger than the observed values. Lower values of a_s could be obtained by choosing different model parameters, but since we work with the simplified model this is not worth while. The essential point is that the simple model does give the qualitatively correct behaviour, and one also finds that the asymmetry increases as T_c is approached from below. If the line shape is calculated at and above T_c, we find that it is flatter at x = 0 than a Gaussian line, so that above T_c the asymmetry parameter a_s should be replaced by a measure for the deviation from a Gaussian line shape.

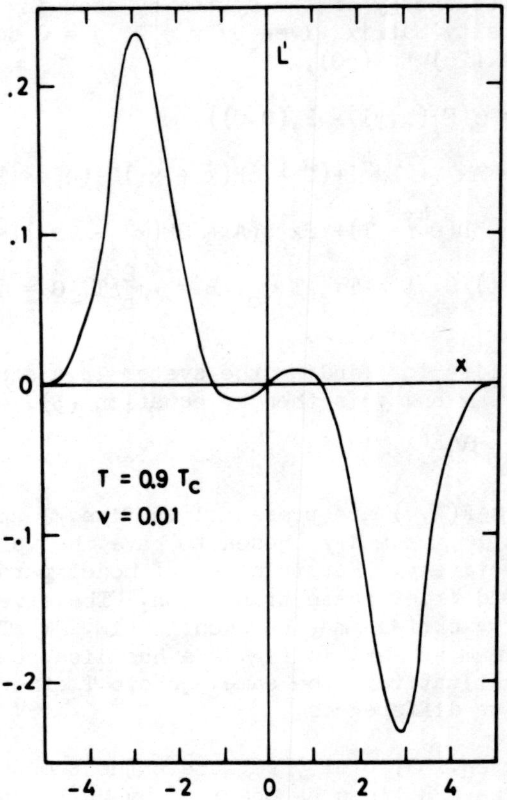

Figure 3. Plot of $L' = d(p_o(x + x_o) + p_o(-(x + x_o)))/dx$ versus x at $T = 0.9\ T_c$. The curve is proportional to the expected derivative of an EPR absorption line measuring the order parameter locally. The model parameters are $\Omega_o^2 = 0.7$, $v_o = 1$, $\gamma = 0.001$, $\nu = 0.01$.

THE CORRELATION FUNCTION

We want to understand the behaviour of the order parameter correlation function defined as

$$\begin{aligned} C_{QQ}(t) &= \langle (Q(o) - Q^o)(Q(t) - Q^o) \rangle \\ &= \int dx'\, dy'\, p_o(x'y') \int dQ'\, \rho(Q';x'y') \cdot \\ &\quad \cdot (Q' - Q^o)\ \langle Q(t) - Q^o \rangle_c \end{aligned} \tag{33}$$

where the conditional average is given by

$$\langle Q(t) - Q^o \rangle_c = \int dx\, dy\, p(xyt|x'y') \cdot$$

$$\cdot \int dQ\, \rho(Q;xy)(Q - Q^o) \tag{34}$$

The integrals all extend from $-\infty$ to $+\infty$. We have for clarity written the fundamental solution of eq.(10) in the form $p(xyt|x'y')$ indicating that

$$p(xyt{=}0|x'y') = \delta(x - x')\, \delta(y - y') \tag{35}$$

The density matrix for our system is given by eq.(16) and has the form

$$(Q;xy) = \frac{1}{\sqrt{2\pi\, \sigma}}\, \exp^{-\frac{1}{2\sigma}(Q-\bar{Q})^2} \tag{36}$$

with $\sigma = \sigma_o\, e^{2y}$, and $\bar{Q} = (x+x_o)\sqrt{\sigma_o}$. The Fokker-Planck equation (10) is two-dimensional for the present case

$$\frac{\partial}{\partial t}\, p(xy;t) = \nabla \cdot \underline{\underline{D}}\, p_o(xy) \cdot \nabla\, \frac{p(xy;t)}{p_o(xy)} \tag{37}$$

and $p_o(xy)$ is given by eq.(32).

The model Hamiltonian is symmetric under the transformation $Q_\ell \to -Q_\ell$, and thus we must expect above T_c that a fluctuation starting at $(x_o y_o)$ decays in the same way as one starting at $(-x_o y_o)$. More precisely let $p_\pm(xy;t)$ be solutions of eq.(37) satisfying $p_\pm(xy;0) = \delta(x_o \pm x)\, \delta(y_o - y)$, then we require that $p_+(xy,t) = p_-(-xy;t)$. Since $p_o(xy) = p_o(-xy)$ it follows from eq.(37) that the diffusion tensor is diagonal, i.e., $D_{xy} = D_{yx} = 0$, whereas $D_{xx} = \mu_x\, kT$, $D_{yy} = \mu_y\, kT$. We thus have arrived at the extreme simplification that the dynamical effects of the bath on the system is contained in only two parameters, the mobilities μ_x and μ_y. In the following discussion we shall assume that $\mu_y \ll \mu_x$, i.e., the center of gravity of the distribution $\rho(Q;xy)$ is much easier to change than its width. We shall see that this assumption will lead to a central peak below T_c, and also above T_c if the nonlinear fluctuations are considered. If one instead chooses $\mu_x \leq \mu_y$ there will be no central peak in the spectrum of the correlation function, $C_{QQ}(\omega)$, either below or above T_c.

We now perform the integrations over Q and Q' in eq.(33) and eq.(34) and find that for $T > T_c$

$$C_{QQ}(t) = \sigma_o\, C_{xx}(t) \tag{38}$$

where we have defined

$$C_{xx}(t) = \int dx'\, dy'\, p_o(x'y')x' \langle x(t)\rangle_c \qquad (39)$$

with

$$\langle x(t)\rangle_c = \int dx\, dy\, p(xyt|x'y')x \qquad (40)$$

Thus the problem of finding the time dependence of the correlation function is reduced to finding the time dependence of $\langle x(t)\rangle_c$. Taking the time derivative of eq.(40) and using the Fokker-Planck equation, we find

$$\frac{\partial}{\partial t} \langle x(t)\rangle_c = \int dx\, dy\, x \nabla \cdot \underline{\underline{D}}\, p_o \cdot \nabla p/p_o \qquad (41)$$

The fundamental solution, $p = p(xyt|x'y')$, starts out as a δ-function at $x'y'$ and approaches the equilibrium distribution $p_o(xy)$. Thus p will vanish with increasing $|x|$ and $|y|$ even faster than p_o, which vanishes exponentially. Thus the surface integrals that arise at $\pm\infty$ when we perform partial integrations in eq.(41) vanish, and we obtain

$$\begin{aligned}\frac{\partial}{\partial t} \langle x(t)\rangle_c &= D_{xx} \int dx\, dy\, p(xyt|x'y') \frac{\partial}{\partial x} \ln p_o \\ &= - \mu_x \left\langle \frac{\partial \Delta F(xy)}{\partial x} \right\rangle_c \end{aligned} \qquad (42)$$

Similarly we find

$$\frac{\partial}{\partial t} \langle y(t)\rangle_c = - \mu_y \left\langle \frac{\partial \Delta F(xy)}{\partial y} \right\rangle_c \qquad (43)$$

$C_{xx}(t)$, whose Fourier transform is the fluctuation spectrum, may be found by solving eqs.(42) and (43) for arbitrary starting values $x_o y_o$, and then averaging over the equilibrium as in eq.(39). Such solutions will be discussed in the following sections.

THE LINEAR CASE AND THE METHOD OF USING A TYPICAL STARTING POINT

In this section we shall discuss linear fluctuation theory where an exact expression for $C_{xx}(t)$ can be obtained. We then show that the exact result may easily be obtained by considering a "test particle" that starts at a "typical starting point" and is driven by the forces $- \nabla \Delta F(x,y)$.

In the linear case one keeps terms only to second order in x and y

$$\beta \Delta F(x,y)_L = \tfrac{1}{2} \omega_x^2 x^2 + C\, xy + \tfrac{1}{2} \omega_y^2 y^2 \qquad (44)$$

where we have used $\omega_x^2 = 8\, Bx_o^2$, $C = 24\, Bx_o$, $\omega_y^2 = 2(1 + 12\, B)$. The linear coupling constant C vanishes above T_c where $x_o = 0$. Eqs. (42) and (43) then become

$$\frac{\partial}{\partial t}\langle x(t)\rangle_c = -\mu_x\omega_x^2\langle x(t)\rangle_c - \mu_x C\langle y(t)\rangle_c$$

$$\frac{\partial}{\partial t}\langle y(t)\rangle_c = -\mu_y C\langle x(t)\rangle_c - \mu_y\omega_y^2\langle y(t)\rangle_c \tag{45}$$

Introducing the relaxation rates $\gamma_x = \mu_x\omega_x^2$, $\gamma_y = \mu_y\omega_y^2$, we find the following solution for $\gamma_y \ll \gamma_x$

$$\langle x(t)\rangle_c = (x(0) - x_1)e^{-\gamma_x t} + x_1 e^{-\gamma_y t} \tag{46}$$

Here we have defined $x_1 = -y(0)C/\omega_x^2$. In the case that γ_y is not much less than γ_x the solution is somewhat more complicated but easily obtained. From eq.(39) we then find the cerrlation function

$$C^L_{xx}(t) = \langle x(0)\ \langle x(t)\rangle_c\rangle = (\langle x(0)^2\rangle - \\ - \langle x(0)x_1\rangle)e^{-\gamma_x t} + \langle x(0)x_1\rangle e^{-\gamma_y t} \tag{47}$$

and using $p_0(xy) \sim \exp(-\beta\Delta F_L)$ we have $\langle x(0)x_1\rangle = \langle x(0)^2\rangle\ C^2/\omega_x^2\omega_y^2$. The spectrum of the order parameter correlation function consists of two Lorentzians

$$C^L_{xx}(\omega) = (\langle x(0)^2\rangle - \langle x(0)x_1\rangle)\frac{2\gamma_x}{\omega^2+\gamma_x^2} + \\ + \langle x(0)x_1\rangle\frac{2\gamma_y}{\omega^2+\gamma_y^2} \tag{48}$$

so that we have a central peak of width γ_y. The ratio of the integrated "intensity" of the central peak to the total integrated intensity is

$$r = \frac{\int_{-\infty}^{\infty} d\omega\ C^{cent}_{xx}(\omega)}{\int_{-\infty}^{\infty} d\omega\ C_{xx}(\omega)} = \frac{\langle x(0)x_1\rangle}{\langle x(0)^2\rangle} \tag{49}$$

which is finite in linear theory as long as the linear coupling constant, C, does not vanish.

In the following we show that the exact result, eq.(47), can be obtained by envisaging a "test particle" that starts at x(0)y(0) and follows eqs.(42) and (43) without averages. In this interpretation the fundamental solution $p(xyt|x(0)y(0))$ is the probability distribution for finding the test particle at x,y at time t if it started at $x(0)y(0)$. Neglecting the averages really means that we neglect the spreading out of $p(xyt|x(0)y(0))$. This will be a good approximation if neighbouring starting points $x(0)y(0)$ have qualitatively the same behaviour.

Thus we consider a "particle" starting at $x(0)y(0)$ which moves with the mobilities μ_x and μ_y driven by the forces $F_x = -\partial\Delta F/\partial x$, $F_y = -\partial\Delta F/\partial y$. Choose a typical starting point $x(0)$. For a given value of $x(0)$ the most probable starting point for y is $y(0)$ obtained from $\partial p(x(0)y)/\partial y = 0$ or $F_y(x(0)y(0)) = 0$, for the linear case at hand we find $y(0) = -x(0)C/\omega_y^2$. The test particle drifts with a relaxation rate γ_x, assumed to be much larger than γ_y, driven by the force F_x until it reaches the point x_1, where the driving force in the x-direction vanishes. Setting $F(x_1 y) = 0$ gives $x_1 = -y(0)C/\omega_x^2 = x(0)C^2/\omega_x^2\omega_y^2$. From then on the particle will move very slowly and proceed along a curve $F_x(x,y) \simeq 0$, with a relaxation rate which is essentially γ_y. Thus we have found that the particle will follow a path given by

$$x(t) = (x(0) - x_1)e^{-\gamma_x t} + x_1\, e^{-\gamma_y t} \qquad (50)$$

The correct correlation function is then obtained by multiplying eq.(50) by $x(0)$ and averaging with $p_o(x)$. Even this last averaging step is avoided if one right away chooses as a typical starting value for x, $x_o = (\langle x(0)^2\rangle)^{\frac{1}{2}}$.

The importance of this last approach is that the correct time dependence of the order parameter correlation function is obtained simply by considering the behaviour of a "test particle" starting at a typical point on the potential surface of $\Delta F(xy)$. Since this approach gave the exact result in the linear case, one may hope that it gives at least qualitatively correct results even in the nonlinear case where an exact solution of eqs.(42) and (43) is impossible.

THE NONLINEAR CASE

If instead of keeping only terms to second order in $\Delta F(x,y)$ one uses the full expression, eq.(26), we can no longer find an exact solution of eqs.(42) and (43) for $\langle x(t)\rangle_c$ or the correlation function. The method of using a typical starting point may, however, still be used. Consider fig. 1.b, where $T = T_c$, and start the test particle at some typical value, say $x(0) = 2$ which is of the order of $\langle x^2\rangle^{\frac{1}{2}} = 1.63$. The most likely starting point for $y = y(0)$, for a prescribed value of $x = x(0)$ is obtained from the equation $F_y(x(0)y(0)) = 0$. We have solved numerically eqs.(42) and (43) without averages using $\mu_y = 0.01\,\mu_x$. The resulting path is seen in fig. 4.a, whereas the time dependence of $x(t)$ is seen in fig. 4.b.

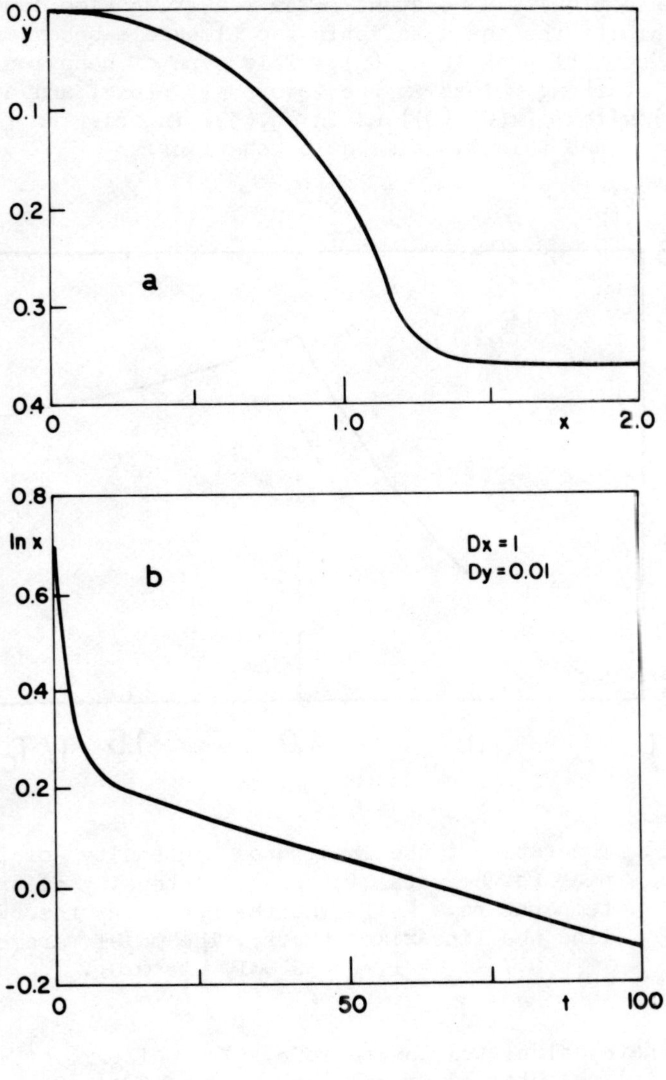

Figure 4. a - Path of test particle starting at (2,- 0.36) at $T = T_c$ and with $\mu_y = 0.01\ \mu_x$.

b - Time dependence of x(t) for the test particle following the path in fig. 2.a.

The test particle indeed moves at first rapidly with a relaxation rate $\gamma_x \gg \gamma_y$, until the curve $F_x \simeq 0$ is reached, from then on $x(t)$ decays much more slowly with a relaxation rate γ_y which is not quite a constant. The resulting time dependence of $x(t)$ exhibits the characteristic two timescale behaviour that implies a central peak in $C_{xx}(\omega)$. This type of behaviour results for all starting points in the lower half plane, and averaging $x(0)x(t)$ with $p_o(x(0)y(0))$ as in eq.(39) in order to find $C_{xx}(t)$, will <u>not</u> erase this two timescale behaviour.

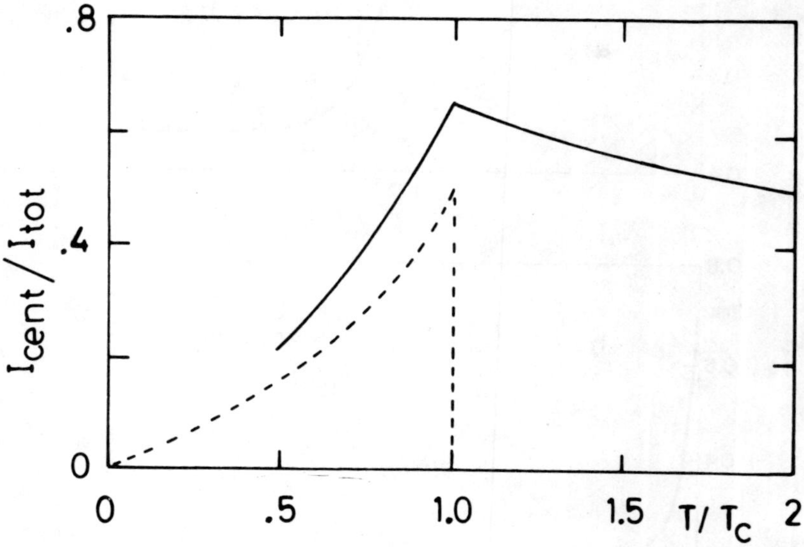

Figure 5. The ratio of the integrated "intensity" of the central peak to the total integrated intensity versus reduced temperature. Full line the nonlinear theory, dashed line the linearized theory. The model parameters are $\Omega_o^2 = 0.7$, $v_o = 1$, $\gamma = 0.001$, $\nu = 0.01$.

We have calculated the ratio $r = I_{cent}/I_{total}$ of eq.(49) using $x_1(y)$ as obtained from $F_x(xy) = 0$. In fig. 5 the resulting r is plotted versus reduced temperature. We have also plotted r as obtained in the linear theory of the previous section. The linear theory not only fails to give a central peak above T_c, it also gives a too small central peak at and below T_c.

Below T_c there is of course a region of linear coupling, but as mentioned above, linearizing the theory gives a too small central peak. Above T_c the central peak intensity decreases since the curve $F_x = 0$ moves further down in the x-y plane as the temperature is increased. Thus, we reach the conclusion that the central peak is expected to vanish at sufficiently high temperatures.

The most important qualitative conclusion of this discussion is that since the $\Delta F(xy)$ surface varies continuously with temperature, one must expect $C_{xx}(t)$ to be a continuous function of temperature. Also the same relaxation processes that give rise to a central peak below T_c - even in the linearized theory - give rise to one above T_c <u>if</u> the nonlinear fluctuations are included. Thus, we are led to interprete the width of the central peak as given by the relaxation rate of the local entropy S_ℓ both below and above T_c.

Quantitatively the simple model is not really appropriate for say $SrTiO_3$ where rather the model Hamiltonian of ref. 13 should be used. Also it seems that choosing a system as small as only one unit cell exaggerates the fluctuations, thus the asymmetry, a_s, of the derivative of the EPR line, fig. 3, was too large, and the intensity of the central peak seems to fall off rather slowly with increasing T above T_c. If one chooses a somewhat larger system say one cell and its nearest neighbours, one would expect the asymmetry to decrease and the central peak to vanish more rapidly - but the nonlinear effects would still remain.

An extension of the previous discussion to the central mode proper - as seen in the inelastic neutron scattering experiments on $SrTiO_3$ - is of course desirable. In this case one measures the spectrum of the correlation function $C_q(t) = \langle x_{-q}(0)x_q(t)\rangle$, where x_q is proportional to the softmode normal coordinate Q_q, with $q = q_o$ - the soft mode wave vector. If the density matrix is factorized in q-space, one obtains the RPA approximation and the system is described by two coordinates,

$$x_q = (Q_q - \bar{Q}_q)/\sqrt{\Delta_q^o}$$

$$y_q = (S_q - S_q^o)/k \tag{51}$$

where

$$S_q = k[(n_q+1)\ln(n_q+1) - n_q \ln n_q] \tag{52}$$

is the entropy of the soft mode coordinate given in terms of the occupation number n_q. In the classical limit we have

$$S_q = \tfrac{1}{2} k \ln(kT\Delta_q) \; ; \quad \Delta_q \simeq kT/\epsilon_q^2 \tag{53}$$

where ϵ_q is the softmode frequency. Having defined the system state by specifying $x_q y_q$, and taking all other phonons to belong to the bath, one again arrives at an expression for the free energy change of the form given in eq.(31), but with different coefficients.

The slow relaxation time of the y-coordinate will then be related to the relaxation time of the soft phonon occupation number, whereas the fast relaxation of the x-coordinate describes the soft mode damping. Thus, both above and below T_c, we find that the central peak is related to a time scale that separates the frequency domain where the phonon occupation number may follow the changes in order parameter from the region where it cannot follow such changes.

We have, however, difficulties with this approach near T_c, since the RPA approximation gives a first order phase transition. The problem is really that both $\Delta^o{}_{q_o}$ and $Q^o{}_{q_o}$ diverge at $T_c (Q^o{}_{q_o} \sim \sqrt{N}, T \leq T_c)$, and it is then difficult to control the approximations necessary in the fourth order term of the Hamiltonian. We hope to discuss these problems in detail in a later publication.

ACKNOWLEDGEMENT

Stimulating discussions with Jens Lothe, Harry Thomas and Erling Pytte are gratefully acknowledged.

REFERENCES

1. For a review see the paper by J.D. Axe, this conference.

2. R.A. Cowley, J. Phys. Soc. Japan, Suppl. 28, S239 (1970).

3. G.J. Coombs and R.A. Cowley, J. Phys. C. Solid St. Phys. 6, 121 (1973).
R.A. Cowley and G.J. Coombs, J. Phys. C. Solid St. Phys. 6, 143 (1973).

4. For a review of the theoretical approaches see the paper by F. Schwabl, this conference.

5. R. Silberglitt, Solid State Commun. 11, 247 (1972).

6. T. Riste, E.J. Samuelsen, K. Otnes and J. Feder, Solid State Commun., 9, 1455 (1971).
J. Feder, Solid State Commun. 9, 2021 (1971).

7. Th. von Waldkirch, K.A. Müller, W. Berlinger and H. Thomas, Phys. Rev. Letters 28, 503 (1972).
Th. von Waldkirch, K.A. Müller and W. Berlinger, Phys. Rev. B 5, 4324 (1972 and 7, 1052 (1973).

8. See for instance P.A.M. Dirac, The Principles of Quantum Mechanics, Oxford University Press, 4th Ed., London 1959).

9. This expression is a slight generalization of the corresponding expression used in L.D. Landau and E.M. Lifshitz, STATISTICAL PHYSICS, Pergamon Press, London 1959, chapter XII.

10. A discussion using such concepts has been given by E.T. Jaynes, Phys. Rev. 108, 171 (1957).

11. For a discussion of Fokker-Planck equations see the review by S. Chandrasekhar, Rev. Mod. Phys. 15, 1 (1943).

12. H. Thomas, in Structural Phase Transitions and Soft Modes, edited by E.J. Samuelsen, E. Andersen and J. Feder, Universitetsforlaget, Oslo, Norway, 1971, p. 15.

13. E. Pytte and J. Feder, Phys. Rev. 187, 1077 (1969).
J. Feder and E. Pytte, Phys. Rev. B 1, 4803 (1970).

14. K.A. Müller and W. Berlinger, Phys. Rev. Letters 29, 715 (1972).

SOME FURTHER OBSERVATIONS OF SOFT PHONON LINE SHAPES IN $KMnF_3$*

S. M. Shapiro, J. D. Axe, and G. Shirane

Brookhaven National Laboratory

Upton, New York 11973 U. S. A.

T. Riste

Research Establishment, Kjeller, Norway

There are two known structural phase transitions in $KMnF_3$ which occur at $T_c^R = 186.5°K$ and $T_c^M = 90°K$.[1] The higher temperature transition is driven by a phonon at the R point (1/2 1/2 1/2) of the Brillouin zone whose frequency approaches zero as $T \to T_c^R$. The dynamics of this transition has been extensively studied and are relatively well understood.[2] Recently a central peak in the soft phonon line shape associated with this phase transition has been observed.[3] The intensity of the central peak diverges as $T \to T_c^R$ and the observed energy linewidth is always resolution limited. As in $SrTiO_3$ only an upper limit to the energy linewidth could be estimated.

In this paper we report on our observation of a central peak associated with the phase transition at $T_c^M = 90°K$ where a phonon at the M point (0 1/2 1/2) becomes soft.[1] We also discuss our attempts to measure a linewidth of the central peak.

Figure 1 shows the inelastic scattering spectra of $KMnF_3$ at an M point for several temperatures above $T_c^M = 90°K$. We observe a sharp peak centered around E = 0 superimposed on the broad overdamped phonon. All higher order contamination is eliminated by the use of a beryllium filter. At T = 140°K the entire

*Work performed under the auspices of the U. S. Atomic Energy Commission and NATO Research Grant.

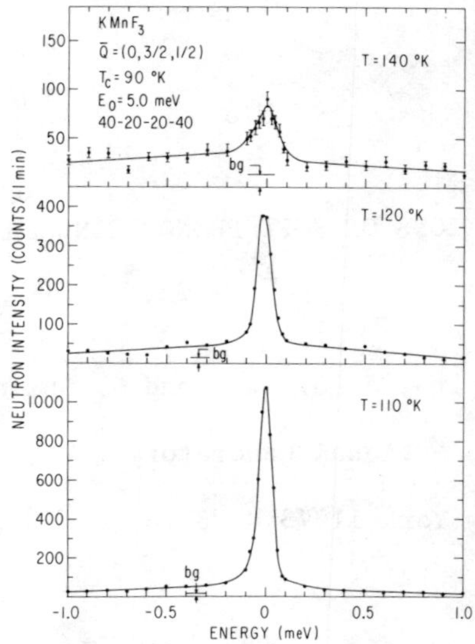

Figure 1. Scattered neutron spectra of $KMnF_3$ at $q = q_M$ at several temperatures. bg represents the level of the room background.

profile of the broad excitation was studied with 13.8 meV incident neutron energy and shows convincingly that the shoulders observed in Fig. 1 represent an overdamped phonon-like excitation. Similar behavior was seen in $KMnF_3$ near T_c^R as shown in Ref. 3. In Fig. 1 we note that the intensity of the central peak diverges as $T \to T_c^M$ and the linewidth remains constant. These characteristics are identical to the central peak observed at T_c^R in $KMnF_3$ and $SrTiO_3$.[3] Unfortunately, the phonon is highly overdamped and as in the case near T_c^R, we are unable to obtain any detailed quantitative values of the parameters used to characterize the power spectrum.[3,4] Nonetheless, we can make some qualitative statements about the central peak deduced from the spectra. Since the sidebands change very little as a function of temperature, we infer that $\omega_\infty^2(T)$ is a weak function of T over the temperature range studied.

The central peak is present at temperatures $T - T_c > 50°K$ and thus is visible over a larger temperature range than the central peak associated with the phonon softening at T_c^R. Also the central peak is observable at relatively large q values, $|q - q_M| = .15Å^{-1}$ at $T - T_c = 10°K$.

In characterizing the central peak in systems that have been

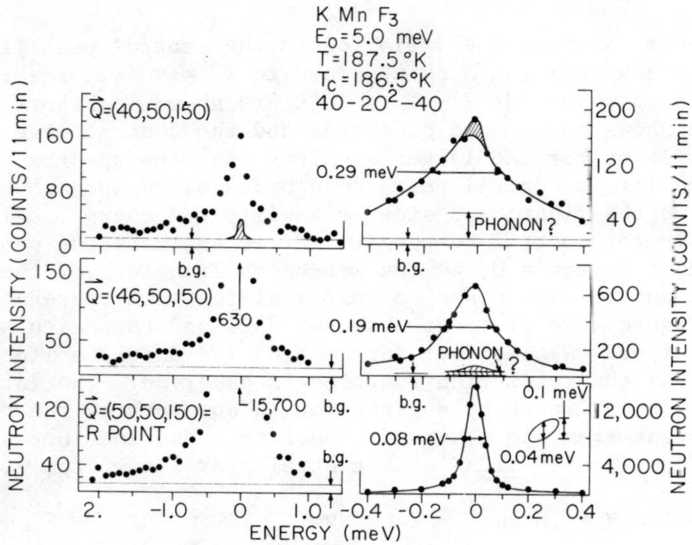

Figure 2. Scattered neutron spectra of $KMnF_3$ for several q near q_R. The hatched area is the incoherent scattering and the estimated phonon level near energy zero. The size of the resolution function is given in the lower right hand corner.

studied to date[4] an important piece of information is lacking: a direct measurement of an intrinsic energy linewidth.[5] We mentioned above that no discernable energy linewidth at either the R or M points was measured at any temperature. However, it is still possible that as q is varied away from the precise soft mode wave vector there will be sufficient q dependence of the decay processes to enable the direct observation of the central peak linewidth.

We first studied the central peak associated with T_c^R at temperatures close to T_c^R. Figure 2 shows the results of varying q from R to M at $T-T_c^R = 1°K$. Although an energy width is apparent, upon closer examination one cannot clearly distinguish the central peak from the overdamped phonon. Thus it is impossible to determine what fraction of the spectra is central peak and what part is phonon. These results are, at most, ambiguous. We also attempted measurements at higher temperatures where the phonon is broader and a two-component spectrum is clearly discernable at the R point. Unfortunately, at these temperatures ($T - T_c^R \sim 10°K$) the central peak away from R is weak and we are unable to observe it for sufficiently large q values to determine an intrinsic linewidth.

We next examined the variation in the central peak linewidth at several q's near the M point close to $T_c^M = 90°K$. The results for several q values at $T - T_c^M = 10°K$ are shown in Figure 3. The left side shows the phonon sidebands and the central peak. It is clear, at least for the larger q values that the spectra can be decomposed into a central peak and a broad overdamped phonon component. On the right hand side we amplify the energy scale and show mainly the central component. If we estimate the phonon intensity near Energy = 0, we can measure a linewidth of the central component and indeed we see an energy width which increases with q. In Figure 4 we plot our observed data and the calculated energy ellipse[6] We see that for q > .05 (a* units) a linewidth greater than the resolution function is observed. The intrinsic width can be estimated by subtracting in quadrature, the resolution height from the observed linewidth. We find that at $T - T_c = 10°K$ and $q = .08a^* = .12Å^{-1}$ the central peak linewidth, γ, is

$$\gamma \sim .1 \text{ meV} = .8 \text{ cm}^{-1} = 2400 \text{ MHz}$$

Although the spectra show a peak around zero energy with a

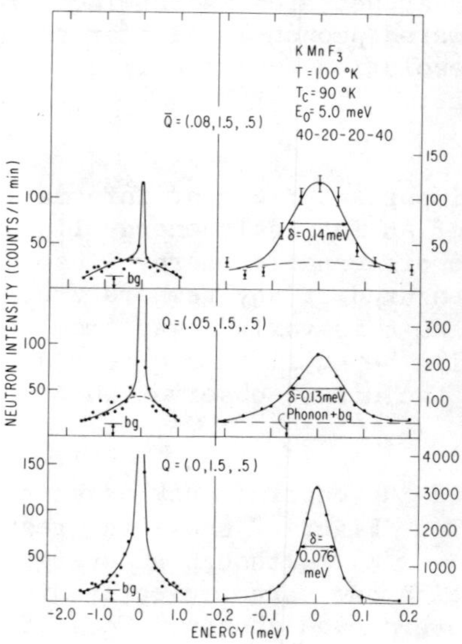

Figure 3. Scattered neutron spectra of $KMnF_3$ for several q near q_M. The incoherent scattering has been subtracted and the dashed line represents the estimated phonon intensity and room background.

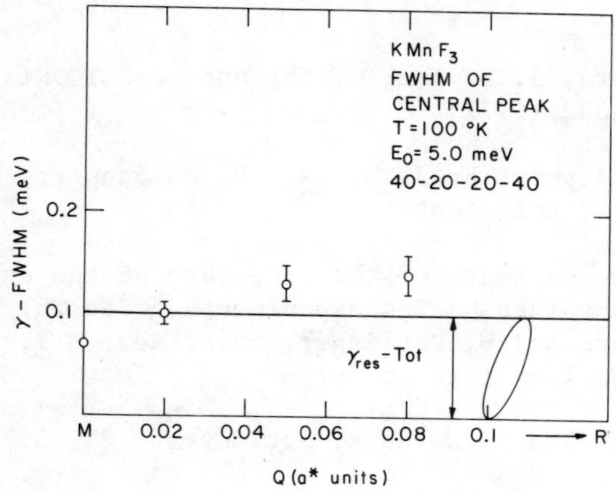

Figure 4. Energy linewidth (γ) of the central peak in $KMnF_3$ vs q (1a* unit = 1.50 Å^{-1}). Also shown is the size of the instrumental resolution function.

linewidth greater than the instrumental width and can be interpreted as the central peak, some words of caution should be added. The phase transition near 90°K has not been extensively studied and the phonon dispersion and its temperature dependence is unknown. Thus, for example, one cannot eliminate the possibility of two overdamped phonon branches near the M point with different linewidths which will give the spectral line shapes observed in Fig. 3. On the other hand, the frequency of such an excitation would require a real part which is anomalously small or an imaginary part anomalously large to give a linewidth as narrow as the central peak observed in Figure 3. To more clearly establish the nature of the central component a more complete knowledge of the phonon dispersion surfaces would be required.

In summary, we have shown that a central peak exists above T_c^M at the M point in the Brillouin zone. Also, we studied its energy linewidth and showed a measurable energy width is observable as q is varied from M to R.

REFERENCES

1. V. J. Minkiewicz and G. Shirane, J. Phys. Soc. Japan 26, 674 (1969); G. Shirane, V. J. Minkiewicz, and A. Linz, Solid State Commun. 8, 1941 (1970).

2. K. Gesi, J. D. Axe, G. Shirane, and A. Linz, Phys. Rev. B 5, 1933 (1972).

3. S. M. Shapiro, J. D. Axe, G. Shirane, and T. Riste, Phys. Rev. B 6, 4332 (1972).

4. See also the paper by J. D. Axe, S. M. Shapiro, G. Shirane, and T. Riste, this conference.

5. There is one estimate of the linewidth of the central peak in $SrTiO_3$ obtained by ESR experiments. See Th. von Waldkirch, K. A. Müller, and W. Berlinger, Phys. Rev. B 7, 1052 (1973) and K. A. Müller, this conference.

6. M. J. Cooper and R. Nathans, Acta Cryst. 23, 357 (1967).

CRITICAL SOUND VELOCITY AND ATTENUATION IN $KMnF_3$

K. Fossheim[x], D. Martinsen[x], and A. Linz[+]

[x]Physics Department, NTH, Trondheim.
[+]Center for Materials Sciences and Engineering, MIT, Mass. 02139.

The critical dynamics of $SrTiO_3$ was first studied by ultrasonic attenuation measurements (1). After the discovery of the central mode (2) an alternative to the original soft mode interpretation (3) has been given by Schwabl (4), suggesting that the critical exponent for attenuation of longitudinal sound along [100] mainly expresses the temperature dependence of the central mode. Thus ultrasonic measurements should be highly relevant in the continuing search for a full description of the central mode.

The present study of $KMnF_3$ was undertaken as an extension of the $SrTiO_3$-work, since the R_{25}-transition in $KMnF_3$ involves the same soft mode. The work was further motivated by the superior crystal quality which could be obtained in $KMnF_3$.

The measurements were made on a single crystal grown in a helium atmosphere by Czochralski technique from a melt containing a few per cent excess of KF. Similar samples have previously shown a mosaic spread of less than 10' of arc. A quality test by neutron scattering performed by Dr. E. Samuelsen at Kjeller indicated vanishing background and very high crystal quality. However, the most sensitive test is probably given by the ultrasonic experiments themselves. During the experiments reported here there were no signs of interference effects in the pulse echo train. Unfortunately the sample was slightly damaged at a later stage, and further results could not be obtained.

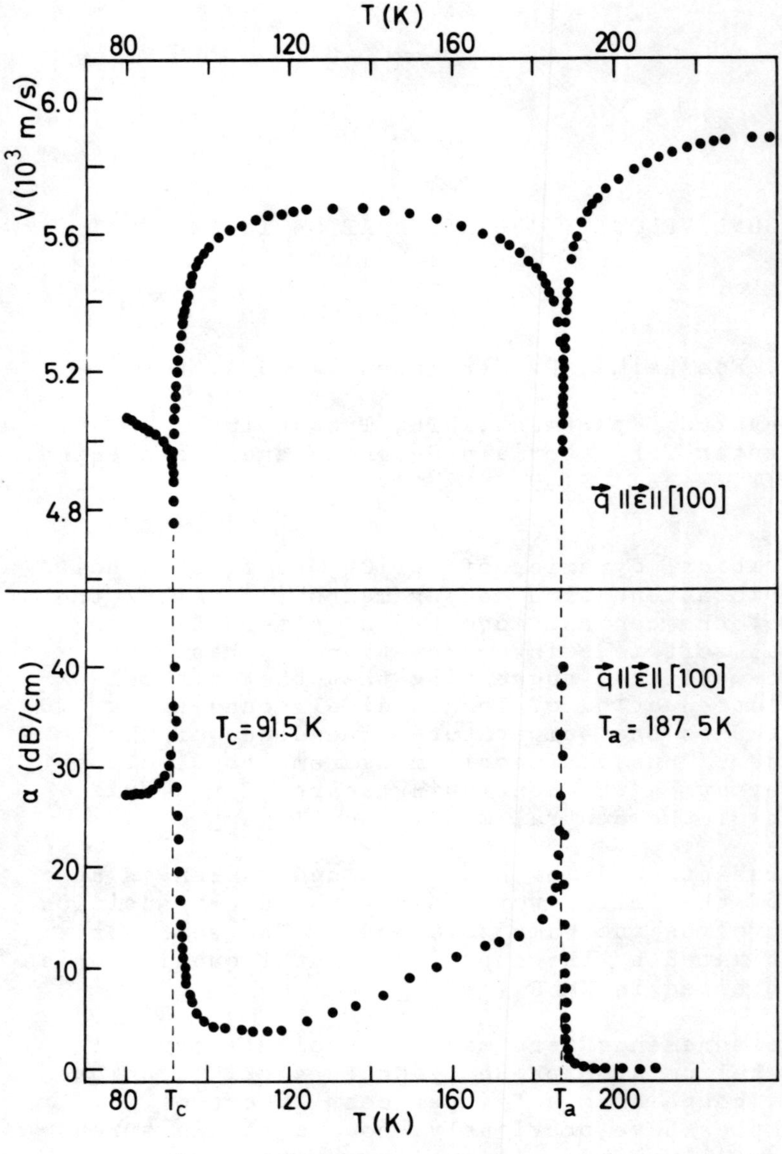

Figure 1. Attenuation and velocity at 11,7 MHz with longitudinal sound along [100] .

Temperature control and resolution in these experiments was 0,02 K over arbitrary time intervals, and the absolute accuracy was 0,1 K by use of copper resistance thermometers. Simultaneous point by point measurements were always made of attenuation and velocity. Before each reading the temperature was kept constant for at least 15 minutes.

The results of velocity and attenuation measurements for 11.7 MHz longitudinal soundwaves along [100] are shown in figure 1.

Above the upper transition the results agree in most respects with independent velocity (5) and attenuation (6) measurements except that a changeover (see below) may be present in our results.

The analysis of critical effects at the R_{25}-transition in $KMnF_3$ is complicated by the 1. order nature of the transition, as evidenced by a hysteresis of about 0,2 K found in our measurements. In plotting the attenuation data we have identified the critical temperature T_a by the minimum in the velocity. This is also in agreement with the maximum of attenuation within our data resolution. A somewhat increased resolution near the attenuation maximum would be desireable to check this further. There is no background attenuation to be corrected for. Thus figure 2 is a direct plot of the data, and contains no parameter adjustments. It is seen that the data indicate two characteristic temperature dependences.

Writing the attenuation as

$$\alpha \sim \omega^n |\varepsilon|^{-\rho}$$

where ω is the sound frequency and ε the reduced temperature, $(T - T_a)/T_a$, we find $\rho = 1,25$ within 1 K above T_a, and $\rho = 1,95$ further away. In Schwabl's theory such exponents arise due to the anisotropic correlations which cause an effective change of dimensionality. Schwabl finds $\rho_3 = \gamma_3 + \alpha_3$ near T_a, and $\rho_2 = 2 - \nu_2\eta_2$ further away. (Exponent indices refer to dimensionality). It should be born in mind that a re-interpretation of the data is possible if the transition temperature is allowed to vary substantially from that value which gives the minimum in the velocity. We believe this possibility ought to be considered, on the basis of more extensive data. Our interpretation is thus only tentative.

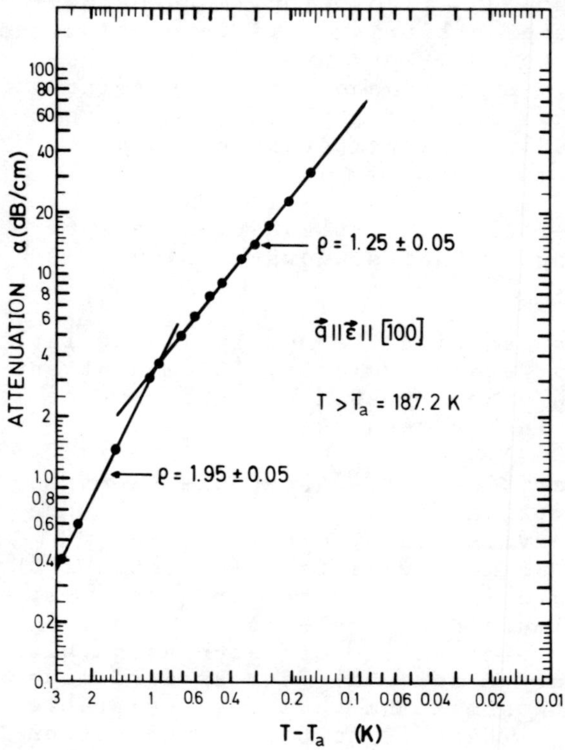

Figure 2. Attenuation of longitudinal waves along [100] above T_a, at 11,7 MHz.

A further complication is found when analyzing the frequency dependence of the attenuation. We do not find the simple ω^2-dependence that is normally expected. Instead we observe a weakly temperature dependent exponent n = n(T) around 1,3-1,4. A similar result was obtained by Furukawa et. al (6). Also, we have just been informed at this meeting of unpublished work by Matsuda et. al (7) supporting this observation. They find n varying from 1,2 to 1,6 when going from T_a to 193 K. The conclusion must be that a simple power law does not hold for longitudinal sound along [100] at frequencies below 100 MHz.

We want to point out one possible clue: In Schwabl' theory one can trace such behaviour back to a violation of the condition $\omega \ll \Gamma_c$, where Γ_c is the central peak width. Space limitation prevents us from discussing this important point further here. However, we would like to suggest that a deviation from ω^2-dependence in the attenuation should also lead to dispersion in sound velocity, a feature not included in Schwabl's theory, and to non-ideal temperature exponents (for instance deviations from $\rho_3 = \gamma_3 + \alpha_3$ etc.). In fact, the whole data analysis has to be carried out in a more complete manner, i.e. in terms of the full T, ω-dependence rather than limiting behaviour. This, however, requires more data than we have at the present. We believe that this analysis could best be carried out in cases of second order transitions, and that when fully performed it will eventually give the relevant parameters that are plugged into the phenomenological structure factor used in neutron scattering (8), and in Schwabl's theory for ultrasonic attenuation.

The data below T_a is to some extent obscured by a shoulder (see figure 1) which we ascribe to domain scattering. The analysis suggests a temperature exponent $\rho = 1,20$ below T_a.

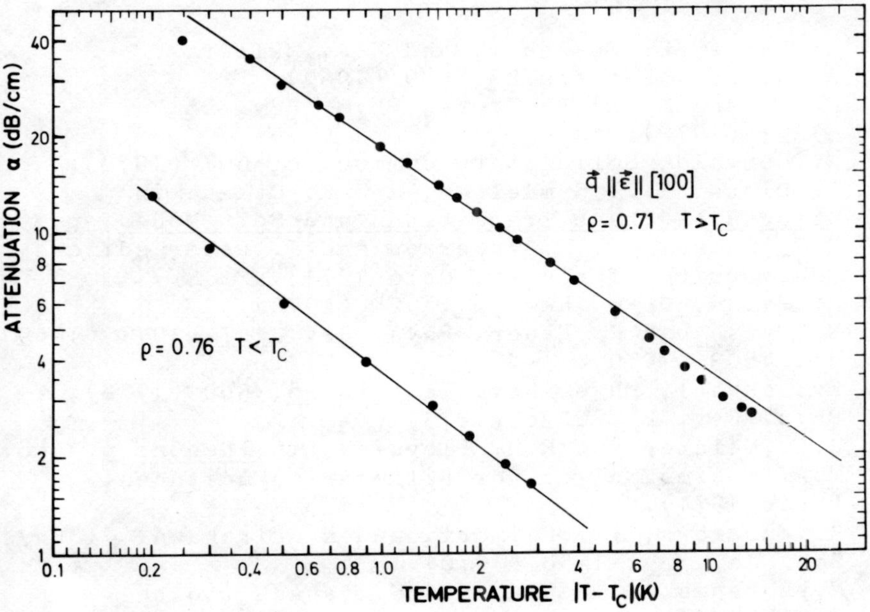

Figure 3. Attenuation of longitudinal waves along [100] above and below T_c, at 11,7 MHz.

The velocity data do not give a unique temperature exponent. There is a gradual decrease in slope as T_a is approached from either side. The minimum is very near symmetric close to T_a, and the limiting exponent is $\mu = 0{,}08 \pm 0{,}03$ out to 0,3 K, again identifying the transition temperature by the minimum in velocity.

The attenuation of longitudinal waves along [100] near the lower transition is plotted on a log-log scale in figure 3. We find that the transition occurs at 91,5 K. The transition is of 2. order, i.e. no hysteresis is observed within our temperature resolution. The attenuation exponents are $\rho = 0{,}76$ for $T < T_c$ and $\rho = 0{,}71$ for $T > T_c$. The velocity exponent is $\mu = 0{,}18 \pm 0{,}03$ for $T > T_c$. Again, some correction may be necessary below T_c due to domain attenuation. No theory has been worked out for this transition yet.

One of the authors (KF) wishes to thank Drs. F. Schwabl and K.A. Müller for discussions, and for communicating their results prior to publication.

REFERENCES

1. B. Berre, K. Fossheim, and K.A. Müller, Phys. Rev. Letters 23, 589 (1969).
K. Fossheim and B. Berre, Phys. Rev. B5, 3929, (1972).
W. Rehwald, Solid State Commun. 8, 607 (1970).
2. T. Riste, E.J. Samuelsen, and K. Otnes, in Structural Phase Transitions and Soft Modes, p.395. E.J. Samuelsen, E. Andersen and J. Feder editors (Universitetsforlaget, Oslo 1971).
3. E. Pytte, Phys. Rev. B1, 924 (1970).
E. Pytte and J. Feder, Phys. Rev. 187, 1077 (1969); B1, 4803 (1970).
4. F. Schwabl, Phys. Rev. Letters 28, 500 (1972); Phys. Rev. B7, 2038 (1973).
5. R.L. Melcher and R.H. Plovnick, in Phonons p. 348. M.A. Nusimovici editor (Flammarion Sciences, Paris 1971).
6. M. Furukawa, Y. Fujimori, and K. Hirakawa, J. Phys. Soc. Jap. 29, 1258 (1970).
7. M. Matsuda, I. Hatta, and S. Sawada (to be published).
8. S.M. Shapiro, J.D. Axe, G. Shirane, and T. Riste. Phys. Rev. B6, 4332 (1972).

RAMAN SPECTRAL STUDY OF THE UPPER PHASE TRANSITION IN $KMnF_3$

D.J. Lockwood and B.H. Torrie

Department of Physics, University of Edinburgh,
Edinburgh, Scotland

INTRODUCTION

Crystals with the perovskite structure, of which $KMnF_3$ is an example, undergo phase transitions as a result of rotations of the anion octahedra about the four-fold axes of the high-temperature cubic phase. The archetype of such transitions is the second order cubic to tetragonal transition in $SrTiO_3$ that results from the condensation of one of the modes associated with the zone boundary R ($\frac{1}{2}$, $\frac{1}{2}$, $\frac{1}{2}$) point. Considerable theoretical and experimental effort has been expended in developing a detailed understanding of this phase transition, as is evident from other papers presented here. $KMnF_3$ undergoes a similar phase transition at T_c = 186K except that in this case the transition is slightly first order in nature, with a hysteresis of a degree or less. In spite of this difference, experimental work to date indicates that the phase transitions in the two crystals can both be described by the same basic theory (1).

We have measured the Raman scattering from a sample of $KMnF_3$ with emphasis on the temperature dependence of both the soft and hard modes in the tetragonal phase. The sample was grown for us by Dr. D.A. Jones of Aberdeen University, and although the sample did show considerable scattering from defects, its quality appears to be better than that of many of the other $KMnF_3$ crystals that have been examined (2).

EXPERIMENT

The Raman spectrum was excited with argon laser light at 476.5 nm so as to avoid fluorescence problems: the laser power was 400 mW. The light scattered at 90° was analysed using a double monochromator with a spectral slit width of 1.7 cm^{-1} for all but the hard mode intensity measurements, where it was 7 cm^{-1}. A digitised data collection system enabled direct computer processing of the results. The $KMnF_3$ crystal was mounted on the cold finger of a continuous-flow cryostat: the crystal temperature could be controlled to within 0.1 K. After changing the sample temperature, a period of at least 15 mins. was spent waiting for thermal equilibrium to be established before running the new spectrum: this period was determined experimentally to be satisfactory. Cycling the temperature below T_c did not affect the results of soft mode measurements. In the hard mode intensity measurements, waiting for an hour at a temperature within 5 K of T_c did not result in any change in the measured intensity. Our experience in this respect agrees with observations made from measurements of the specific heat of $KMnF_3$ (3).

Typical Raman spectra for $T<T_c$ are shown in Fig. 1.

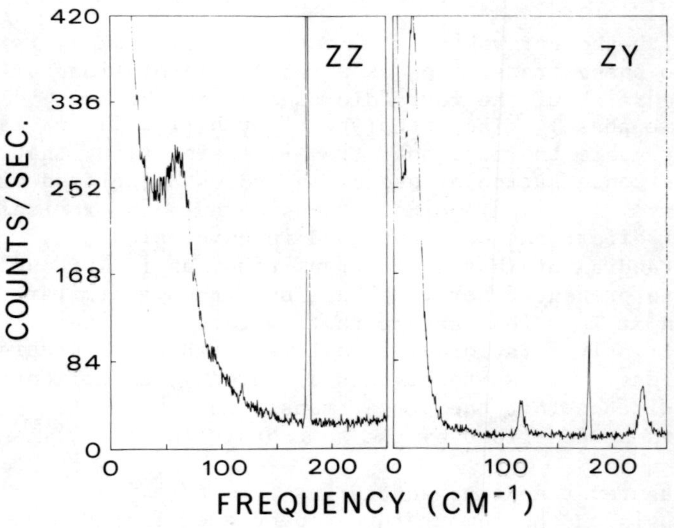

Figure 1. Polarized Raman spectra of $KMnF_3$ at 125 K. The sharp feature at 178 cm^{-1} is a plasma line.

The spectrum with (ZZ) polarization exhibits a soft mode of A_{1g} symmetry and the (ZY) spectrum shows the other soft mode of E_g symmetry. Five other Raman active modes are predicted for the tetragonal phase and these have all been observed. Full details of the group theory and observations for both the tetragonal and orthorhombic (T<92 K) phases will be published elsewhere.

SOFT MODE TEMPERATURE DEPENDENCE

The soft-mode spectra were analysed with a computer using a least-squares routine. The E_g mode was fitted to a damped simple harmonic oscillator (DSHO) model of the form

$$\left[S\omega_o^2\gamma^2\omega(\bar{n}(\omega)+1)\right]/\left[(\omega_o^2-\omega^2)^2+\gamma^2\omega^2\right],$$

where S is the strength, ω_o the frequency, and γ the width of the oscillator; $\bar{n}(\omega)$ is the usual Bose population factor. Because of the clean background spectrum, good fits were obtained to within 8 cm^{-1} of the exciting line. The values of ω_o and γ obtained for the E_g mode at various temperatures are shown in Fig. 2. According to this data, the line becomes overdamped at a frequency $\omega_o = \gamma/\sqrt{2} \approx 16$ cm^{-1}. For an overdamped mode, the parameters S, ω_o and γ are highly correlated, and this is reflected in the large standard deviations near T_c indicated by the error bars in Fig. 2. At low temperatures, the standard deviations are less than the size of the symbols used in the figure. The A_{1g} mode was also fitted to a DSHO model, but this time the fitting procedure was more complicated because the (ZZ) spectrum contains a background component that goes as ω^{-2} approximately, and which is of comparable intensity to the soft mode. Good fits could not be obtained for temperatures greater than about 160 K. The A_{1g}-mode ω_o and γ show a temperature dependence of the same form as the E_g soft mode, and at 105 K $\omega_o = 71$ cm^{-1} and $\gamma = 17$ cm^{-1}. It is interesting to note that in the low-temperature limit the anisotropy in the soft mode frequencies for $KMnF_3$ is the same as that for $SrTiO_3$ (4).

INTENSITIES OF THE HARD MODES

The temperature dependent parts of the Raman cross-section are $(\bar{n}+1)\,|P_{\alpha\beta}|^2$, where $P_{\alpha\beta}$ is a component of the polarizability tensor. Expanding $P_{\alpha\beta}$ in terms of normal mode coordinates A(qj) gives

$$P_{\alpha\beta} = P_{\alpha\beta}^{o} + \sum_{qj} P_{\alpha\beta}\begin{pmatrix} q \\ j \end{pmatrix} A(qj) + \sum_{\substack{q_1 j_1 \\ q_2 j_2}} P_{\alpha\beta}\begin{pmatrix} q_1 q_2 \\ j_1 j_2 \end{pmatrix} A(q_1 j_1) A(q_2 j_2) + \ldots$$

Normal first order scattering is associated with the second term in this expansion and second order scattering with the third term. If the crystal has inversion symmetry and is cubic, then there is no first order scattering. For T close to T_c the main contribution to the second order scattering will come from the

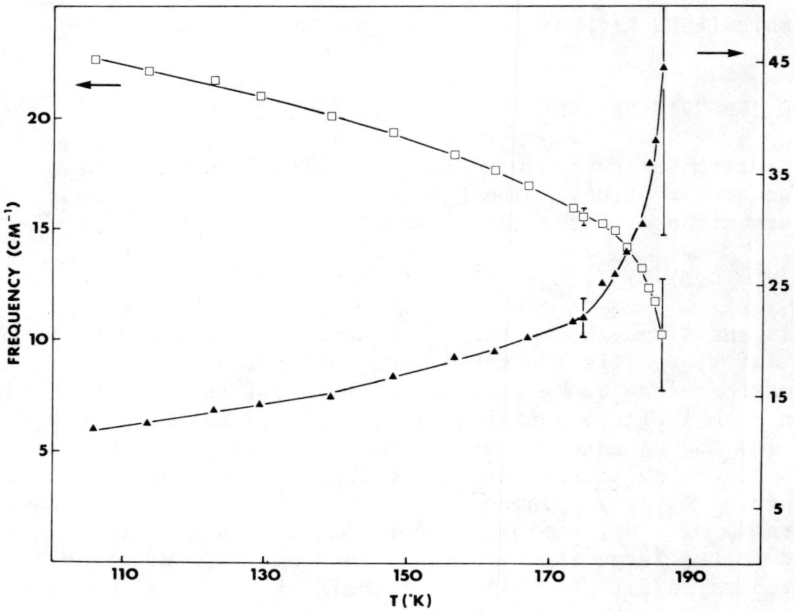

Figure 2. E_g soft mode frequency and width versus temperature.

soft mode because of its low frequency, so that

$$P_{\alpha\beta} \simeq \sum_{q_1 j_1} P_{\alpha\beta}\begin{pmatrix} q_1 q_s \\ j_1 j_s \end{pmatrix} A(q_1 j_1) A(q_s j_s),$$

where $A(q_s j_s)$ can be divided up into a static part and a dynamic part. The static part is proportional to the order parameter ϕ and gives rise to scattering which could be regarded as first order for T below T_c. When the static part is much larger than the dynamic part, the intensity of the hard modes should be proportional to ϕ^2 and hence to ω_o^2. As $(T_c-T)\rightarrow 0$, the dynamic part will become more important, and above T_c there is no static displacement so that the total contribution to the intensity will come from the dynamic part of $A(q_s j_s)$.

We have measured the intensities of several hard modes, but here we concentrate on the intensity of the E_g mode at 227 cm^{-1}, which has a good signal to noise ratio. Scans of this peak were made at several temperatures and the integrated intensities of the observed peaks are shown in Fig. 3. Note that there is a discontinuity in the curve in Fig. 3 at T_c as expected for a first order transition and that the integrated intensity does not go to zero for $T>T_c$; the residual weak peak disappears gradually as T is increased until it is lost in the background for $T>230$ K.

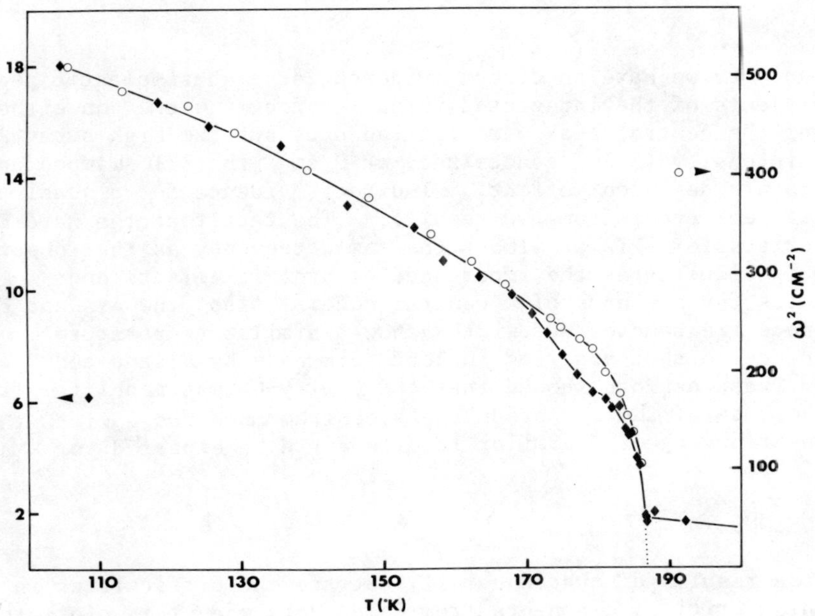

Figure 3. The temperature dependence of the E_g (227 cm^{-1}) mode intensity and the E_g soft mode frequency squared.

This observation is in qualitative agreement with the theory given above and indicates that dynamic effects are important in the range T_c to T_c + 40 K.

Below T_c, the results are again in qualitative agreement with the theory, but the accuracy is not good enough to justify the extraction of temperature exponents. Suffice it to say that a straight line, I versus $(T-T_c')$, can be fitted to short regions of the curve provided that an appropriate choice of $T_c'>T_c$ is made. The dip shown in Fig. 3 between 175 and 180 K is reproducible and may be partially caused by the difficulty in separating the static and dynamic parts of the scattering. The dynamic scattering could add to the observed peak near T_c and increase its intensity, which would leave an apparent dip in the intensity at lower temperatures. Note that a similar, but smaller dip, is also found in the birefringence and NMR results (5,6). In Fig. 3, the intensity measurements have been scaled to fit ω_o^2 for the E_g mode. There is general agreement between the intensity and ω_o^2 values for T away from T_c, as predicted above, and the disagreement for T close to T_c is not significant because of the uncertainties in ω_o and I discussed earlier.

CENTRAL MODE

As yet, we have no direct evidence for a quasi-elastic peak. Measurements of the integrated intensity from ~ 6 cm^{-1} on either side of the central peak were carried out, but the high stray light intensity (3×10^9 counts/sec. at 0 cm^{-1} in (ZZ)) swamped any temperature dependent effect. Indirect evidence for a quasi-elastic peak comes from two results. The fact that the hard E_g mode exists for $T>T_c$ at almost the same frequency as that observed below T_c illustrates the importance of dynamic effects and indicates the presence of a central mode. Also, the A_{1g} and E_g soft mode frequencies and widths show a similar temperature dependence to that observed in lead germanate by Hisano and Ryan (7). These authors showed that the Cowley-Coombs model (8) fitted their data, thereby producing a central mode and a width that no longer diverged. Similar results could be expected for $KMnF_3$.

CONCLUSIONS

The results discussed here illustrate the difficulties in obtaining critical exponents from Raman data when T is close to T_c. In the case of $KMnF_3$ the difficulties are accentuated by the first order nature of the transition, which limits the range over which temperature dependences can be measured and makes it impossible to obtain two decades of temperature dependent behaviour. Our results show that the intensity of a hard mode is proportional to the square of the soft mode frequency for temperatures away from T_c. Indirect evidence was found for the existence of a central component.

We wish to acknowledge helpful discussions with R.A. Cowley and A.D. Bruce, and thank D.A. Jones for providing the crystal.

REFERENCES

1. S.M. Shapiro, J.D. Axe, G. Shirane and T. Riste, Phys. Rev. B<u>6</u>, 4332 (1972) and references therein.
2. Yu.A. Popkov, V.V. Eremenko and V.I. Fomin, Sov. Phys. Solid State <u>13</u>, 1701 (1972) and also private communications.
3. V.G. Khlyustov, I.N. Flerov, A.T. Silin and A.N. Sal'nikov, Sov. Phys. Solid State <u>14</u>, 139 (1972).
4. J.M. Worlock and D.H. Olson, in <u>Light Scattering in Solids</u>, edited by M. Balkanski (Flammarion, Paris, 1971), p.410.
5. S. Hirotsu and S. Sawada, Solid State Comm. <u>12</u>, 1003 (1973).
6. F. Borsa, Phys. Rev. B<u>7</u>, 913 (1973).
7. K. Hisano and J.F. Ryan, Solid State Comm. <u>11</u>, 1745 (1972).
8. R.A. Cowley and G.J. Coombs, J. Phys. C. <u>6</u>, 143 (1973).

SOFT MODE AND CRITICAL OPALESCENCE IN $SrTiO_3$

E.F. Steigmeier, H. Auderset and G. Harbeke

Laboratories RCA Ltd., Zurich - Switzerland

ABSTRACT

The soft mode frequency is determined from spectral light scattering measurements in $SrTiO_3$ and found to be proportional to the static order parameter over the whole range of the tetragonal phase. In the second part, measurements of the critical opalescence near the structural phase transition in $SrTiO_3$ are reported. The total scattering intensity is measured as a function of temperature and angle (3, 90 and 177 deg) and found to behave according to an Ornstein-Zernike-like expression. This quasi-elastic scattering is interpreted as of dynamic origin. It is concluded that (at least above T_c) the coupling of the light is directly to the phonon density fluctuations (central mode) and not indirectly via the soft mode.

Light scattering studies attract growing interest for the study of structural phase transitions since they can contribute considerably to the understanding of the dynamics involved in it, as is wellknown for the liquid-gas transition (1). The 108 K transition in $SrTiO_3$, of second order, consists in a doubling of the unit cell as a result of alternate static rotations of neighboring TiO_6 octahedra about the cubic axis. The mean value of the rotational angle represents the order parameter, and the dynamic rotation about this mean value the soft mode.

It is the purpose of this paper to present further results on the behaviour of the soft mode frequency, a study of which was reported recently (2), and mainly to provide more complete results on the critical opalescence in $SrTiO_3$ as a function of temperature and angle (3). It will be shown that the latter effect is quite substantial in $SrTiO_3$. The light is found to couple predominantly directly to the phonon density fluctuations and not via the soft mode.

The crystals used for this work are cut from a boule which had been carefully annealed at about 2200 K for about two hours (4). The samples, cut to the size of 6 x 2 x 1.5 mm^3, are (110)-oriented to reduce the number of domains (5). They are cooled extremely carefully, i.e. with only 0.06 deg/min, through the transition. At each temperature, controlled to 0.01 deg, ample time is allowed for equilibration (0.5 h far from T_c and up to 2 h near T_c) while constantly monitoring the soft mode frequency (2), before a data point is taken. The results are then highly reproducible over the whole temperature range (provided that T_c is not crossed).

The experimental details of the Raman measurements which were used to determine the soft mode frequency have been given previously (2). A comparison, which is given in Fig. 1, of the results with the measurements of the static order parameter (6) can be summarized as

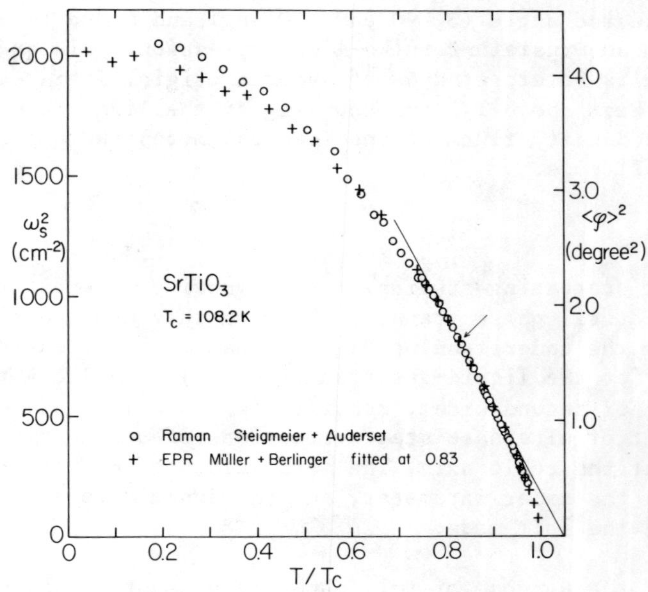

Figure 1: Square of the soft mode frequency and of the static order parameter (ref. 6) versus temperature. The arrow indicates where the order parameter curve is adjusted to the other curve.

follows: After proper data reduction (7) we find that the soft mode frequency ω_s and the static order parameter $\langle\varphi\rangle$ are proportional to each other over the whole range of the tetragonal phase to within 3 %. This is considered to be a remarkable empirical result which, we feel, deserves further theoretical study. In the Landau limit ($0.75 < T/T_c < 0.9$) this proportionality is expected theoretically (2), but the extension into the critical regime ($0.9 < T/T_c \lesssim 1.0$) seems by no means trivial.

The experimental setup used for the measurements of the critical opalescence is given in Fig. 2. HeNe laser light of about 5 mW im-

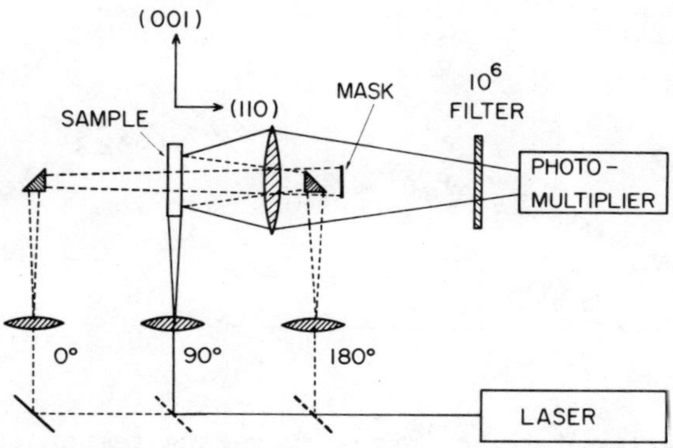

Figure 2: Experimental arrangement to measure the angle dependent opalescence in $SrTiO_3$.

pinges onto the crystal in one of three alternative directions which are selected at each temperature by using the corresponding mirror. The scattered light is collected under a small solid angle (about 5 deg inside the crystal) and, after appropriate attenuation, is detected integrally by a photomultiplier. The measurements were performed on two different crystals one of which (#1-8) had been used previously (2) for the high accuracy soft mode frequency measurements. The opalescence results obtained on it were reported by Steigmeier et al. (3). The more detailed results obtained on the second crystal (#1-9) are given here; they support the previously drawn conclusions.

Fig. 3 gives the total (i.e. frequency integrated) scattering intensity versus temperature for three different scattering angles, obtained on the second crystal. Constant background intensities are already subtracted in Fig. 3. They were determined at 60 K and 160 K for the 3 deg curve and amount to 1314 counts per second (cps) and

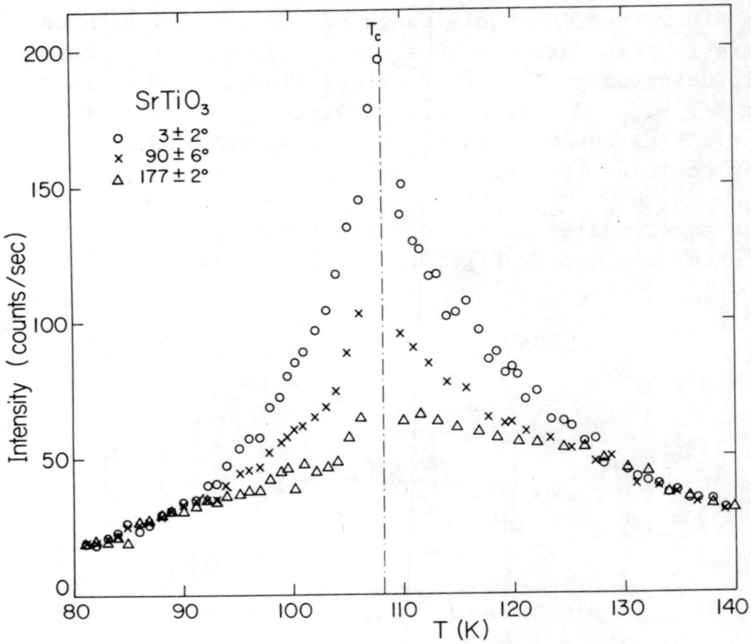

Figure 3: Scattered light intensity versus temperature for three scattering angles. Background subtracted (see text). T_c = 108.2 K.

1223 cps, respectively. It is assumed that this background consists of scattering due to crystalline imperfections and possibly to non-critical phonon density fluctuations. The 90 and 177 deg curves are fitted to the 3 deg curve at 82 K and 138 K. It should be pointed out that the 3 and 177 deg data in Fig. 3 have been corrected for internal reflection; the high refractive index ($n \cong 2.40$) induces 21 % internal backscattering for the forward scattering case and vice versa. Note that the observed total intensity is several orders of magnitude larger then the Raman (3) and Brillouin intensity (8).

Contrary to the widespread belief that 'opalescence' in solid state transitions is entirely caused by domain scattering (8,9) we find that the results of Fig. 3 indicate true critical opalescence since they are reminiscent of an Ornstein-Zernike expression (10).

We have attempted (3) to use such an expression for the analysis of our data. The total intensity would then be given by

$$I = \frac{A\,T\,\chi^*(\epsilon)}{1 + q^2\xi^{*2}(\epsilon)} \qquad (1)$$

where A is a constant containing the coupling coefficients, χ^* is the generalized response function, $\epsilon = (T-T_c)/T_c$ is the reduced temperature, ξ^* the generalized correlation length and q the scattering wavevector, given by $|q| = 2\,|k_i|\sin(\Theta/2)$ with Θ the scattering angle. χ^* and ξ^* may or may not be given by simple power laws of ϵ since the generalized response function χ^* may in principle contain single phonon terms (representing the order parameter susceptibility χ) but also terms to second order in the phonon coordinates (corresponding to collective or phonon density excitations (11,12).

Several phenomenological and microscopic theories to describe a central mode have become available in the last years (13). In many of them, the central mode results from a frequency dependent damping of the soft mode, and the coupling of the light is to the latter only. According to our results (Fig.3) this description is not appropriate for $SrTiO_3$ above T_c since the light cannot couple to the soft corner mode in the cubic phase. We therefore are led to the conclusion that only direct coupling of the light to the phonon density fluctuation occurs, a mechanism which was proposed in a theory of Wehner and Klein (11,12). This means that above T_c the collective or phonon density part is the only contribution to the generalized response function χ^*, the contribution from the order parameter susceptibility being zero. Below T_c the indirect coupling via the soft mode is in principle permitted. In this context, it is interesting to note that the corresponding effect, namely the direct coupling of light to the temperature fluctuations, has also been observed in liquids (14) where however it is in general smaller than the indirect coupling via the adiabatic pressure fluctuations.

Fig. 4 shows a plot of 1/T versus ϵ for $T > T_c$. We note that for $\epsilon > 0.1$ the data can be represented by using a temperature exponent γ close to 4/3. A value $\gamma = 1.29$ is expected to hold for the exponent of the order parameter susceptibility (15), but the agreement might be somewhat fortuitous since the connection between χ^* and χ (11,12) will still have to be clarified. For an angle of 3 deg, where one presumes to be in the limit $q^2\xi^{*2} \ll 1$, we observe a saturation effect in Fig. 4 which can be described with a phenomenological form $\epsilon^{-1.33}/(1+0.02\epsilon^{-1.46})$. One possible cause for this is that the coupling coefficient (which essentially is the temperature derivative of the refractive index (11)) depends on ϵ. There is independent evidence for this (16) from the fact that the refractive index is very strongly temperature dependent near T_c in the cubic phase.

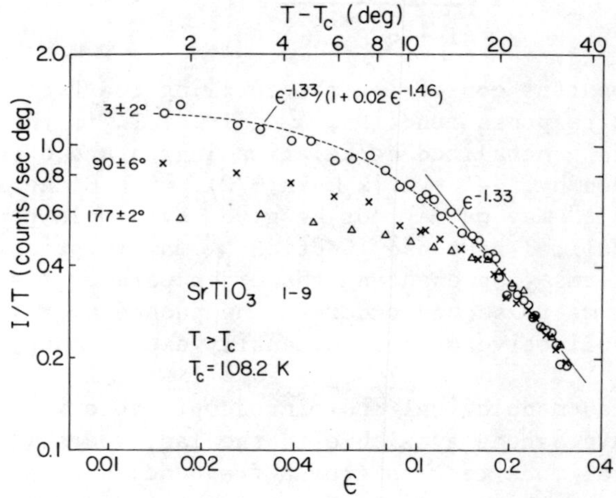

Figure 4: Scattered light intensity divided by temperature versus $\epsilon = (T-T_c)/T_c$ for $T > T_c$. The full line shows a dependence $I/T \propto \epsilon^{-1.33}$.

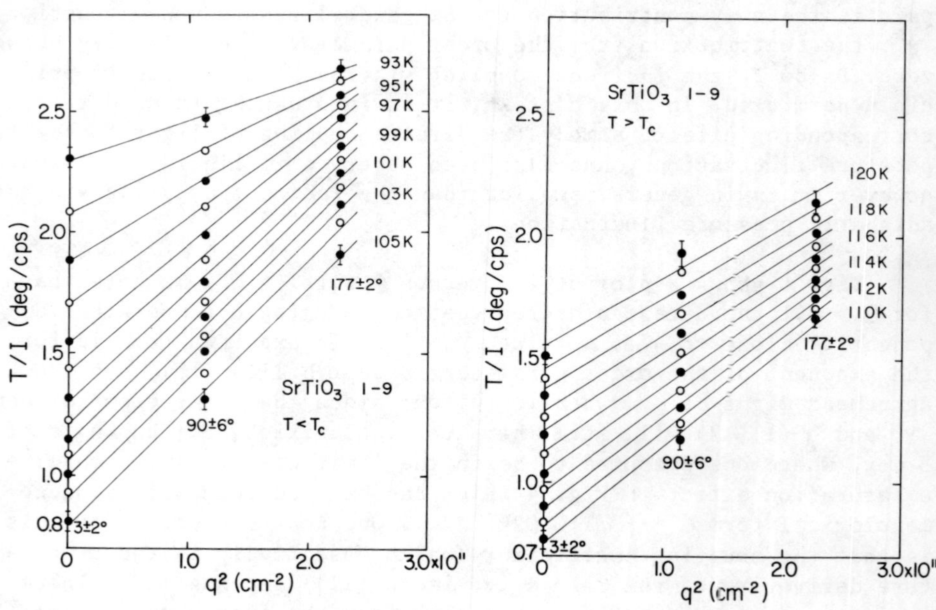

Figure 5: Ornstein-Zernike plots: T/I versus square of wavevector for different temperatures below and above T_c.

A check for the compatibility of our results with an angular dependence according to eq.(1) is made in Fig. 5 which shows Ornstein-Zernike plots T/I versus q^2 for different temperatures. Within error limits we find that our data are compatible with an angular dependence according to Ornstein and Zernike. From the slope and the intercept with the ordinate, one can determine the parameter ξ^* at each temperature. We find values increasing from 4 to 8 x 10^{-6} cm on approaching T_c.

It should be pointed out that the critical opalescence reported here is believed to be of a dynamic origin (i.e. due to phonon density excitations) and not caused by static strain for the following reasons :

(a) In our crystals we do not observe microscopically any crystallographic domains in the cubic phase or any other defects which change with temperature (3). Below T_c the number of domains and their wall thickness do not change between 106 K and 80 K.

(b) There is evidence that crystals of better quality produce a higher opalescence. Crystal 1-8 reported previously (Fig. 1 of ref. 6) shows a higher efficiency than crystal 1-9 reported here (Fig. 3) and it definitely was of better quality (judging from the strain birefringence observed microscopically and from the extension of the Rayleigh wing in the Raman spectrum).

(c) Strain induced first order Raman scattering is not observed above T_c.

(d) We further note that the relation $\omega_s^2 \propto \chi^{-1}$ finds an apparent breakdown if one uses the exponents $\beta = 1/3$ obtained for the soft mode frequency (2) and $\gamma = 1.29$ for the susceptibility (15). This can only be cured by renormalizing the frequency ω_s (17), which in turn implies a dynamic nature (18) of the central mode in neutron and light scattering.

All these points speak against a strain-caused or enhanced effect.

In conclusion, the present results give evidence for critical opalescence in $SrTiO_3$ which is considered to be of dynamic origin. The angular dependence which is observed near T_c is compatible with an Ornstein-Zernike-like expression. The light couples directly to the phonon density fluctuation (or central mode) and not indirectly via the soft mode. The coupling coefficient given by the thermooptic coefficient seems to be temperature dependent near T_c.

We gratefully acknowledge stimulating discussions with R. Klein, W. Rehwald and R.K. Wehner.

REFERENCES

(1) See e.g. review paper by H.Z. Cummins, Proc. Intern. School of Physics on Critical Phenomena, Varenna 1970 (Academic Press 1971) p. 380 and this volume.

(2) E.F. Steigmeier and H. Auderset, Solid State Comm. 12, 565 (1973).

(3) E.F. Steigmeier, H. Auderset and G. Harbeke, Solid State Comm. 12, 1077 (1973).

(4) The crystals used for ultrasonic studies (W. Rehwald, Phys. kond. Materie 14, 21 (1971)) originated from the same boule.

(5) K.A. Müller, W. Berlinger, M. Capizzi and H. Gränicher, Solid State Comm. 8, 549 (1970).

(6) K.A. Müller and W. Berlinger, Phys. Rev. Lett. 26, 13 (1971).

(7) i.e. using the individually determined T_c values. Note that T_c = 108.2 K, used for the light scattering data, is obtained from the extrapolation of the cube of the soft mode frequency versus temperature, and the same value also from the peak in the critical opalescence. T_c = 105.5 K was obtained for the crystals in the EPR work (ref. 6).

(8) A. Laubereau and R. Zurek, Z. Naturf. 25a, 391 (1970).

(9) See e.g. S.M. Shapiro and H.Z. Cummins, Phys. Rev. Lett. 21, 1578 (1968).

(10) L.S. Ornstein and F. Zernike, Proc. Acad. Sci. Amsterdam 17, 793 (1914); Physik. Z. 19, 134 (1918).

(11) R.K. Wehner and R. Klein, Physica 62, 161 (1972).

(12) For details see also R. Klein, this volume.

(13) See references cited in ref. 3.

(14) C.L. O'Connor and J.P. Schlupf, J. Chem. Phys. 47, 31 (1967). We thank Profs. Cummins and Benedek for pointing this out to us.

(15) See e.g. K. A. Müller, this volume, and references cited in ref. 2.

(16) E. Courtens, Phys. Rev. Lett. 29, 1380 (1972), Fig. 2c.

(17) S.M. Shapiro, J.D. Axe, G. Shirane and T. Riste, Phys. Rev. B 6, 4332 (1972).

(18) We thank Dr. T. Schneider for bringing this point to our attention.

PHONON TRANSPORT THEORY AND THE CENTRAL MODE

R. Klein

Fachbereich Physik, Universität Konstanz
D-775 Konstanz, Germany, and Laboratories RCA,
CH-8048 Zürich, Switzerland

ABSTRACT

The response function of a dielectric crystal is calculated from anharmonic lattice dynamical theory, including explicitely phonon transport. The phonon self-energy, in general given by the vertex part, is determined from a Boltzmann type transport equation, expressing the coupling of the phonon to space and time-dependent fluctuations of the phonon occupation number density. These fluctuations can be "thermodynamic" or "dielectric", and both can give rise to quasi-elastic scattering. Their contributions to the difference between adiabatic and isothermal responses are discussed. The theory described here is mainly applied to light scattering experiments and it is suggested that direct coupling of light to phonon density fluctuations must be taken into account in addition to ordinary one-phonon scattering in which a central mode appears through anharmonic coupling.

I. INTRODUCTION

For most problems in lattice dynamics the anharmonic interactions among the phonons of a dielectric solid can be considered to be a weak perturbation with the effect of a slight shift of the harmonic frequencies to a temperature dependent value and the introduction of a finite lifetime of the phonons. Lorentzian line shapes for the response functions, measured in various scattering experiments, are usually excellent descriptions of the result of these measurements; the reason being that the frequencies Ω, one is interested in, are large compared to the inverse relaxation times of the system of interacting thermal phonons.

At $\Omega\tau < 1$, however, low-order perturbation theory breaks down, the summation of perturbation theory leads to a Boltzmann type integral equation (1) - (7). The self-energy Π of a phonon has under these circumstances no longer the simple form which leads to Lorentzian line shapes. Instead, additional structure in the response function arises from fluctuations in space and time of the number density of thermal phonons. The singularities of $\Pi(\underline{Q}\Omega)$ as $\underline{Q}$ and Ω go to zero are important in making the connection of the many-body theory of an interacting phonon system with the macroscopic thermodynamic and elastic theories (4), (8), (9). As an example we mention that the difference between adiabatic and isothermal elastic constants c_S and c_T results from the different behavior of the terms in $\Pi(\underline{Q}\Omega)$ due to phonon transport if one allows first Ω to go to zero and then $\underline{Q}$ or vice versa (10). At the gas-liquid phase transition a central peak arises due to entropy fluctuations, the intensity of which is just given in terms of the difference between adiabatic and isothermal. The behavior of $\Pi(\underline{Q}\Omega)$ in the case of the interacting phonon system at small Ω does also contain a term giving rise to quasielastic scattering (11). But besides these so-called "thermodynamic" fluctuations the varying phonon density consists in general also of "dielectric" fluctuations, as was first pointed out by Cowley and Coombs (12). These consist of fluctuations of the phonon density away from local thermodynamic equilibrium. They do not contribute to the difference between adiabatic and isothermal and it can be shown that they can be related to a frequency dependent viscosity.

This paper reviews phonon transport theory in the form as developed by Klein and Wehner (6) and starts from the problem of light scattering from a dielectric solid, in which scattering from phonon density fluctuations is included (13). Since we are not only interested in the "thermodynamic" fluctuations we generalize the theory developed for light scattering from second sound (13) by using an improved solution of the transport equation. The result is an expression for the self-energy which can be interpreted in terms of a coupling of the phonon state to fluctuations of local thermodynamic equilibrium (entropy fluctuations in the hydrodynamic limit) and to fluctuations away from local equilibrium. Both modes give rise, in general, to quasi-elastic response (14). Finally, it is pointed out that the possibility of direct coupling of light to phonon density fluctuations exists, a mechanism which was studied in detail in Ref. (13) for entropy fluctuations. There it was shown, that this direct process can be quite important for solids.

II. LIGHT SCATTERING IN LIQUIDS AND IN SOLIDS

Instead of discussing phonon transport theory in a formal way, we will illustrate the need for such a theory by considering the example of light scattering. Since this method has been very successful for the investigation of the liquid-gas transition and is also used for structural phase transformations (15), it seems worthwile to start from the spectrum of scattered light, which is proportional to the Fourier transform of the susceptibility auto-correlation function

$$(2.1) \qquad J(\underline{Q}\Omega) = \int_{-\infty}^{\infty} dt\, e^{i\Omega t} \left\langle \delta\chi(\underline{Q},t)\,\delta\chi(\underline{Q},0)\right\rangle .$$

Here, $\Omega = \omega_i - \omega_s$, $\underline{Q} = \underline{q_i} - \underline{q_s}$ denote the transfer of energy and quasi-momentum, respectively, of the incident light (index i) to the system (index s refers to scattered light, and $\overline{\underline{Q}} \equiv - \underline{Q}$). The transferred wave vector $\underline{Q}$ is given by the scattering angle Θ according to $|\underline{Q}| = 2\, q_i \sin(\Theta/2)$, so that the maximum wave vector of an excitation, which can be looked at in a light scattering experiment, is of the order of $10^{-5}\mathrm{cm}^{-1}$.

Let us start by considering first a simple liquid, described by the two independent thermodynamic variables density ρ and temperature T. In this case

$$(2.2) \qquad \delta\chi(r,t) = \left(\frac{\partial\chi}{\partial\rho}\right)_T \delta\rho(r,t) + \left(\frac{\partial\chi}{\partial T}\right)_\rho \delta T(r,t)$$

Since the temperature dependence of the susceptibility at constant density is very small in most fluids (16), susceptibility fluctuations are very well described in terms of density fluctuations. Working with this approximation, as was first done by Einstein in his theory of critical opalescense, one can decompose $\delta\rho$ into adiabatic pressure fluctuations, which is ordinary (first) sound, and isobaric entropy fluctuations, which are called second sound, if they are propagating:

$$(2.3) \qquad \delta\chi(r,t) \approx \left(\frac{\partial\chi}{\partial\rho}\right)_T \left\{ \left(\frac{\partial\rho}{\partial p}\right)_S \delta p(r,t) + \left(\frac{\partial\rho}{\partial S}\right)_p \delta S(r,t) \right\}$$

Note that the coupling of light is via the density dependence of χ only, and that the entropy fluctuations appear through their coupling (proportional to thermal expansion α) to first sound.

In order to calculate the spectrum, it is neccessary to compute correlation functions such as $\left\langle \delta p(\underline{Q}t)\,\delta p(-\overline{\underline{Q}}0)\right\rangle$. These are obtained from the dynamic equations of the fluid, such as the Navier - Stokes equation, the heat flow equation and the continuity equation. Therefore, the dynamical behavior determines the light scattering spectrum. The result of the calculation is the well-known three-peak spectrum, consisting of emission and absorption of propagating sound waves and of a central peak of overdamped second sound or heat diffusion (17).

The static properties of the system are contained in the integrated intensities. In the Einstein approximation the ratio of scattered intensities of second to first sound is given by the Landau-Placzek ratio

$$R_{LP} = I_2/2I_1 = (C_p/C_v) - 1 = (C_S/C_T) - 1 \tag{2.4}$$

If one decomposes $\delta\chi$ directly into pressure and entropy fluctuations, the Landau-Placzek ratio gets modified by a term including the temperature dependence of the susceptibility:

$$R = R_{LP}\left[1 + \frac{C_p}{\alpha C_V}\,\frac{(\partial\chi/\partial T)_\rho}{(\partial\chi/\partial\rho)_S}\right]^2 \tag{2.5}$$

If one could seperate the scattered intensity from the entropy fluctuations into a part arising from the indirect mechanism leading to the Landau-Placzek ratio and a rest which arises from direct coupling, the result for the latter would be

$$I'(Q) = \frac{kT^2}{VC_v}\,2\,\frac{c_S}{c_T}\left(\frac{\partial\chi}{\partial T}\right)_\rho^2 \tag{2.6}$$

The coupling constants which appear in light scattering (for instance Raman tensor, elasto-optical constants or $(\partial\chi/\partial\rho)_T$) are usually considered to be constants. Considering the term analogous to (2.6) for solids it must be mentioned that in cases like $SrTiO_3$ and $BaTiO_3$ it is known (18),(19) that the refractive index shows an unusual temperature dependence near T_c. The experiments by Courtens on optical birefrigence reveal that Δn is different from zero above T_c in a small temperature interval , where at least one of the refractive indices n_c or n_a must change rather rapidly with temperature. But it is just $(dn/dT)^2$ which is proportional to the total scattered intensity arising from the direct coupling of light to the thermodynamic fluctuation.

In dielectric crystals the fluctuations of the susceptibility arise in general from the atomic motions and can therefore be represented by an expansion into phonon normal coordinates $A(\lambda)$, where $\lambda = (\underline{q}j)$ represents phonon wave vector $\underline{q}$ and polarization index j:

$$(2.7)\qquad \delta\chi(\underline{Q}) = \frac{\hbar}{V}\Big\{\sum_{j} P_1(\Lambda)A(\Lambda)+\sum_{\lambda_1\lambda_2} P_2(\lambda_1\lambda_2)A(\lambda_1)A(\lambda_2)\Big\}\ ,$$

where $\Lambda = (\underline{Q}j)$. The coupling coefficients P_1 and P_2 are simply related to the elasto-optical coefficients p_{ik} or the Raman tensor and the temperature dependence of the susceptibility at constant volume (13).

If the expansion, eq. (2.7), is used in eq. (2.1), we find the spectrum of scattered light being expressed by correlation functions of phonon normal coordinates, which can be written as imaginary parts of corresponding retarded Green functions. As long as only the one-phonon terms in eq. (2.7) are used ($P_2 \equiv 0$), the spectral density function is given by

$$(2.8)\qquad J_1(\underline{Q}\Omega) = \frac{2\hbar^2}{V^2}(1 + n(\Omega))$$

$$\sum_{j_a j_b} \text{Im}\ P_1(\Lambda_a)P_1(\overline{\Lambda}_b)G_\Omega(A(\Lambda_a)\,|\,A(\Lambda_b))$$

with $\Lambda_{a,b}= (\underline{Q},j_{a,b})$, $n(\Omega)$ the Bose-Einstein function and the one-phonon Green function

$$(2.9)\qquad G_\Omega(A(\Lambda_a)\,|\,A(\Lambda_b)) = i\int_0^\infty dt e^{i\Omega t}\Big\langle\big[A(\Lambda_a,t),A(\overline{\Lambda}_b,0)\big]\Big\rangle$$

We have to calculate the one-phonon Green function for the phonon of wave vector $\underline{Q}$ equal to the transferred momentum. This is a rather long-wavelength phonon which due to the anharmonicities of the lattice is coupled to the large number of thermal phonons, having much larger wave vectors.

The spectrum $J_1(\underline{Q}\Omega)$ represents conventional first-order Brillouin or Raman scattering, depending on whether the branch indices j_a and j_b refer both to the same acoustic or optical branch. If $j_a \neq j_b$, the expression (2.8) can be used for coupled phonon mode scattering as observed by Steigmeier et al.(20), Katiyar et al.(21) and Fleury and Lazey (22). But from now on we will concentrate on the simpler case of a one-branch model, symbolically represented in Fig. 1a.

The Green function in eq.(2.8) corresponds to a phonon of wave vector $\underline{Q}$ interacting with thermal phonons of wave vectors $\underline{q}_1= \underline{q} + \frac{1}{2}\underline{Q}$ and $\underline{q}_2= -\underline{q} + \frac{1}{2}\underline{Q}$, which are nearly equal and opposite and have practically the same energy. The self-energy for the phonon $\underline{Q}$ can in this case not be calculated by simple perturbation theory, as was clearly demonstrated by Sham (3). Instead, we are in a situation where the wavelength of the susceptibility fluctuation is large compared to the mean free path of the thermal phonons and where one period of the fluctuation is long compared to the lifetime of thermal phonons due to scattering among themselves. Under these circumstances it is clear that we have to use phonon transport theory. We have a situation which in many respects resembles hydrodynamics. The calculation of the phonon self-energy and the derivation of phonon transport equations has been discussed by a number of methods by various authors (1)-(7). The Hamiltonian for the system of interacting phonons is given by

$$H = H_o + V \tag{2.10}$$

$$H_o = \sum_{\lambda} \hbar\omega_\lambda a^+(\lambda)a(\lambda) \tag{2.10a}$$

$$V = \hbar \sum_{\lambda_1\lambda_2\lambda_3} V_3(\lambda_1\lambda_2\lambda_3)A(\lambda_1)A(\lambda_2)A(\lambda_3) + \ldots \tag{2.10b}$$

$$A(\lambda) = a(\lambda) + a^+(\bar{\lambda}) . \tag{2.11}$$

The second part of $\delta\chi$, eq. (2.7), corresponds to two-phonon scattering such that $\underline{q}_1 + \underline{q}_2 = \underline{Q}$, summing $\underline{q}_1$ and $\underline{q}_2$ over the whole Brillouin zone. We are interested here in the spectrum for small Ω, where only the difference processes remain in which $\underline{q}_1$ is a phonon thermally present in the system. Writing again $\underline{q}_{1,2}=\pm\underline{q} + \underline{Q}/2$ we see that the light couples to two phonons with nearly equal but opposite wave vectors. Using (2.11), the second part of (2.7) includes terms like

$$n_{\underline{q}}(\underline{Q}) = a^+\left(\underline{q} - \frac{\underline{Q}}{2}\right)a\left(\underline{q} + \frac{\underline{Q}}{2}\right), \tag{2.12}$$

which is known as the Wigner operator and is for $|\underline{Q}| \ll |\underline{q}|$ a representation for fluctuations of wave vector $\underline{Q}$ of the phonon occupation numbers for the phonons of mode $\underline{q}$ (6). Therefore, the terms proportional to P_2 in eq. (2.7) include the possibility of a direct coupling of light to phonon density fluctuations, see Fig.1b. It should also be noted that taking into account $P_2 \neq 0$, leads to phonon Green functions with three or four operators $A(\lambda)$, for details we refer to ref. (13). In "ordinary" second order Raman scattering one is interested in much higher frequencies Ω, where phonon density fluctuations are unimportant.

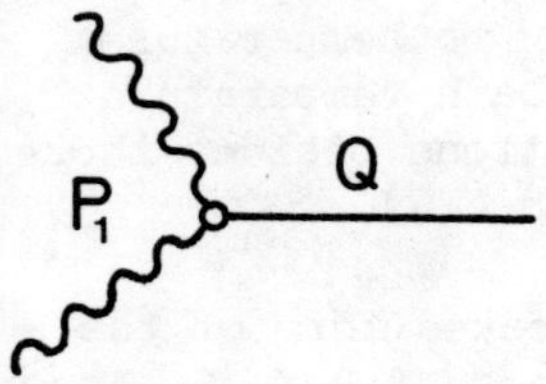

Fig. 1a

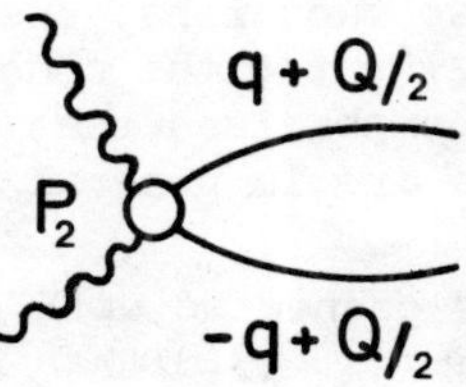

Fig. 1b

III. RESULTS OF PHONON TRANSPORT THEORY

As we have seen in eq. (2.8) the one-phonon Green function determines one important part of the spectrum. Neglecting polarization mixing (j = j') this function has the form

$$G(\underline{Q}\Omega) = \frac{2\omega_Q}{\omega_Q^2 - \Omega^2 + 2\omega_Q \Pi(\underline{Q}\Omega)} \tag{3.1}$$

where the self-energy

$$\Pi(\underline{Q}\Omega) = \Delta(\underline{Q}\Omega) - i\Gamma(\underline{Q}\Omega) \tag{3.2}$$

arises from all the interactions of the mode ω_Q with all the other phonons. The real part of $\Pi(\underline{Q}\Omega)$ renormalizes the harmonic frequency ω_Q^2 to $\tilde{\omega}_Q^2$, which is temperature dependent. The imaginary part of $\Pi(\underline{Q}\Omega)$ represents lifetime effects. It is an odd function of the frequency, and it is usually sufficient to write

$$\text{Im } \Pi(\underline{Q}\Omega) = \Omega\Gamma_Q/\omega_Q \ , \tag{3.3}$$

where Γ_Q is a constant. In this case, which is a very good approximation for thermal phonons, the spectrum consists of two Lorentzians, centered at $\Omega = \pm\tilde{\omega}_Q$, of half-widths 2 Γ_Q. To get this result from perturbation theory, it is sufficient to consider the lowest-order processes, diagrammatically represented by the bubble diagram. The frequency of the excitation, Ω, is assumed to be large compared to the inverse lifetime τ_q of those phonons with which it interacts. This is the collision-less regime, $\Omega\tau_q >> 1$.

Investigating quasi-elastic scattering, one is however, interested in the opposite situation, known as the collision-dominated regime, where one period of the excitation is long compared to the phonon lifetimes, so that local equilibrium in the phonon density can be established. Under these circumstances, ordinary perturbation theory is inapplicable and one has to sum up the so-called ladder diagrams or vertex corrections (3). If we assume for the moment that the phonon density fluctuations are just fluctuations in the local temperature then $\delta n_{\underline{q}}(\underline{Q}\Omega)$ is just the difference of two Bose functions, one corresponding to temperature $T(\underline{r},t) = T_o + \delta T(\underline{r},t)$, the other having the bath temperature T_o. But it must be emphazised that other fluctuations besides those of the temperature are in general included in $\delta n_{\underline{q}}(12)$.

How $\delta n_{\underline{q}}$ is connected with a more complicated form of the self-energy than eq. (3.3) has been the subject of Ref. 6. According to the general Green function theory of anharmonic phonon systems (23), one considers instead of the retarded Green function eq. (2.9), the time-ordered Green function $G^{\xi}(A(\lambda)|A(\lambda')) = TA(\lambda t)A(\lambda',0)$, which in turn is given by dynamical correlations $f_2(\lambda t,\lambda' t')$. The equations of motion of these latter functions are studied and the retarded quantities, which we are interested in, follow as boundary values of the time-ordered functions in the upper half plane of the complex frequency. In this way one finds the Dyson equation for the one-phonon Green function

$$G_{jj'}(\underline{Q}\Omega) = G^o_j(\underline{Q}\Omega)\,\delta_{jj'} - \sum_{j''} G^o_j(\underline{Q}\Omega)\,\Pi_{jj''}(\underline{Q}\Omega)\,G_{j''j'}(\underline{Q}\Omega). \tag{3.4}$$

The self-energy $\Pi_{jj'}(\underline{Q}\Omega)$ is the retarded boundary value of the time-ordered quantity

$$\Sigma(\Lambda t,\Lambda' t') = 3\sum_{\lambda_1\lambda_2} V_3(\Lambda\lambda_1\lambda_2)\quad f_2(\lambda_1 t,1')f_2(\lambda_2 t,2') \times \Gamma_3(1',2',\Lambda' t') \tag{3.5}$$

where $\Lambda = (\underline{Q}j)$, $\Lambda' = (\underline{Q}j')$ and a variable $k = (\lambda_k,t_k)$ implies summation over λ_k and integration over t_k. In diagrams:

(3.6) $\Sigma =$

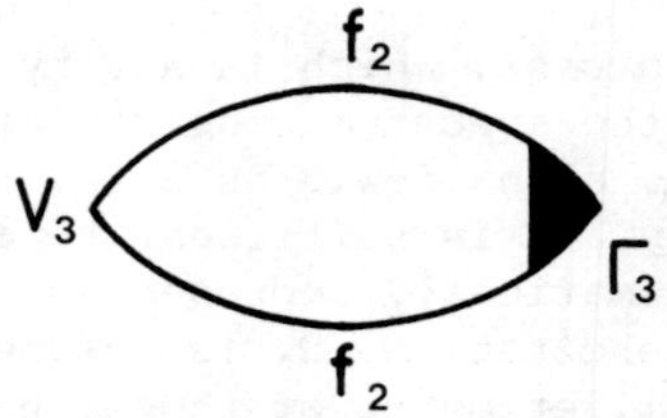

The quantity Γ_3 is the vertex part and is the solution of an inteтral equation

$$\Gamma_3(1,2,3) = \Gamma^o_3\ (1,2,3) + \Gamma^o_3\ (1,4',5)\,f_2(4',4'') \times \Gamma^o_3\ (2,4'',6)\,f_2(5,5')\,f_2(6,6')\,\Gamma_3(5',6',3) \tag{3.7}$$

(3.8) $\Gamma_3{}^{o}(1,2,3) = -\ i\ 6\ V_3(\lambda_1\lambda_2\lambda_3)\delta(t_1 - t_3)\delta(t_2 - t_3)$.

In diagrams:

(3.9) 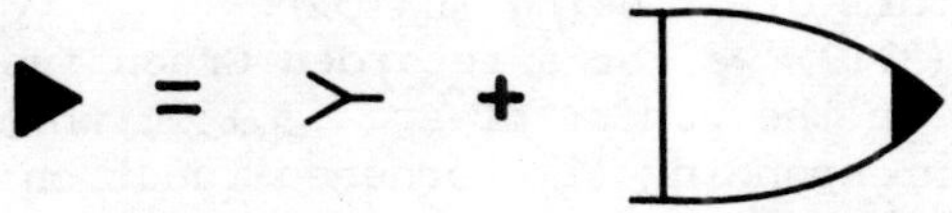

Using the iterations of this integral equation in the diagram for Σ gives the infinite series of bubble diagrams with an increasing number of ladder rungs

(3.10) 

At this point it should be mentioned that 4-phonon processes are neglected in the present treatment. Including terms of quartic anharmonicities in the Hamiltonian (2.10b) adds to the right hand side of (3.9) the socalled chain diagrams (6)

(3.11) 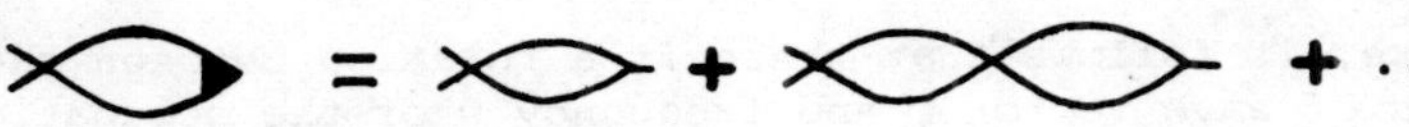

This class of diagrams has been considered by Enz (24) for the case of $SrTiO_3$. Beck and Meier (25) use both classes, (3.11) and (3.9) for a treatment of the same substance.

If we can solve the integral eq. (3.7) for Γ_3, the self-energy can be calculated. Before discussing the results, we want to point out the connection of Γ_3 with phonon density fluctuations δn_q. Having an operator representation for the varying phonon number density, eq. (2.12), we ask for the response of the phonon occupation numbers to an external perturbation, using Kubo's response theory. One adds to the anharmonic Hamiltonian, eq. (2.10), a term representing the coupling of an external perturbation $\xi(\underline{Q}\Omega)$ to the phonon normal coordinates

(3.12) $$H'_t = \sum_{\Lambda} A(\bar{\Lambda})\xi(\Lambda)e^{-i\Omega t}e^{\varepsilon t}.$$

The time-dependent expectation value of the phonon density is then given by

$$(3.13)\quad \left\langle n_{\underline{q}}(\underline{Q})\right\rangle_t \equiv \delta n_{\underline{q}}(\underline{Q}t) = \frac{1}{i\hbar}\int_{-\infty}^{\infty} \Theta(t-\tau)\left\langle\left[n_{\underline{q}}(\underline{Q},t-\tau),H'_{\tau}\right]\right\rangle d\tau$$

the time Fourier transform of it being $\delta n_{\underline{q}}(\underline{Q}\Omega)$.
Using eqs. (2.12) and (3.12) we get a retarded Green function of three phonon operators on the r.h.s. of eq. (3.13), which can be calculated from the corresponding time-ordered function

$$(3.14)\quad G_3(1,2,3) = \left\langle T\, A(\lambda_1 t_1)A(\lambda_2 t_2)A(\lambda_3 t_3)\right\rangle .$$

But this function is again given in terms of the dynamical correlations f_2 and the vertex part Γ_3:

$$(3.15)\quad G_3(1,2,3) = f_2(1,1')\, f_2(2,2')\, f_2(3,3')\, \Gamma_3(1',2',3'),$$

in diagrams:

(3.16) $G_3 =$

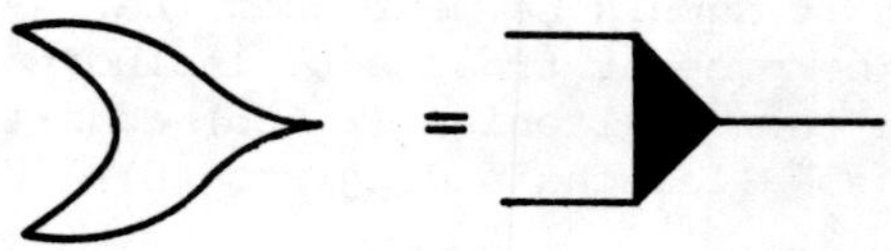

Therefore, it follows that there is a relation between the fluctuations of wave vector $\underline{Q}$ and frequency Ω of the occupation number density of the phonon mode $\underline{q}$ and the vertex corrections. The integral equation (3.7) for Γ_3, which we can consider as an integral equation for the variable phonon density, is therefore the transport equation of the interacting phonon system. This integral equation is of central importance, since it represents the analog of the heat transport equation in the case of fluids.

After this general outline of the phonon transport theory, the results of it are given here in the form which was used in Ref. 13, where the equation for Γ_3 is rewritten in terms of a function D, defined by

$$(3.17)\quad \Gamma_3(1,2,3) = D(1,2;1',2')\,\Gamma_3^{\,o}(1',2',3).$$

Going over to the corresponding retarded quantities, it is shown, that the function

$$(3.18)\quad Y_{\lambda\lambda'}(\underline{Q}\Omega) = \frac{1}{\Omega-\underline{Q}\cdot\underline{v}_{\lambda}+2i\Gamma_{\lambda}} \times$$
$$\times \int_{-\infty}^{\infty} dt_1 e^{i\omega_1 t_1}\int_{-\infty}^{\infty} dt_2 e^{i\omega_2 t}\; D(\lambda_1 t_1,\lambda_2 t_2;\lambda_3 0,\lambda_4 0)$$

$$\lambda_{12} = \left(\pm\underline{q} + \frac{1}{2}\underline{Q}, j\right) \quad , \quad \lambda_{34} = \left(\mp\underline{q}' - \frac{1}{2}\underline{Q}, j'\right)$$

$$\omega_1 = \frac{1}{2}\Omega + \omega_\lambda + i\eta \ , \ \omega_2 = \frac{1}{2}\Omega - \omega_\lambda + i\eta \ ; \ \underline{v}_\lambda = \frac{\partial\omega(\lambda)}{\partial\underline{q}}$$

satisfied an equation very similar to a phonon Boltzmann equation

$$(3.19) \quad (\Omega - \underline{Q}\cdot\underline{v}_\lambda)\ Y_{\lambda'\lambda}(\underline{Q}\Omega) = \delta_{\lambda'\lambda} - \frac{i}{n_{\lambda'}(n_{\lambda'}+1)} \sum_{\lambda''} \mathcal{L}_{\lambda'\lambda''} Y_{\lambda''\lambda}(\underline{Q}\Omega)$$

where $\mathcal{L}$ is the collision operator first introduced by Peierls (26) for three-phonon processes.

It is the solution of this transport equation which determines the self-energy

$$(3.20) \quad \Pi_{jj'}(\underline{Q}\Omega) = -36i \sum_{\lambda\lambda'} V_3(\bar{\Lambda}\lambda_1\lambda_2)\ N_{\lambda\lambda'}(\underline{Q}\Omega)\ V_3(\lambda_3\lambda_4\Lambda')$$

and the fluctuating phonon occupation numbers

$$(3.21) \quad \delta n_\lambda(\underline{Q}\Omega) = -6i \sum_{j\lambda'} N_{\lambda\lambda'}(\underline{Q}\Omega)\ V_3(\lambda_3\lambda_4\Lambda)\ A_j(\underline{Q}\Omega)$$

where

$$(3.22) \quad N_{\lambda\lambda'}(\underline{Q}\Omega) = i\beta\hbar n_\lambda(n_\lambda + 1)\left(-\delta_{\lambda\lambda'} + \Omega Y_{\lambda\lambda'}(\underline{Q}\Omega)\right)$$

and $A_j(\underline{Q}\Omega)$ denotes the strain introduced by the external perturbing field $\xi(\underline{Q}j)e^{-i\Omega t}$.

IV. SOLUTION OF THE TRANSPORT EQUATION

The method to solve the Boltzmann type transport equation for $Y_{\lambda\lambda'}(\underline{Q}\Omega)$, eq. (3.19), makes use of some basic properties of the collision operator $\mathcal{L}$ (27). Neglecting the back-scattering part (6) of $\mathcal{L}$ would bring us back to lowest-order perturbation theory, where the response function has essentially Lorentzien form. But if $\underline{Q}$ and Ω are small in (3.19), the collision term is important. Splitting $\mathcal{L}$ up into a part $\mathcal{L}^N$, which conserves quasi-momentum (Normal-processes) and the rest $\mathcal{L}^U$, containing Umklapp-processes, one uses the fact that the following eigenfunctions belonging to eigenvalue zero are known

$$(4.1) \qquad \mathcal{L}_{\lambda'\lambda''}\omega_{\lambda''} = 0$$

$$(4.2) \qquad \mathcal{L}^{N}_{\lambda'\lambda''} \underline{q}'' = 0$$

The first property implies that the total collision operator leaves the distribution, corresponding to a local temperature, unchanged:

$$(4.3) \qquad n_\lambda(\underline{r},t) = \left[\exp\left(\frac{\hbar\omega_\lambda}{kT(\underline{r},t)}\right) - 1\right]^{-1}$$

$$(4.4) \qquad n_\lambda(\underline{Q}\Omega) \approx n_\lambda + \beta\hbar n_\lambda (n_\lambda + 1)\,\omega_\lambda \frac{\delta T(\underline{Q}\Omega)}{T} .$$

From eq. (4.2) it follows that Normal processes cannot change a local equilibrium distribution, drifting with a local drift velocity $\underline{v}(\underline{r},t)$

$$(4.5) \qquad n_\lambda(\underline{r},t) = \left[\exp \frac{\hbar(\omega_\lambda - \underline{v}(\underline{r},t)\cdot\underline{q})}{kT(\underline{r},t)} - 1\right]^{-1}$$

$$(4.6) \qquad n_\lambda(\underline{Q}\Omega) \approx n_\lambda + \beta\hbar n_\lambda(n_\lambda + 1)\left\{ \omega_\lambda \frac{\delta T(\underline{Q}\Omega)}{T} + \underline{v}(\underline{Q}\Omega)\cdot\underline{q} \right\}$$

Since the eigenfunctions ω_q and q_i $(i = 1,2,3)$ have a direct physical meaning and since they dominate the solution of the Boltzmann equation in the low frequency regime, one expands the function $Y_{\lambda\lambda'}$ in terms of the four eigenfunctions mentioned above and a rest, which belongs to eigenvalues different from zero. Specializing to a one-branch model, we write

$$(4.7) \qquad \sqrt{n_{q'}(n_{q'}+1)}\; Y_{q'q}(\underline{Q}\Omega) \equiv Z_{q'q}(\underline{Q}\Omega)$$

$$= a^{o}_{\underline{q}}(\underline{Q}\Omega)\,\eta^{o}_{\underline{q}'} + \sum_{i=1}^{3} a^{i}_{\underline{q}}(\underline{Q}\Omega)\,\eta^{i}_{\underline{q}'} + \tilde{Z}_{q'q}(\underline{Q}\Omega)$$

where

$$(4.8) \qquad \eta^{o}_{q'} = \mu_o \hbar\beta\sqrt{n_{q'}(n_{q'}+1)}\;\omega_{q'}$$

$$(4.9) \qquad \eta^{i}_{q'} = \mu_i \hbar\beta\sqrt{n_{q'}(n_{q'}+1)}\; q'_i$$

and μ_o, μ_i are normalizing factors. Comparing (4.7) with (4.6) it is seen that in a drifting local equilibrium we have $\tilde{Z} = 0$ and the coefficients a^o and a^i are given by the local temperature fluctuation and the local drift velocity, respectively. If one assumes that Normal processes are very fast, it follows that local equilibrium is established quickly, so that $\tilde{Z}$ can be neglected. Under these circumstances, phonon density fluctuations are essentially fluctuations of the local temperature, as treated in Ref. 13 for the case of light scattering from second sound. The "non-thermodynamic" fluctuations, contained in $\tilde{Z}$, however, can also be important.

Since all the other eigenvalues and corresponding eigenfunctions of $\mathcal{L}^N$ and $\mathcal{L}^U$ besides those mentioned above are not known one usually introduces a relaxation time approximation in the following way

$$(4.10)\quad \sum_{q''} (n_{q'}(n_{q'}+1)n_{q''}(n_{q''}+1))^{-1/2}\mathcal{L}^N_{q'q''}\, Z_{q''q} = \frac{1}{\tau_N}\,\tilde{Z}_{q'q}(\underline{Q}\Omega)$$

$$(4.11)\quad \sum_{q''}\left(n_{q'}(n_{q'}+1)n_{q''}(n_{q''}+1)\right)^{-1/2}\mathcal{L}^U_{q'q''}\,\tilde{Z}_{q''q} = \frac{1}{\tau_U}\left(\sum_{i=1}^{3} a^i_q \eta^i_{q'} + \tilde{Z}_{q'q}\right).$$

Using eq. (4.7) together with these definitions of Normal and Umklapp relaxation times τ_N and τ_U in the transport equation (3.19), one finds its solution in the form

$$(4.12)\quad Z_{q'q}(\underline{Q}\Omega) = \frac{1}{\Omega - \underline{Q}\cdot\underline{v}_{q'} + i/\tau}\left\{\sqrt{n_{q'}(n_{q'}+1)}\;\delta_{q'q} + \frac{1}{\tau_N}\sum_{i=1}^{3} a^i_q\eta^i_{q'} + \frac{i}{\tau}a^o_q\eta^o_{q'}\right\}$$

$$\frac{1}{\tau} = \frac{1}{\tau_U} + \frac{1}{\tau_N}$$

where the coefficients a^i_q $(i = 0,1,2,3)$ are still unknown. They can be determined from a system of linear algebraic equations, which is obtained from the Boltzmann equation by multiplication with $\eta^o_{q'}$ and $\eta^i_{q'}$, respectively, and a summation over $\underline{q}'$. Using the orthogonality of the decomposition (4.7), we have

$$\Omega a_q^o - \underline{c}\cdot\underline{Q}\, a_q^i = \sqrt{n_q(n_q+1)}\,\eta_q^o \tag{4.13}$$

$$-\underline{c}\cdot\underline{Q}\, a_q^o + \left(\Omega + \frac{i}{\tau_U}\right)a_q^i - \sum_{q'} \eta_{q'}^i \underline{Q}\cdot\underline{v}_{q'} \widetilde{Z}_{q'q} = \sqrt{n_q(n_q+1)}\,\eta_q^i \tag{4.14}$$

Here,

$$\underline{Q}\cdot\underline{c} = \underline{Q}\cdot\sum_{q'} \eta_{q'}^o \underline{v}_{q'} \eta_{q'}^i , \tag{4.15}$$

which, in a Debye model of constant sound velocity v, is equal to $\omega_2 = vQ/\sqrt{3}$. Expressing $\widetilde{Z}$ in eq. (4.14) by eq. (4.7), the expansion coefficients a_q^i are given by:

$$\frac{a_q^o(\underline{Q}\Omega)}{\sqrt{n_q(n_q+1)}} = \frac{\eta_q^o\left(\Omega + i\gamma(\underline{Q}\Omega)\right) + \omega_2\eta_q^i\left(1 + \dfrac{\underline{Q}\cdot \underline{v}q}{\Omega - \underline{Q}\cdot\underline{v}_{\underline{c}} + i/\tau}\right)}{\Omega^2 - \frac{i}{\tau}M_{11}(\underline{Q}\Omega)\,\omega_2^2 + i\Omega\gamma(\underline{Q}\Omega)} \tag{4.16}$$

$$\frac{a_q^i(\underline{Q}\Omega)}{\sqrt{n_q(n_q+1)}} = \frac{\frac{i}{\tau}M_{11}\omega_2\eta_q^o + \Omega\eta_q^i\left(1 + \dfrac{\underline{Q}\cdot \underline{v}q}{\Omega - \underline{Q}\cdot\underline{v}_q + i/\tau}\right)}{\Omega^2 - \frac{i}{\tau}M_{11}(\underline{Q}\Omega)\,\omega_2^2 + i\Omega\gamma(\underline{Q}\Omega)} \tag{4.17}$$

where

$$\gamma(\underline{Q}\Omega) = \frac{1}{\tau} - \frac{1}{\tau_N}\left(\Omega + \frac{i}{\tau}\right)M_{11}(\underline{Q}\Omega) \tag{4.18}$$

$$M_{11}(\underline{Q}\Omega) = \sum_{q'} \eta_{q'}^1\eta_{q'}^1 \frac{1}{\Omega - \underline{Q}\cdot\underline{v}_{q'} + i/\tau} \tag{4.19}$$

It is interesting to note that in the limits where $\Omega\tau \ll 1$ and $\underline{Q}\cdot\underline{v}_q\tau \ll 1$, we have $iM_{11}/\tau \to 1$ and $(\Omega + i/\tau)M_{11} \to 1$, so that $\gamma \to 1/\tau_U$.

In this way it is easily seen that for $\tau_N \to 0$ the solution reduces to the one derived in Ref. (13).

V. THE SELF-ENERGY

In eq. (3.20) the self-energy was related to the solution of the transport equation. The latter consists of various terms; the last two terms in eq. (4.12) represent essentially fluctuations of a local thermodynamic equilibrium, whereas the other terms correspond to fluctuations away from this local equilibrium. The importance of such terms with regard to the dielectric behaviour near a structural phase transition was clearly pointed out by Cowley and Coombs (12), who first showed, using a less general solution than eqs. (4.16) and (4.17) that the coupling of a one-phonon state to these multiphonon excitations is quite different in the two classes of fluctuations.

This property can already be demonstrated in the simple case where only one branch of acoustic phonons is considered. Therefore, we specialize our general solution of the transport equation to this case in order to avoid lengthy expressions. In this acoustic approximation the anharmonic coupling parameters V_3 can be related to Grüneisen parameters γ_q by (10)

$$V_3(\underline{Qqq}) = \frac{1}{6}\left(\frac{\hbar}{2\omega_{\underline{Q}}\rho V}\right)^{1/2} Q\omega_q \gamma_q . \tag{5.1}$$

Similar quantities can be introduced in the more general case, where phonon $\underline{Q}$ is not a long-wavelength acoustic phonon.

The resulting self-energy can be written in various forms, either by separating the thermodynamic fluctuations completely from the "dielectric" fluctuations, or in a form which makes the transition from the high frequency to the low frequency response particulary transparent. Choosing the latter possibility, we have

$$\begin{aligned} \Pi(\underline{Q}\Omega) = \frac{Q^2 T C_V}{2\omega_{\underline{Q}}\rho} \Big\{ & -\overline{\gamma^2} + \bar{\gamma}^2 + \frac{\tau\Omega}{VC_V}\sum_q c_q\gamma_q(\gamma_q - \bar{\gamma}) \times \\ & \times \frac{(\Omega - \underline{Q}\cdot v_q)\tau}{1 + (\Omega - \underline{Q}\cdot \underline{v}_q)^2\tau^2} - i\,\frac{\Omega\tau}{VC_V}\sum_q c_q\gamma_q(\gamma_q - \bar{\gamma})\frac{1}{1 + (\Omega - \underline{Q}\cdot\underline{v}_q)^2\tau^2} \\ & + \gamma^2\,\frac{\omega_2^2(\Omega + i/\tau)M_{11}(\underline{Q}\Omega)}{\Omega^2 - \frac{i}{\tau}M_{11}(\underline{Q}\Omega)\omega_2^2 + i\Omega\gamma(\underline{Q}\Omega)} \Big\} \end{aligned} \tag{5.2}$$

$$C_V V\overline{\gamma^n} = \sum_q c_q\gamma_q^n \qquad (n = 1,2) \tag{5.3}$$

C_V and c_q are, respectively, the specific heats at constant volume per volume and per mode q.

This generalizes our early result (13) in two ways. First we get contributions to the imaginary part of the self-energy at very low frequencies from two different types of terms, and secondly, the real part of this expression has the $(\underline{Q}\Omega)$ dependence which is needed to renormalize the harmonic phonon frequency for all values of $\Omega\tau$.

In order to discuss this result, it is first pointed out that one introduces a wave vector and frequency-dependent elastic constant (9), (10)

$$c(\underline{Q}\Omega) = \frac{\rho}{Q^2}\left(\omega_Q^2 + 2\omega_Q \Pi(\underline{Q}\Omega)\right) \tag{5.4}$$

As was mentioned in the introduction the limit of $\omega_Q \Pi(\underline{Q}\Omega)/Q^2$ as $\underline{Q},\Omega \to 0$ depends on which quantity approaches zero first. As discussed in detail in Ref. 10, the isothermal and the adiabatic elastic constants are obtained as

$$c_T = \lim_{Q\to 0} \lim_{\Omega\to 0} c(Q\Omega) \tag{5.5}$$

$$c_S = \lim_{\Omega\to 0} \lim_{Q\to 0} c(Q\Omega) \tag{5.6}$$

Taking these two limits in our expression (5.2) for the self-energy gives

$$c_T = c^{(0)} - TC_v\overline{\gamma^2} \quad ; \quad c^{(0)} = \rho v^2 \tag{5.7}$$

$$c_S = c^{(0)} - TC_v(\overline{\gamma^2} - \overline{\gamma}^2) \tag{5.8}$$

The difference between adiabatic and isothermal depends on $\overline{\gamma}$. All terms in the expression for the self-energy which are proportional to γ originate from phonon transport, meaning that they come from diagrams other than the simple bubble diagram, whose contribution is proportional to $\overline{\gamma}^2$.

Therefore the terms $\overline{\gamma^2} - \overline{\gamma}^2$ in (5.2) renormalize ω_Q^2 to the adiabatic frequency ω_S^2. This means that part of the anharmonicities change the phonon into an adiabatic sound wave (first sound). But this sound wave is coupled to two entities, one arising from the dielectric fluctuations, the other from thermodynamic ones. The first do not contribute to the difference between adiabatic and isothermal, since their contribution to $c(Q\Omega)$ is zero in both limits. But the last term in (5.2) gives $-\overline{\gamma}^2$ in the isothermal limit, so that the renormalized frequency $\tilde{\omega}(Q\Omega)$ is equal to the isothermal value ω_T in the static limit. At high frequencies, however, the thermodynamic fluctuations die out, as expected, but from the dielectric fluctuations we get

$$\tilde{\omega}^2(Q\Omega) = \omega_T^2\left(1 - \frac{Q^2 TC_V}{\rho\omega_Q^2}\overline{\gamma^2}\right) \qquad (\Omega \to \infty) \tag{5.9}$$

This result shows also clearly that phonon transport is negligible in the high-frequency regime.

The more important property of the result (5.2) is that its imaginary part consists of two terms. There is a generalization to all $\Omega\tau$ of what was earlier derived as a propagator for entropy fluctuations. The response of the local entropy density to the deformation $u(Q\Omega)$ is

$$(5.10) \qquad \frac{\delta S(Q\Omega)}{C_v} = \frac{\omega_2^2(\Omega + i/\tau)M_{11}(Q\Omega)}{\omega_2^2 \cdot \frac{i}{\tau} M_{11}(Q\Omega) - \Omega^2 - 2i\Omega\gamma(Q\Omega)} \bar{\gamma}u(Q\Omega)$$

The coupling of these thermodynamic fluctuations to the one-phonon state, described by the Green function $G(\underline{Q}\Omega)$, is via $\bar{\gamma}^2$. If the entropy fluctuations get overdamped, a central peak in the spectral density arises, the width of which is given by the thermal diffusivity times the transferred momentum squared.

In addition, there is the a term in the self-energy (5.2), which is a weighted sum of Lorentzians centered around $\underline{Q}\cdot\underline{v_q} = \omega_Q \cos\Theta$, where Θ is the angle between $\underline{Q}$ and $\underline{q}$, varying from 0 to π. Here $\underline{v_q}$ is the group velocity of phonon $\underline{q}$. In the simplified model employed to derive (5.2), which is the one-branch Debye model, this term gives rise to a rather broad "central peak" which extends out to the phonon peaks, thus giving rise to a spectral density where the phonon "sits on the shoulder" of the quasi-elastic contribution. Another important difference to the thermodynamic fluctuations is that the coupling to these "dielectric" fluctuations is partly via γ^2. This part arises from the bubble diagram (first term in eq. (3.10)). If the acoustic approximation (5.1) is not made, the internal lines of the bubble diagram can represent phonons from another branch. Then, the group velocity $\underline{v_q}$ in the denominator of the dielectric propagator corresponds to this branch. In case of an optical branch, $\underline{v_q} \approx 0$, and under this condition the central component would be much narrower and a clear separation of it from the phonon peaks is found. It must also be mentioned that the dependence of the width of this central component on $\underline{Q}$ is different from the one of the entropy peak; if $\underline{v_q}=0$ the width is independent of $\underline{Q}$. Therefore, it should in principle be possible to discriminate between the two cases experimentally.

As can be seen from eq. (5.2) the coupling of the two fluctuations to the one-phonon state is different. Applying this difference to the case of Nb_3Sn, which has a soft transverse acoustic mode exhibiting a central peak (28), we have $\bar{\gamma}^2 = 0$, but $\gamma^2 \neq 0$, so that in this case the thermodynamic fluctuations do not appear above T_c in the neutron scattering from the TA mode. There is no coupling of a transverse acoustic phonon to the temperature. The only remaining contribution comes from the bubble diagram.
Making the approximation of putting $\underline{v_q}\cdot\underline{Q} = 0$ in the self-energy and having $\bar{\gamma} = 0$ (TA mode) the spectrum can be written in the form used be Shirane and Axe (28). The renormalized soft phonon fre-

quency is given by

$$(5.11) \qquad \widetilde{\omega}^2(Q\Omega) = \omega_Q^2 - \frac{Q^2 T C_V}{\rho}\,\overline{\gamma^2}\,\frac{1}{1+(\Omega\tau)^2}$$

Identifying $\omega_Q = \omega_\infty$, $\widetilde{\omega}(\Omega=0) = \omega_0$ and $\delta^2 = Q^2 T C_V \overline{\gamma^2}/\rho$, we have $\omega_0^2 = \omega_\infty^2 - \delta^2$. The spectrum is then given by

$$(5.12) \quad \frac{1}{\Omega}\,\mathrm{Im}\,G(Q\Omega) \sim \frac{\dfrac{\delta^2\tau^{-1}}{\Omega^2+\tau^{-2}}}{\left(\omega_Q^2 - \dfrac{\delta^2\tau^{-2}}{\tau^{-2}+\Omega^2} - \Omega^2\right)^2 + \Omega^2\left(\dfrac{\delta^2\tau^{-1}}{\Omega^2+\tau^{-2}}\right)^2} \; .$$

Finally, we would like to mention that the part of the self-energy arising from fluctuations away from local thermodynamic equilibrium can be interpreted as a frequency dependent viscosity.

VI. DIRECT COUPLING OF LIGHT TO PHONON DENSITY FLUCTUATIONS

So far we have discussed the interaction of a one-phonon state with fluctuations in the phonon density, $\delta n_{\underline{q}}(\underline{Q}\Omega)$. The idea behind it was an experim ent, in which the external probe couples to this one-phonon state, but since this state is coupled anharmonically (via $\overline{\gamma}^2$ and/or $\overline{\gamma^2}$) to phonon density fluctuations, the spectrum of scattered light (or neutrons) exhibits these latter fluctuations indirectly. As discussed in Sect. II there is also the direct coupling of light to phonon density fluctuations arising from the terms of eq. (2.7) proportional to P_2. As a result, in the general expression for the spectrum, higher-order Green functions appear. These functions have to be calculated by approximations which are consistent with those for the one-phonon Green function. Therefore, one has to go beyond a simple Hartree factorization. As shown in Ref. (13), two types of terms appear, a pure P_2^2 term and interference effects P_1P_2 which one can interpret as a renormalization of the coupling to the one-phonon state P_1^2. Furthermore, both contributions can be reduced to expressions which are determined by the four-point function D(1,2,1',2') of eq. (3.17), which is directly related to the solution of the transport equation.

The central peak in the light scattering spectrum which we had discussed in Section V appears because of anharmonic coupling: Schematically we draw the following diagram

(A)

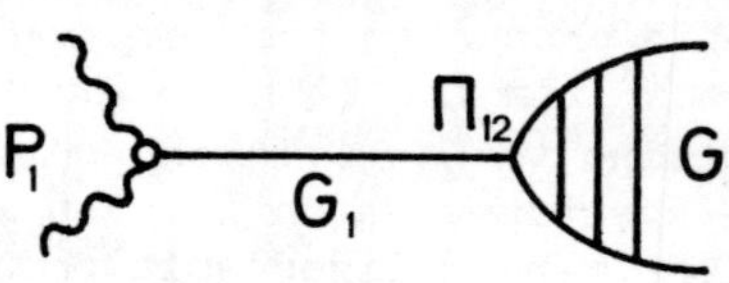

Light couples via P_1 to the one-phonon state of wave vector equal to the transfered momentum. The propagator G_1 for this phonon has been calculated in the preceeding sections. A self-energy expression results which can be interpreted as consisting of a coupling (of strength π_{12} given by weighted sums over $V_3(\underline{Q}\ \underline{q}\ \overline{\underline{q}})$) to phonon density fluctuations $\delta n_{\underline{q}}(\underline{Q}\Omega)$ (indicated in the diagram by the ladder) described here for simplicity by a propagator G_2. This mechanism to detect a central peak can be called indirect.

If in addition P_2 is included in eq. (2.7), one also takes the direct coupling of light to phonon density fluctuations into account, which can be illustrated by

(B)

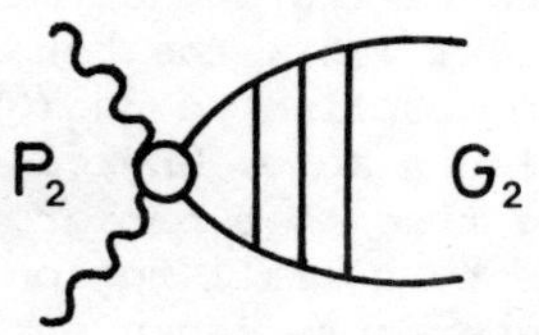

The total intensity of the central peak is in general not simply given by the sum of the two processes (A) and (B), since there are interference effects between the direct and the indirect process, proportional to P_1P_2:

(C)

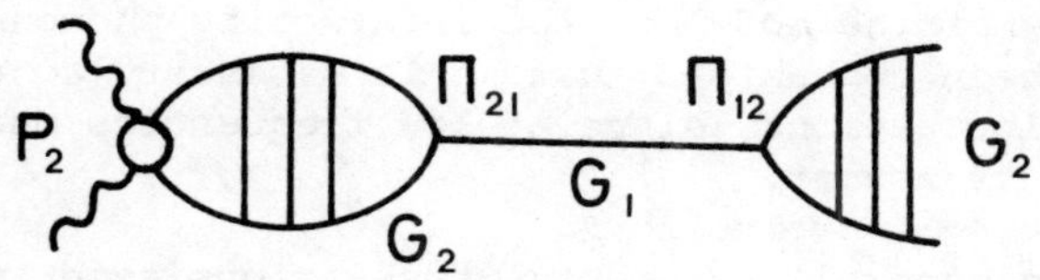

They lead to a renormalization of P_1 to

$$\tilde{P}_1 = P_1 + P_2G_2\pi_{21} \quad .$$

Since light scattering is proportional to $\tilde{P}_1^2$, we get a term $P_1P_2G_2\pi_{21}$, which can be negative. In this way destructive interference becomes possible as well as constructive interference (29). The possibility of so-called antiresonances should also be mentioned. Examples of these effects have been calculated in Ref. 13 for the case, in which G_2 is the "thermodynamic" propagator of entropy fluctuations.

The situation discussed here can be put into the form of anharmonically coupled oscillators. One of them is the adiabatic sound wave $\underline{Q}$, and in addition there are the "thermodynamic" and the "dielectric" oscillators. All these oscillators are different from the phonons which are used to define the Hamiltonian, from which we started. More or less collective character is included in their definitions (30).

Within this picture of coupled oscillators light scattering from phonon density fluctuations by the indirect process arises from coupling the light into the first-sound oscillator. But since the latter is coupled to G_2 the response function of the coupled system will also exhibit effects arising from G_2. In addition, there is also the possibility of direct coupling of the light to the collective fluctuations G_2.

This latter mechanism might be particulary important in cases like $SrTiO_3$ where the soft-phonon state is not a zone center mode. The observation by Steigmeier et al.(31) of critical opalescence in this material even at $T > T_c$ suggests the presence of a process like (B). Since first-order Raman scattering is not allowed and the Brillouin peaks behave non-critical (32), the total integrated intensity must arise from the direct coupling to $\delta n_q(Q)$, where q is to be summed over the whole Brillouin zone. Whereas for most q the scattering is uncritical, giving rise to a background which is practically temperature independent, a small region around q_R has to give rise to the strongly temperature dependent contribution.

VII. CONCLUSION

The appearance of a quasi-elastic peak in the low-frequency response of a dielectric crystal has here been investigated from the point of view of the model of the interacting phonons. In particular, it has been emphazised that it is necessary to consider phonon density fluctuations, since at low frequencies one has to allow for phonon transport.

Besides these general statements, the various experimentally observed phase transitions, have to be considered seperately, since the symmetry properties of the modes involved determine which mechanism is operative. In this connection is should be mentioned that for the case of $SrTiO_3$ and similar transitions Enz (24) and Beck and Meier (25) have used a rather special property of the phonon dispersion curve to derive quasi-elastic neutron scattering as a sideband to the soft mode at the R corner of the Brillouin zone. However, it does not seem to be necessary to consider this mechanism, since the generalization (12) of the four-phonon diagram which was first considered in this context by Silberglitt (33) to include phonon transport can in principle describe the observations.

It should also be mentioned that besides the anharmonic lattice dynamical theories discussed here, phenomenological models have been studied (34), (35) to account for the effects of the quasi-elastic peak in terms of frequency-dependent damping functions (36). These theories have been developed quite far and experiments like neutron scattering, electron spin resonance and ultrasonic attenuation near the phase transitions are discussed (34), (37).

Many helpful discussions with Drs. W. Czaja, G. Harbeke, W. Rehwald, E. F. Steigmeier, J. Jäckle and N. Theodorakopoulos and, in particular, the collaboration of Dr. R. K. Wehner and of W. Hess are gratefully acknowledged.

VIII. REFERENCES

1. C. Horie and J. A. Krumhansl, Phys. Rev. 136, A 1397 (1964)

2. P. C. Kwok and P. C. Martin, Phys. Rev. 142, 495 (1966)

3. L. J. Sham, Phys. Rev. 156, 494 (1967)

4. W. Götze and K. H. Michel, Phys. Rev. 157, 738 (1967)

5. C. P. Enz, Ann. Phys. 46, 114 (1968)

6. R. Klein and R. K. Wehner, Phys. kondens. Materie 8, 141 (1968)
10, 1 (1969)

7. G. Niklasson, Phys. kondens. Materie 14, 138 (1972) and references given therein.

8. W. Götze and K. H. Michel, Phys. Rev. 156, 963 (1967);
Z. Phys. 223, 199 (1969)

9. R. A. Cowley, Proc. Phys. Soc. 90, 1127 (1967)

10. R. K. Wehner and R. Klein, Physica 52, 92 (1971)

11. R. Klein, J. de Physique 33, C2 - 11 (1972)

12. R. A. Cowley and G. J. Coombs, J. Phys. C: Solid St. Phys.
6, 121, 143 (1973)

13. R. K. Wehner and R. Klein, Physica 62, 161 (1972)

14. Experimental evidence for a central mode in connection with solid state phase transitions was reported at the first Geilo conference (Structural Phase Transition and Soft Modes, edited by E. J. Samuelsen et al., Oslo 1971) by Riste et al. for neutron scattering from $SrTiO_3$ and by Steigmeier et al. for light scattering from SbSI. Since then direct or indirect observations of a central mode in a number of materials have been reported, some of them are Nb_3Sn, $KMnF_3$, KH_2AsO_4 and CsH_2AsO_4.

15. See, for instance, Light Scattering in Solids, edited by M. Balkanski (Flammarion, Paris, 1971).

16. C. L. O'Connor and J. P. Schlupf, J. Chem. Phys. 47,31 (1967)

have shown that in H_2O the term $(\partial\chi/\partial T)_\rho$ must be taken into account in order to explain their measurements. The present author thanks Prof. Cummins for bringing this reference to his attention.

17. I. L. Fabelinski, Sov. Physics-Uspekhi 63, 474 (1957); R. D. Mountain, Rev. Mod. Phys. 38, 205 (1966)

18. E. Courtens, Phys. Rev. Letters 29, 1380 (1972)

19. R. Hofmann, Dissertation No. 4009, ETH Zürich (Juris Druck Verlag Zürich, 1968); M. G. Cohen, M. DiDomenico, Jr., and S. H. Wemple, Phys. Rev. B 1, 4334 (1970).

20. E. F. Steigmeier, G. Harbeke and R. K. Wehner, Ref. 15, p. 396

21. R. S. Katiyar, J. F. Ryan and J. F. Scott, Ref. 15, p. 436.

22. P. A. Fleury and P. D. Lazey, Phys. Rev. Letters 26, 1331 (1971

23. R. K. Wehner, phys. stat. sol. 22, 527 (1967)

24. C. P. Enz, Phys. Rev. B 6, 4695 (1972) and to be published.

25. H. Beck and P. F. Meier, Helv. Phys. Acta (to be published).

26. R. E. Peierls, Quantum Theory of Solids, Oxford University Press,(London, 1955)

27. J. A. Krumhansl, Proc. Phys. Soc. 85, 921 (1965)

28. G. Shirane and J. D. Axe, Phys. Rev. Letters 27, 1803 (1971)

29. D. W. Pohl and S. E. Schwarz (Phys. Rev., to be publ.) have measured the thermo-optic and elasto-optic constants of NaF and therefore P_1 and P_2. The results indicate destructive interference which has been found subsequently by D. W. Pohl, S. E. Schwarz, and V. Irniger (to be published) by forced Rayleigh scattering.

30. For the case of adiabatic sound waves and entropy fluctuations collective normal coordinates have been used in Ref. 13 and in R. Klein and R. K. Wehner, Phonon Scattering in Solids, edited by H. J. Albany, Paris, 1972, p. 18.

31. E. F. Steigmeier, H. Auderset and G. Harbeke, Solid State Comm. (to be published), and contribution to this volume.

32. A. Laubereau and R. Zurek, Z. Naturf. 25 a, 391 (1970)

33. R. Silberglitt, Solid State Comm. 11, 247 (1972)

34. F. Schwabl, Phys. Rev. Letters 28, 500 (1972); Z. Phys. 254, 57 (1972) and contribution to this volume.

35. T. Schneider, Phys. Rev. B 7, 201 (1973)

36. S. M. Shapiro, J. D. Axe, G. Shirane and T. Riste, Phys. Rev. B 6, 4332 (1972)

37. F. Schwabl, to be published.

PHASE TRANSITIONS IN SOLID CD_4

Werner Press and Alfred Hüller

Institut für Festkörperforschung der
Kernforschungsanlage Jülich, Germany

1. INTRODUCTION

Talking of order-disorder type transitions in solids, one either thinks of binary alloys (e.g. β-brass) or of magnetic systems. This leaves out a large group of phase transitions concerning order or disorder of molecular orientations. Orientationally disordered phases are characterized by a symmetry at the molecular sites, which is higher than the symmetry of the molecule itself. Numerous examples of orientationally disordered solids reveal strongly plastic behaviour (the entropy change with melting being smaller than or comparable to that at possible orientational transitions). The counter part of plastic crystals are the liquid crystals, which are ordered with respect to orientations and disordered with respect to the molecular centres. Orientational ordering means a reduction of the site symmetry at the molecular centre to the symmetry of the molecule, or, in many cases, to a subgroup of the molecular symmetry.

Orientational phase transitions ideally occur without change of the centre-of-mass structure. There are two prominent examples representative of such orientational phase transitions and they appear to represent two extremes. The one is the family of the ammonium halides, especially NH_4Cl/ND_4Cl, which will be the topic of another lecture. The other is solid methane. What is the difference between the two cases? In phase II of NH_4Cl we have a strong "built in" double minimum potential (more precisely: two groups of 12 equivalent minima) which allows for two orientations of the NH_4^+-ions. The deep potential well is caused by the negatively charged Cl^--ions which occupy the corners of cubes surrounding each tetrahedron. The NH_4^+-ions thus are strongly localized in either one of the minima and the transition to phase III means an increase in the population of one of the two orientations without a strong effect on the depth of the potential well (only an asymmetry is introduced). In the high-temperature phase of solid methane (phase I), however, we find a situation with extremely shallow angle-dependent potentials (time-averaged). The reason for this is that the cubic crystalline field in the case of methane is not caused by electrostatic potentials, but by van der Waals forces between the disordered molecules. Though the orientational distribution in phase I of solid methane is largely isotropic, there are no free rotations occurring in this phase. The short time behaviour is characterized by strong local fluctuations of orientation dependent potentials. This must be seen on the background of an orientation dependent interaction energy (phase transition at 27 $^\circ$K for CD_4) which is of the same order of magnitude as the rotational energy spacing of the molecules (7.5 $^\circ$K for $2\hbar^2/2\theta_{CD_4}$). The transition to phase II means the ordering of most of the molecules and for this orientational localization to take place fairly high potential barriers must develop.

In the following, we will be primarily concerned with the high-temperature phase transition of solid CD_4. It will be necessary, first, to note some important facts in connection with the phase diagram and the phase transitions of solid methane.

Methane solidifies at 90 $^\circ$K. At normal pressure, CH_4 has just one phase transition at 20.4 $^\circ$K, while CD_4 has two transitions at 27.0 $^\circ$K and 22.1 $^\circ$K, respectively. First evidence of these phase transitions was obtained in specific heat measurements (1, 2). These showed λ-type anomalies at the transition points. Simultaneously, however, they seemed to indicate a small first-orderness at all transitions, as hysteresis effects of a few hundredth of a degree were found (2). A phase-diagramm for CH_4, determined via measurements of the dielectric constant under pressure (3), is shown in Fig. 1. In this work, we restrict ourselves to normal pressure and to CD_4, mainly.

The question of the structure in the various methane phases was an unsolved problem for a long time. X-ray measurements in CH_4 and CD_4 (4-6) indicated a fcc centre-of-mass structure at all temperatures. Consequently, both

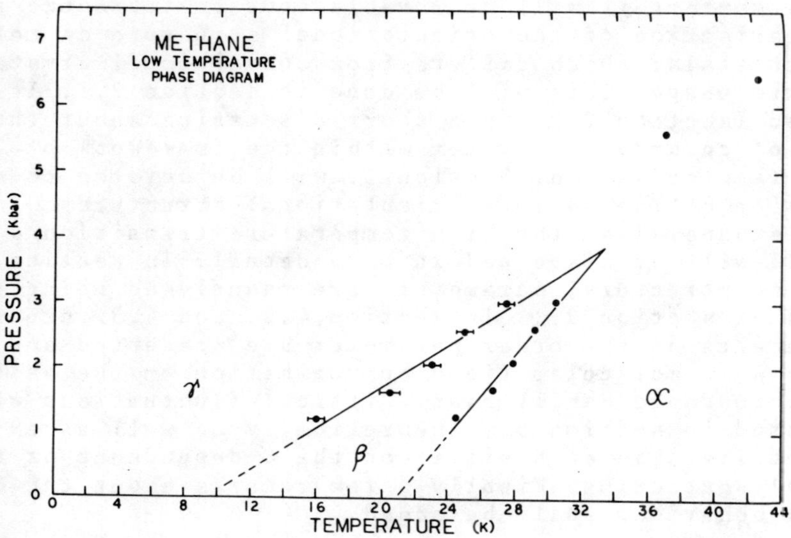

Fig. 1. Phase diagram of solid CH_4, obtained by measurements of the dielectric constant (ref. (3)).

phase transitions were believed to have strictly orientional character. Recently we have learned that this is not exactly what happens (7, 8). In its low temperature phase (phase III at normal pressure has no analogue in CH_4) CD_4 undergoes a 1.2 % tetragonal distortion (8). Rigorous evidence on the orientational structure of CD_4 can be obtained by neutron diffraction only (9-11).

It is not the place here for a detailed discussion of the physical properties in all the various phases of solid methane. Also, only few of the experimental studies so far published were directly concerned with the phase transitions. Therefore, the reader is referred to a recent and comprehensive review of experimental results (12) - not including the more recent neutron measurements - which also may serve as a source of further references. Considerably more attention to the topic "phase transitions" has been paid by theoreticians, especially in the "classic" paper by James and Keenan (13), by the "Kyoto group" (14) and more recently by Alexander and Lerner-Noar (15).

Our main purpose is the discussion of data as obtained by neutron scattering, especially by coherent neutron scattering from single crystals of solid CD_4.

In this context it will be advantageous to introduce a parameterisation of the orientational structure of molecular crystals, which differs from conventional crystallographic usage. This will be done in section 2.1. It is followed (section 2.2) by a short discussion about the notion of an order parameter within the framework of this parameterisation. Section 3. will be devoted to a discussion of the various orientational structures of solid methane. Then the high temperature transition CD_4-I/II will be discussed in more detail. In section 4.1. some structural parameters are reanalysed using the results of section 2.1. In section 4.2. and 4.3. neutron measurements of the order parameter are presented and discussed in molecular field approximation on the basis of a microscopic Hamiltonian. Critical fluctuations will be treated in section 5., theoretically as well as experimentally. The emphasis is on the $\vec{Q}$-dependence of the critical scattering. Finally a few remarks about the dynamical behaviour shall be added.

2. TEMPERATURE DEPENDENT STRUCTURE ANALYSIS IN MOLECULAR CRYSTALS

2.1. Concept of a rotational form factor

A structure factor generally may be written as

$$F(\vec{Q}) = \int_{\text{all space}} \tilde{a}(\vec{r}\,')\exp(i\vec{Q}\vec{r}\,')d\vec{r}\,' \qquad (2.1)$$

where $\vec{Q}$ is the scattering vector and $\tilde{a}(\vec{r}\,')$ is a scattering length density, i.e. a density distribution weighted with the coherent scattering length of the respective atoms. In general, calculations start out from a Gaussian distribution in direct space and the Fourier transform then is a bilinear expression in Q-space, namely

$$F(\vec{Q}) = \sum_i b_i \exp(i\vec{Q}\vec{R}_i^o)\exp(-\vec{Q}\overleftrightarrow{B}_i\vec{Q}) \qquad (2.2)$$

$\vec{R}_i^o$ denotes equilibrium positions, $\overleftrightarrow{B}_i$ the Debye-Waller tensor. For most crystalline solids this is a reasonable approach. In molecular crystals with large librational amplitudes, however, remarkable deviations from Gaussian distributions may occur. This has been accounted for by more sophisticated expressions, basing either on (i) mechanistic models (16, 17) or (ii) the use of higher cumulants in Eq. (2.2) (16), that is the formal generalisation of this expression beyond bilinear terms.

(i) is based on expansions, which tend to break down above some limiting value for the librational amplitude, while (ii) in general demands many additional parameters.

In case of rigid molecules (internal modes of vibration have much higher eigenfrequencies than the external or lattice modes) an alternative approach (18, 19) seems to be promising. Instead of considering single atoms, we simultaneously look at all atoms situated at the same distance from the molecular centre. For simplicity, take just one spherical shell. Furthermore we shall assume that the orientational motion is uncorrelated with the translational or centre-of-mass motion. We then may write, with $\vec{r}' = \vec{R} + \vec{r}$

$$F(\vec{Q}) = \sum_j \exp(i\vec{Q}\cdot\vec{R}_j^o)\exp(-W_j(\vec{Q}))\cdot F_j^{rot}(\vec{Q}) \qquad (2.3)$$

where

$$F_j^{rot}(\vec{Q}) = \int_{cell} a_j(\vec{r})\exp(i\vec{Q}\cdot\vec{r})d\vec{r} \qquad (2.4)$$

is a rotational form factor and the Debye-Waller factor concerns the centre-of-mass motion only. The summation in Eq. (2.3) runs over the unit cell; a generalisation to several shells in one molecule is straightforward. In neutron diffraction (index j is dropped) $a(\vec{r}) = \delta(r-\rho)\cdot b(\theta,\phi)/r^2$, where ρ is the radius of the shell under consideration and (θ,ϕ) are the polar angles of $\vec{r}$. $b(\theta,\phi)$ will now be expanded into a system of symmetry-adapted harmonics; in the case of cubic symmetry, the cubic harmonics $K_{\ell m}(\theta,\phi)$ (20) will be the appropriate system.

$$b(\theta,\phi) = \sum_{\ell,m} b_{\ell m}\, K_{\ell m}(\theta,\phi) \qquad (2.5)$$

One advantage of this expansion is, that for high site symmetry only few coefficients $b_{\ell m}$ are non-zero. Tetrahedral symmetry ($\bar{4}3m$) may serve as an example: Up to the order $\ell = 11$ there is at most one harmonic in each order, which-contributes in Eq. (2.4). The first few terms of an expansion are

$$b(\theta,\phi)=b_{01}+b_{31}\cdot K_{31}(\theta,\phi)+b_{41}\cdot K_{41}(\theta,\phi)+b_{61}K_{61}(\theta,\phi)+\ldots \qquad (2.6)$$

Explicit expressions for the harmonics invariant under the operations of the tetrahedral group are listed in Table I. The meaning of the expansion coefficients may be illustrated as follows: b_{01} represents an angle-independent distribution and is related with the sum over

$$K_{01}(\Omega)=1$$
$$K_{31}(\Omega)=\sqrt{105}xyz$$
$$K_{41}(\Omega)=(5\sqrt{21}/4)(x^4+y^4+z^4-\tfrac{3}{5})$$
$$K_{61}(\Omega)=(231\sqrt{26}/8)(x^2y^2z^2+\tfrac{1}{22}[K_{41}]-1/105)$$
$$K_{71}(\Omega)=\tfrac{11}{4}\sqrt{1365}xyz(x^4+y^4+z^4-5/11)$$
$$K_{81}(\Omega)=\tfrac{65}{16}\sqrt{561}(x^8+y^8+z^8-\tfrac{28}{5}[K_{61}]-\tfrac{210}{143}[K_{41}]-\tfrac{1}{3})$$

Table I Cubic harmonics , which are invariant under the operations of the tetrahedral group (up to the order $\ell = 8$). $\Omega = (x, y, z)$, with $x^2 + y^2 + z^2 = 1$, denotes polar coordinates. Functions in square brackets indicate functions with normalization factors omitted.

the scattering amplitudes of all atoms on the spherical shell. As $\int K_{\ell m}(\theta,\phi)d\phi d\cos\theta = 0$ for $\ell \geq 1$, the higher order terms represent a modulation of the constant distribution (Fig. 5). To calculate the rotational form factor, $\exp(i\vec{Q}\cdot\vec{r})$ is expanded into the same set of orthonormal functions:

$$\exp(i\vec{Q}\cdot\vec{r}) = \sum_{\ell'm'} 4\pi i^{\ell'} j_{\ell'}(Qr)K_{\ell'm'}(\Omega_Q)K_{\ell'm'}(\Omega) \quad (2.7)$$

and the integration in Eq. (2.4) is performed. The result is

$$F^{rot}(\vec{Q}) = 4\pi\sum_{\ell m} i^{\ell} j_{\ell}(Q\rho)b_{\ell m}K_{\ell m}(\Omega_Q) \quad (2.8)$$

where Ω_Q denotes the polar coordinates of the scattering vector Q and Q is its modulus. The $b_{\ell m}$ are the expansion coefficients of $b(\Omega)$. $j_{\ell}(Q\rho)$ is a spherical Bessel function of order ℓ, $\Omega = (\theta,\phi)$.

The method is especially advantageous if the diffraction pattern is determined by a few expansion coefficients only. Such a situation will arise: (i) if the orientational localization is weak, (ii) if the radius of the molecular shell is small ($\rho \lesssim 1$ Å). It is clear, that the expansion in Eq. (2.8) converges the more rapidly the less the molecules under consideration are localized orientationally. This manifests itself in two extremes: A constant (isotropic) distribution in direct space leads to just one (Fourier) coefficient, while a δ-peak in direct space requires an infinite number of coefficients.

The aforesaid is illustrated in Fig. 2a). Gaussian peaks (related by tetrahedral symmetry) on a spherical

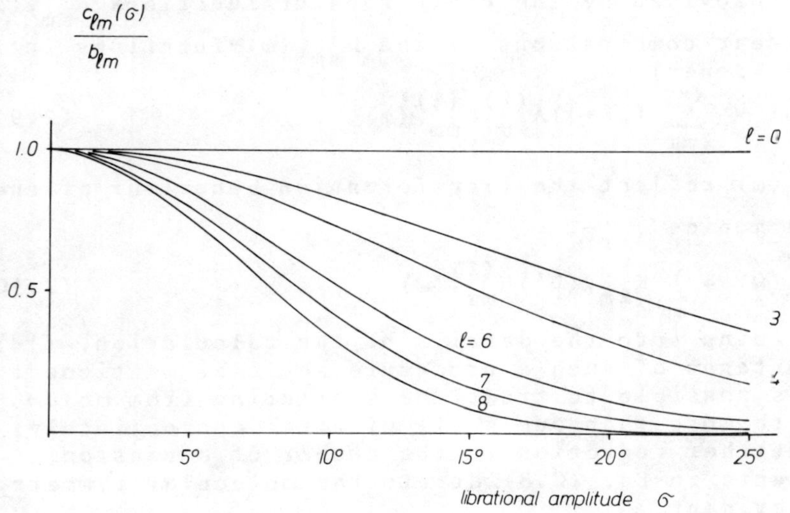

Fig. 2.(a) Expansion coefficient of four Gaussians, related by tetrahedral symmetry, in cubic harmonics. The expansion coefficients $b_{\ell 1}(\sigma)$ are normalized by $b^{o}_{\ell 1}$, their value at $\sigma = 0$.

surface have been expanded into cubic harmonics. Reduced expansion coefficients $b_{\ell m}/b^{o}_{\ell m}$ as a function of the half width at half maximum (librational amplitude σ) are presented. The assumption of a Gaussian distribution certainly is arbitrary.

The spherical Bessel functions $j_{\ell}(Q\,\rho)$ serve as weighting functions. (ii) is due to the fact that for small radii ρ they will suppress the contribution of higher order harmonics within the Q-values accessible in a scattering experiment.

Additional features of an expansion into cubic harmonics become evident in a generalized treatment. It starts by first evaluating coefficients $b^{o}_{\ell m}$ in a coordinate system fixed in the individual molecules and then to transform to the crystal system. Molecular orientations are properly described by the set of Euler angles, here collectively denoted by ω. By analogy with what has been done above, a distribution function $f(\omega)$ is defined, which gives the probability of finding a molecule in the orientation ω with respect to the crystal system. $f(\omega)$ is expanded into a complete and orthogonal set of functions, which is provided by the cubic rotator functions $U^{(\ell)}_{mm'}(\omega)$ (13) (linear combinations of the $D^{(\ell)}_{mm'}(\omega)$-functions introduced by Wigner)

$$f(\omega) = \sum_{\ell mm'} (2\ell+1) A^{(\ell)}_{mm'} U^{(\ell)}_{mm'}(\omega) \tag{2.9}$$

The $U^{(\ell)}_{mm'}(\omega)$ reflect the transformation behaviour of the cubic harmonics

$$K_{\ell m}(\Omega) = \sum_{m'} K_{\ell m'}(\Omega') U^{(\ell)}_{mm'}(\omega) \tag{2.10}$$

Without going into the details of the calculation, (18) two advantages of such a procedure shall be mentioned: (i) it is possible to treat the scattering from molecules with more than one shell of atoms appropriately; (ii) a further reduction in the number of expansion coefficients in Eq. (2.8) due to the molecular symmetry becomes evident.

A formalism based on Euler angles and the functions introduced here will be needed again in the context of the orientation dependent interaction between pairs of molecules (section 4.3.).

2.2. Concept of an orientational order parameter

A phase transition from an orientationally disordered to an ordered phase means that a whole set of expansion coefficients $b_{\ell m}$ (see proceeding section), which is zero in the disordered phase, will become symmetry-allowed. In case of anti-ferro order there will be superlattice reflections governed by these additional coefficients, which may easily be determined in a neutron diffraction experiment.

What is the order-parameter in an orientational transition? Evidently, an orientational order-parameter is linked with the expansion coefficients that become symmetry-allowed. Three examples may help to illustrate this in more detail.

(i) The situation is especially simple in o-H_2 and p-D_2. These are quantum systems and as the orientational coupling is weak, the rotational quantum number J=1 is a good quantum number. In the ordered phase the quantization axes of the molecules are aligned along body diagonals. Typically for systems with quadrupole-quadrupole interaction a superstructure occurs, with four sublattices on which different body diagonals are distinguished (21). The order parameter η has been defined as the projection of the density distribution onto a distinguished body diagonal, i.e. $\eta \sim \langle 3\cos^2\theta - 1 \rangle$ (22). Correspondingly in our formalism there is just one expansion coefficient of order $\ell = 2$ (Eqs. 2.5 and 2.8) which becomes non-zero and it represents the order parameter within a constant factor. The particularity of o-H_2 lies in the fact that the density distribution $\rho(\Omega)$ may be evaluated directly from $\rho(\Omega) \sim \psi_{1,0}(\Omega) \cdot \psi^*_{1,0}(\Omega)$ where $\psi_{1,0}(\Omega)$ is the wave function of a hydrogen molecule with J = 1, M = 0.

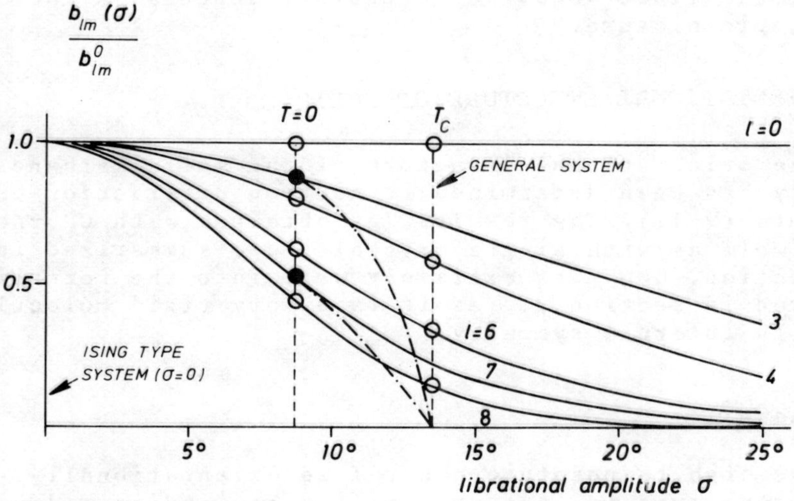

Fig. 2.(b) Change of the expansion coefficients in an orientational phase transition.

(ii) In cases like the ammonium halides, an Ising model may be a useful approximation (23). There the order parameter is defined by $\eta_{ISING} = (n_+ - n_-)/(n_+ + n_-)$; n_+ and n_- denote the occupation of two alternative orientations. In Fig. 2b this corresponds to a librational amplitude $\sigma = 0$, where all expansion coefficients which become symmetry-allowed rise to 1 (in reduced units) in the ordered phase. There is no difference between η_{ISING} and any one of these expansion coefficients, for example b_{31}. Away from $\sigma = 0$ the Ising model is an approximation only, which is the better, the stronger the orientational localization. On the other hand, an expansion may not be tractable in a region close to $\sigma = 0$, since too many parameters will contribute.

(iii) The general case, in between Ising-type systems and orthohydrogen, may best be visualized with the help of Fig. 2b again. A phase transition starts out from a certain orientational localization and leads to an increased localization in the sense that (a) additional expansion coefficients $b_{\ell m}$ become non-zero (order/lower symmetry) and (b) the $b_{\ell m}$ already present in the disordered phase increase in value. The order parameter will be a function of coefficients $b_{\ell m}$ (with $b_{\ell m} = 0$ in the disordered phase). In general the lowest order coefficient (of the symmetry-adapted harmonic or of the corresponding rotator functions) will represent the leading term and the others will be secondary quantities. The exact behaviour certainly depends on the microscopic picture.

3. ORIENTATIONAL STRUCTURE OF SOLID CD_4

The orientational structure of the solid methanes recently has been determined in neutron diffraction experiments (9-11). The results, as obtained with CD_4-powder as well as with single crystals, are summarized in this section. CD_4 fits extremely well into the formalism presented in section 2., as it is a very rigid molecule with high internal symmetry.

3.1. Phase I

The high temperature phase I is orientationally disordered, that is, the probability of finding a deuterium atom on a spherical surface with radius $\rho_{C-D} = 1.093$ Å (around the carbon atoms) is very nearly constant. The space group is Fm3m (9, 10) with just one molecule in the primitive cell (lattice constant $a_o = 5.85$ Å at 35 °K). Slight deviations from complete orientational disorder must be due to the cubic crystalline field. In phase I the rotational formfactor may be written (19)

$$F^{rot}(\vec{Q}) = \left\{ b_C + 4b_D \left[j_o(Q\rho) + b_{41}^{cub} \cdot j_4(Q\rho) \cdot K_{41}(\Omega_Q) + \ldots \right] \right\} \quad (3.1)$$

The experimental information on b_{41}^{cub} is not very precise: only $b_{41}^{cub} \lesssim 0.25$ (in reduced units) may be deduced. This is sufficient, however, to conclude that the assumption of complete disorder for phase I is a reasonable approximation.

3.2. Phase II

In the diffraction pattern of phase II superlattice reflections appear, which indicate an enlargement of the cubic cell by a factor of eight. The space group (9, 11) is Fm3c with a lattice constant a_o = 11.64 Å at 24.5°K. As we know that the lattice of the carbon atoms is fcc both above and below the transition, the phase change must be related to the orientational structure, that is to the onset of orientational order. The arrangement of the methane molecules in phase II is shown in Fig. 3. It may be characterized as follows:

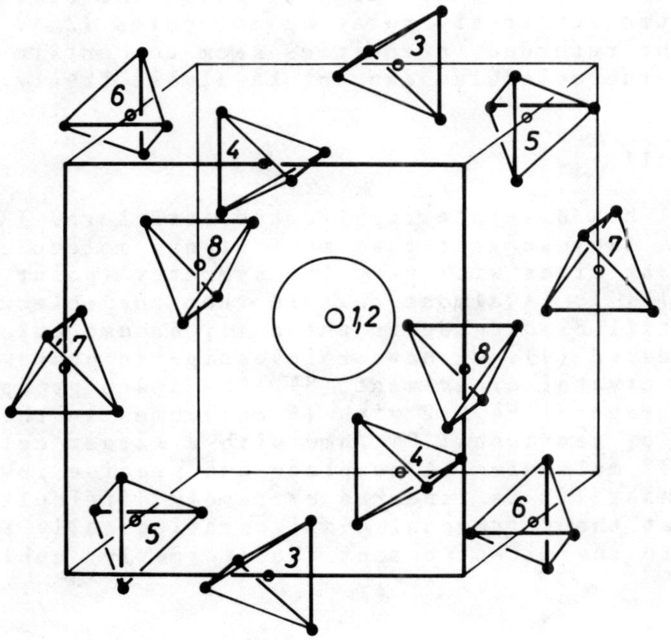

Fig. 3. Arrangement of the eight sublattices in CD_4-II. The figure shows the "octahedral" symmetry in this phase (space group Fm3c) with a disordered molecule in the symmetry centre.

(i) six of eight molecules order in an anti-ferro way (antiferrorotational phase). The orientations on each of the six ordered sublattices are different and the site symmetry is tetragonal $\bar{4}2m$.

(ii) Due to this antiferrorotational ordering the

symmetry at the two remaining sublattice sites is octahedral. The molecules at these sites are orientationally disordered: Octahedral symmetry 432 contains symmetry elements which are not contained in tetrahedral symmetry $\overline{4}3m$. The orientation dependent potential is rather weak and probably not much different from that present in phase I. The structure analysis again does not give very precise information about b_{41}^{dis}, which is the first higher order expansion coefficient. In contrast to the behaviour in phase I, fluctuations of the potential should be very weak, since each disordered molecule is surrounded by a shell of twelve ordered nearest neighbours. In the light of this it is not too surprising, that nearly undamped rotational excitations have been found in CH_4 at $4^{o}K$, with energies shifted only slightly below the corresponding values for freely rotating molecules (25). In addition, the recorded intensities seem to confirm that only 1/4 of the molecules can rotate almost freely.

3.3. Phase III

CD_4-III has a rather complicated structure. IR-measurements (26) suggest that most of the molecules are centred at sites with very low symmetry (point symmetry 1 or $\overline{1}$). It is almost certain that the molecules which are still disordered in phase II, become orientationally ordered (9). We have reinvestigated phase III in a single crystal experiment (8). The space group probably is tetragonal $P4_22_12$ with 16 molecules in the primitive cell or tetragonal $P4_2/nmc$ with a larger cell containing 32 molecules. A complete and precise investigation of phase III is rendered extremely difficult by the fact that the tetragonal c-axis statistically is pointing into the directions of the (formerly) cubic axes.

3.4. Nature of the phase transitions

In contrast to specific heat measurements, microscopic quantities (like the order parameter) may be investigated directly in neutron diffraction experiments. The following results concerning the nature of the phase transitions are obtained. Fig. 4 shows the temperature dependence of a superlattice reflection (310 reflection, when indexing in a face-centred cell) occurring below the II/III transition together with the tetragonal distortion of phase III. Intensities change rather abruptly from zero (above $T_c = 22.1^{o}K$) to some finite value below

T_c. The transition clearly is a first order transition. The behaviour at the I/II transition is shown in Fig. 6. From the temperature dependence of superlattice reflections no discontinuity can be detected. In addition, no hysteresis effects have been found ($\Delta T \lesssim 0.025\,^{o}K$). A small first order contribution cannot be excluded, however.

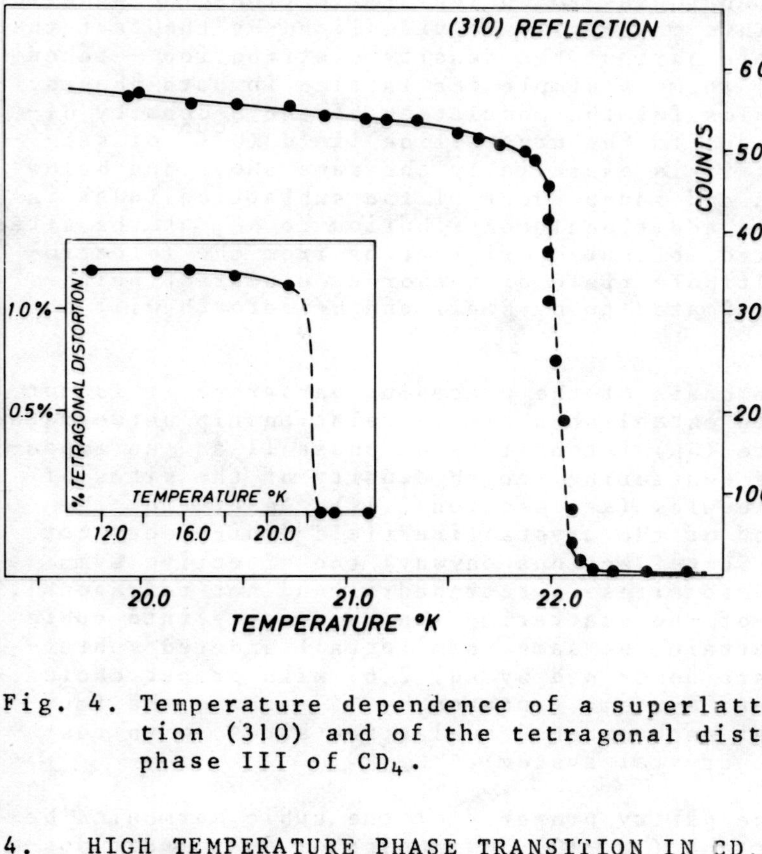

Fig. 4. Temperature dependence of a superlattice reflection (310) and of the tetragonal distortion in phase III of CD_4.

4. HIGH TEMPERATURE PHASE TRANSITION IN CD_4: ORIENTATIONAL ORDER PARAMETER

4.1. Rotational form factors in phase II

From the last section it is clear that the high temperature phase transition of solid CD_4 deserves most interest. It resembles to an antiferromagnetic phase transition, as does solid orthohydrogen (27). Phase I corresponds to a paramagnetic phase, while phase II has its analogue in antiferromagnetic order. Due to this anti-ordering which occurs on six of the eight sublattices the temperature dependence of the expansion coefficients (see section 2) may be studied directly at the superlattice reflections of phase II. In the following isotropic and identical (translational) Debye-Waller factors for all molecules in phase II shall be assumed. Then the carbon atoms and the isotropic part of the D-density distribution contribute to centre-of-mass reflections only, that is to reflections present in phase I already (analogous to nuclear reflections in magnetic systems). This may easily be visualized by the fact that the isotropic part of the density distribution - taken by itself - forms a simple fcc lattice in both phases. The same holds for the modulation of the D-density distribution due to the crystalline field (b_{41}^{cub} of section 3.1.): it is essentially the same above and below $T_c = 27.0^{o}K$ and independent of the sublattice index in phase II. An additional contribution to b_{41} at the sites of disordered molecules originating from the (electrostatic) multipole field of the ordered nearest neighbours is estimated to be small and henceforth will be neglected.

On the basis of the preceding paragraph it is not difficult to establish a simple relationship between the superlattice (SL) intensities of phase II and an expansion of the scattering length density at the sites of ordered molecules (see section 2.1.). Apart from the contribution of the crystalline field (which does not affect the SL-reflections anyway) the effective symmetry at these sites is tetrahedral and not tetragonal. Expansions of the scattering length density into cubic harmonics attain the same form for all ordered sublattices and are described by Eq. 2.6. with proper choice of coordinate systems (rotated by 45° around the fourfold axis at each ordered sublattice sites, with respect to the crystal system).

Now the parity property of the cubic harmonics becomes important (18, 19): Harmonics of even order possess inversion symmetry (positive parity) and odd order harmonics do not (negative parity). As has been done before with the isotropic distribution, the harmonics of even order ($\ell \geq 4$) may be grouped alone. It is found that this part of the D-density distribution forms a primitive cell (P-cell) containing four sublattices only. Even order harmonics consequently do not contribute to general reflections of the large face-centred cell, with indices $h_1h_2h_3$ all odd. These general superlattice reflections are solely due to harmonics $K_{\ell m}(\Omega)$ with odd order ℓ. The situation can be visualized without great difficulty, but considering pairs of ordered molecules, related by inversion symmetry (Fig. 5). In addition it can be shown that the rotational form factor arising from each pair, is $F^{rot} \sim \{ 1 + (-1)^{h_i+\ell} \}$ and $(h_i + \ell)$ must be even to yield a non-zero result. The centre-of-mass positions generate a phase factor $(-1)^{h_i}$ and the parity of the $K_{\ell m}(\Omega_0)$ is introduced by the inversion operation. Hence, the contributions of the two groups of harmonics are well-separated. SL-reflections with indices all even are due to even order harmonics only and harmonics with ℓ odd contribute to SL-reflections with indices all odd.

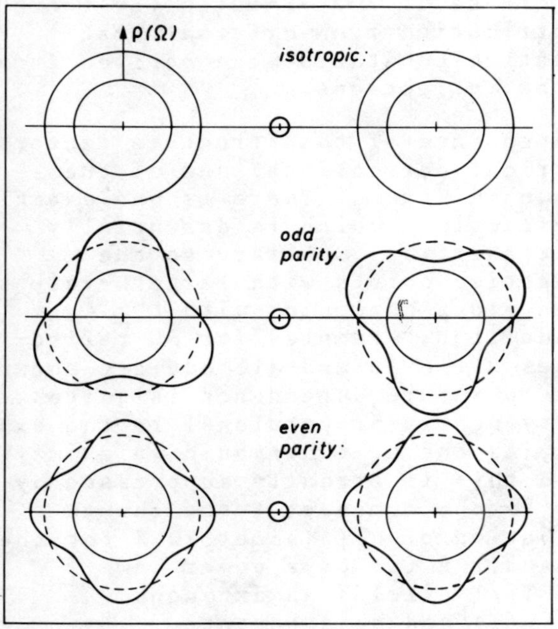

Fig. 5. Schematic diagram of density distributions on a sphere: the contours stand for even and odd parity harmonics and isotropic distributions, always related by inversion symmetry.

Reflections $h_1h_1h_2$ all odd are systematically absent in the space group Fm3c and the first non-zero reflection has indices 531. Experimentally, the systematic absences demand the choice of rather inconvenient zones, e.g. [112].

4.2. Measurement of superlattice reflections in phase II

The measurements (28) were performed with a two axis spectrometer using a Ge-(311) monochromator and operating at λ = 1.829 Å (calibrated with Al_2O -powder). A set of eight reflections in a [1$\bar{1}$2]-zone of a single crystal was recorded as a function of temperature ($h_1h_2h_3$ = 440, 622, 804; 753, 531, 731; 660, 642) by scanning the scattering angle. Gaussians were fitted to the measured peaks and integrated intensities as well as peak positions extracted from the fit. Due to a large mosaic (35 min) of the sample used, extinction corrections turned out to be unnecessary. Also, probably for reasons of the steep increase of the SL-intensity there was only negligible contribution from critical scattering. In addition, lattice constants were derived from the peak positions of the reflections.

The temperature dependence of the structure factors (after the usual geometrical corrections) and of the lattice constant is shown in Fig. 6. There is one example for a centre-of-mass reflection, which is essentially unchanged by the phase transition. Contrary to the example in Fig. 6, at lattice points with large Q-values the contribution of even order harmonics with $\ell \geq 4$, becomes important. In addition, examples for SL-reflections with Miller indices all even and all odd are shown. We observe that their temperature dependence is quite different. $|F|_{odd}$ is very nearly proportional to the expansion coefficient b_{31}, as the next harmonic which may contribute is $K_{71}(\Omega)$ and this is strongly suppressed by $j_7(Q\rho)$. For $|F|_{even}$ b_{41} is the dominant term, though there is a small contribution of b_{61} (about 10 % for the reflections presently used). A fit of a power law $|F| \sim \varepsilon^{\beta}$ with $\varepsilon = (T_c - T)/T_c$ yields an exponent $\beta = 0.170 \pm 0.010$ for $b_{31}(T)$ and an exponent $\beta' = 0.24_8 \pm 0.012$ for $b_{41}(T)$. In Table II the exponents as obtained for the different reflections are listed. We also have attempted to fit all data simultaneously with temperature dependent structural parameters. The resultant R-factor is 7.1 % with only slightly changed exponents. Generally speaking it may be necessary to repeat almost complete structure determinations at a sequence of temperatures, to account for various temperature-dependent parameters. In the case of methane the parity of the $K_{\ell m}(\Omega)$ helps to separate the contributions and this simplifies the problem.

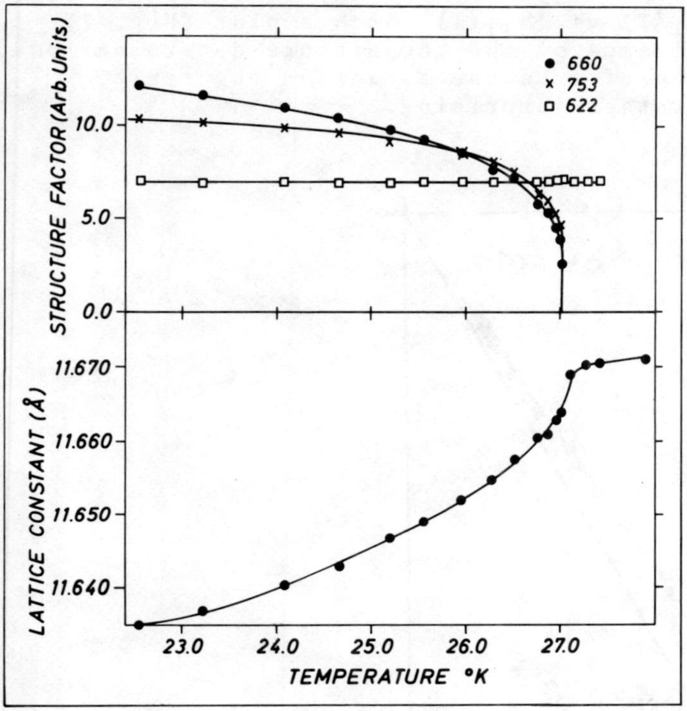

Fig. 6. Superlattice reflections (660, 753) and centre-of-mass reflection (622) in phase II of CD_4; in the lower part of the figure, the temperature dependence of the lattice parameter is shown.

$h_1h_2h_3$	β	$h_1h_2h_3$	β'
753	0.171 (4)	660	0.257 (5)
731	0.176 (3)	642	0.238 (6)
531	0.159 (6)		
	0.170 ± 0.010		$0.24_8 \pm 0.01_2$

Table II "Exponents" β and β' as obtained for the different superlattice reflections in solid CD_4.

The ratio $\beta'/\beta = 1.47 \pm 0.06$ may be obtained directly, plotting $b_{41}(T)$ vs. $b_{31}(T)$. Such a plot (Fig. 7) removes the influence of the temperature determination. On the background of a Landau expansion the ratio $\beta'/\beta = 1.47$ is rather surprising.

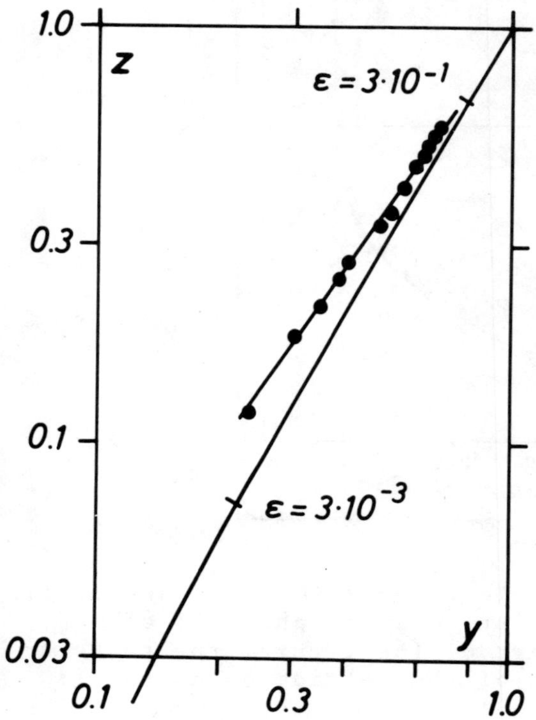

Fig. 7. Comparison of y and z, calculated from the solution of the self-consistency equation and determined experimentally. From the graph a theoretical ratio $\beta'/\beta = 1.65$ is obtained, whereas the experiment gives $\beta'/\beta = 1.47$.

4.3. Molecular field approximation (MFA)

How can the experimental results be understood? We base the discussion (28) on the James and Keenan model Hamiltonian (13) for solid CD_4. Therein no kinetic energy term is contained. Hence, quantum effects as for example introduced by spin statistics (three spin species in solid methane) are neglected. Also, the influence of the cubic crystalline field is omitted. The (electrostatic) multipolar interaction between two molecules depends on their instantaneous orientation, which may be expressed in terms of the Euler angles ω. For a general multipolar interaction, we have

$$H=1/2\sum U^{(\ell)}_{\mu\mu'}(\omega_i)\bar{C}^{\ell\ell'}_{\mu\nu\mu'\nu'}{}^{ij}U^{(\ell')}_{\nu\nu'}(\omega_j) \qquad (4.1)$$

where the $U^{(\ell)}_{\mu\mu'}(\omega)$ are the rotator functions, introduced in section 2.1.; $\bar{C}^{\ell\ell'}_{\mu\nu\mu'\nu'}{}^{ij}$ is a purely geometrical tensor and only depends on (i) the relative position of two molecules which interact and (ii) a coupling constant $I_\ell \cdot I_{\ell'}/R^{\ell+\ell'+1}$ (I_ℓ is the ℓth multiple moment). James and Keenan (13) retain octupole octupole interaction between nearest neighbour molecules only:

$$H_{JK}=I_3^2/(2R^7)\cdot\sum U^{(3)}_{1\mu}(\omega^i_a)C^{ij}_{ab\,\mu\nu}U^{(3)}_{1\nu}(\omega^j_b) \qquad (4.2)$$

Here a, b denote the eight sublattices in phase II of CD_4; $C^{ij}_{ab\,\mu\nu}$ is a 7 x 7 matrix and R the nearest neighbour distance. The approximations of James and Keenan are justified by the fact that (i) I_3 is the leading term in an (electrostatic) multipole expansion and (ii) the octupole octupole interaction decreases with R^{-7}. A formal analogy to the Heisenberg (Ising) Hamiltonian is evident.

In the MFA the distribution function (over Euler angles) of the molecules is given by

$$f_a(\omega^i_a)=G\cdot\exp\left[-(\beta\cdot I_3^2/R^7)\sum_\mu U^{(3)}_{\mu 1}(\omega^i_a)\sum_{\nu j}C^{ij}_{ab\,\mu\nu}\langle U^{(3)}_{\nu 1}(\omega^j_b)\rangle\right] \qquad (4.3)$$

G is a normalization factor. Eq. 4.3 is a self-consistency equation, as $\langle U^{(3)}_{\nu 1}(\omega^i_b)\rangle = \int d\omega f_b(\omega^j_b)U^{(3)}_{\nu 1}(\omega^j_b)$.

It has been solved by James and Keenan (13) and leads to the correct prediction of the structure of phase II. Their result will be used in the following. Rotating the coordinate system by $\pm\,\pi/4$ around fourfold axes at the sites of ordered molecules (see section 4.1.) all distribution functions defined in the new (primed) systems look the same and an expansion of $f(\omega)$ (see Eq. 2.9.) yields:

$$f(\omega')=1/8\pi^2+7\cdot y\cdot U^{(3)}_{11}(\omega')+9\cdot z\cdot U^{(4)}_{11}(\omega')+\ \dots \qquad (4.4)$$

y stands for the octupolar component of the ordering and is the order parameter. Both y and z are zero above the phase transition and reach finite values below - due to the breaking of symmetry. The relationship with the scattering length density of the deuterium atoms is

$$a(\Omega')=b^{o}_{01}+y\ b^{o}_{31}K_{31}(\Omega')+z\ b^{o}_{41}K_{41}(\Omega')+\ldots \qquad (4.5)$$

with $b^{o}_{01} = 1$, $b^{o}_{31} = \sqrt{35/9}$ and $b^{o}_{41} = -\sqrt{7/3}$ in units of $4\cdot b_D/\sqrt{4\pi}$, where b_D is the coherent scattering length of a deuterium atom.

As usually we now look at an expansion of the free energy F=E-TS in terms of the order parameter y.

$$F(y;T) = a(T)y^2 + b(T)y^4 + \ldots \qquad (4.6)$$

The internal energy is obtained by replacing the $U^{(3)}_{1\mu}$ in the Hamiltonian by their expectation values; E contributes to a(T) only. The entropy, finally, is $S = -Nk\sum_a\int d\omega f_a(\omega)\ln(8\pi^2 f_a(\omega))$ (summation over all "magnetized" sublattices). kT_c is defined by $a(T_c) = 0$, which is fulfilled at $kT_c = (51/4)I_3^2/R^7$. b(T) is proportional to $3\overline{u^2}^2 - \overline{u^4} = (429-427)/7007$ (convention $\overline{u^n} = \int U^{(3)}_{\mu}{}^{n}(\omega)d\omega$) and hence is very small. One is close to $b(T_c) = 0$, where the transition would change its nature from second to first order. In this context two things are important: (i) the cancelling of the terms contributing to b(T) is a purely geometrical effect; the addition of terms neglected in the Hamiltonian may well increase b(T) or even reverse its sign; (ii) it is our feeling that the most important multipolar interaction neglected in the Hamiltonian stems from the crystalline field (14). As it is certainly impossible to apply multipolar fields conjugate to secondary quantities in the case of CD_4, the only method is to change the crystalline field by application of pressure (this influences the octupole-octupole interaction as well).

The result $\beta = 1/2$ (MFA) is valid only in an extremely narrow range below T_c. To obtain the temperature dependence of y and z in the range accessible to the experiment, the exact solutions of the self-consistency equation must be considered:

$$y = \int U^{(3)}_{11}(\omega')f(\omega')d\omega' \quad \text{and} \quad z = \int U^{(4)}_{11}(\omega')f(\omega')d\omega'$$

The calculation does not provide a straight line in a double logarithmic presentation, that is, the order parameter may not be written as a simple power law $y \sim \varepsilon^{\beta}$.

In such a description the "exponent" β would be temperature dependent. In the interesting temperature range ($3 \cdot 10^{-3} \leq \varepsilon \leq 3 \cdot 10^{-1}$) it changes from 0.295 to 0.235, that is an "exponent" of roughly 1/4 obtains. On the other hand, the ratio of β'/β can be evaluated and turns out to be 1.65 in MFA (28).

To compare these theoretical results with the experiment we have performed a magnetothermo-mechanical correction (29). An effective phase transition temperature $kT_c^*(T) = (51/4)I_3{}^2/R(T)^7$ is introduced and its value determined from the measured temperature dependent lattice spacing. This procedure defines an effective reduced temperature $\varepsilon^* = (T_c^* - T)/T_c^*$. We should like to point out that in contrast to magnetic systems, where such corrections have been applied (30), the R-dependent "exchange interaction" is well known and can be calculated directly in our example. New values for the exponents are obtained, namely $\beta = 0.21 \pm 0.01$ and $\beta' = 0.31 \pm 0.015$, which brings experiment and theory into closer agreement. We thus have a fairly good understanding, at least a qualitative understanding, of the - at the first glance - strange temperature dependence of the superlattice intensities in phase II of CD_4.

We would conclude from the above reasoning that b_{31} is the true order parameter, while the other quantities have secondary character and are induced by the octupolar ordering. In the present classical approach, at $T \to 0^o K$ $f(\omega)$ would become a sum over delta functions, thus giving rise to an infinitely good orientational localization: hence all expansion coefficients $b_{\ell n}$ would become finite.

In addition it may be mentioned that principally hexadecapolar (2^4-pole) ordering is possible without simultaneous ordering of 2^3-poles: the octupolar ordering breaks symmetries, which need not be broken with 2^4-pole ordering, while the converse is not possible. This may be compared with a (simpler) magnetic system composed by spins S with $S \geq 1$ (31). There a restriction to dipolar and quadrupolar magnetic order has been made. thus allowing for $\langle S_z \rangle$ and/or $\langle S_z^2 - S(S+1)/3 \rangle$ different from zero. The following Hamiltonian has been studied:

$$H = 1/2 \sum J_{ij} s_i^z s_j^z + 1/2 \sum K_{ij} (s_i^{z^2} - S(S+1)/3)(s_j^{z^2} - S(S+1)/3.$$

There are two special cases contained in the general solution: (i) if $J_{ij} = 0$ only quadrupolar ordering occurs; (ii) if $K_{ij} = 0$ the order parameter is $\langle S_z \rangle$, but there is also an induced quadrupolar ordering.

5. CRITICAL FLUCTUATIONS IN CD_4-I

Looking at the coherent scattering in phase I of CD_4 the ensemble averaged part of the density distribution gives rise to Bragg-intensity proportional to $|F(Q)|^2$. In addition, there is disorder scattering, originating from the D_4-tetrahedra with an intensity $S(\vec{Q})_{DIS} = \langle F^2(\vec{Q})\rangle - \langle F(\vec{Q})\rangle^2$, the average means an ensemble average. In the limit of high temperatures ($T \gg T_c$) correlations between the orientations of molecules are negligible and the ("incoherent") scattering between the reciprocal lattice points is described by $S(\vec{Q})_{DIS}$, which depends on $|Q|$ only. With approaching T_c a short range orientational order is developing which gives rise to critical scattering. This intensity which is fairly localized in reciprocal space, is growing at the expenses of the disorder scattering and diverges with approaching T_c. Whereas the "magnetization" in phase II depends on $\langle U_{l\mu}^{(\ell)}(\omega)\rangle$ on the various sublattices, the critical scattering (phase I) caused by the orientational fluctuations is due to the correlation function $\langle U_{l\mu}^{(3)}(\omega_i)U_{l\nu}^{(\ell)}(\omega_j)\rangle$. The intensity peaks at positions which become superlattice reflections in phase II and hence may be directly observed with neutron diffraction.

5.1. Experimental

We have performed a scattering experiment (32) using a large single crystal with a mosaic spread of 8^{min} only. (A different crystal was used for the-measurement of SL-intensities in phase II.) Quasielastic intensities were recorded with a three-axis spectrometer operating at $\lambda = 1.829$ Å (in order to eliminate $\lambda/2$-neutrons, a Ge-(311) monochromator was used). The instrument was operated without and with use of an analyser unit, being set to elastic scattering in the latter case. The energy resolution was 1.0 meV then and hence was large compared to the inelasticity of the scattering (quasielastic linewidth $\Gamma \lesssim 0.25$ meV, with the approximate sign valid far away from T_c). Especially close to T_c an effective integration over energy is taking place in both modes of operation.

The Q-resolution typically was 0.015 Å^{-1} (FWHM) horizontally and 0.12 Å^{-1} vertically. So far, no resolution corrections have been performed. As in the study of superlattice intensities, the analysis took place in a [1$\bar{1}$2]-zone of the crystal.

5.2. Temperature dependence of the critical scattering

Fig. 8 shows the temperature dependence of the critical scattering (peak intensity $S(\vec{Q} = \vec{Q}_L)$ and the correlation length $\xi(T)$, as determined at a zone-boundary L-point. Due to the lack of resolution corrections, no critical indices have been extracted yet (without corrections: $\gamma \approx 1$). In Fig. 8 a small discontinuity of the transition is indicated, as the curves extrapolate to a temperature about 0.4^{o} below T_c. We believe that this is largely due to resolution effects: (i) the low vertical collimation means an effective integration in the vicinity of T_c; a correction, which assumes a complete integration increases $\xi(T)$ by a factor $\sqrt{3}$; (ii) very close to T_c the inverse correlation length ξ^{-1} decreases to values, which are of the same magnitude as the horizontal resolution.

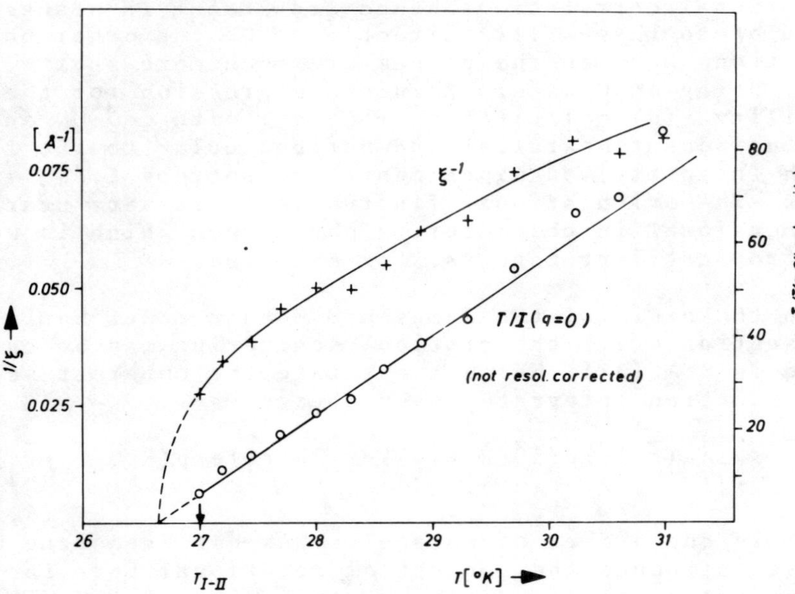

Fig. 8. Inverse static susceptibility and inverse correlation length as a function of temperature (data are not resolution corrected).

5.3. q-dependence of the critical scattering

The distribution of the critical scattering in reciprocal space is related with the nature of the orientational interaction. It has been studied at two symmetry-related L-points in the $[1, \bar{1}, 2]$-zone, namely at $(5, 3, \bar{1})/2$ and $(5, \bar{1}, \bar{3})/2$. Fig. 9 shows the intensity contours at these two zone boundary points: evidently the contours look rather different. The critical scattering is strong along the $[1\bar{1}\bar{1}]$-direction connecting $(53\bar{1})/2$ with the adjacent zone centres (220) and $(31\bar{1})$. The equivalent $[1\bar{1}1]$-direction for the other L-point intersects the zone of scattering at an angle of 70.5^{o}. Consequently we have intensity contours in two cuts, which are very nearly perpendicular to each other and we may gain rather complete information about the q-distribution of the scattering. The observed strong anisotropy is not in controversy with the point symmetry $\bar{3}m$ at the L-points (with the $\bar{3}$ axis directed along $\langle 111 \rangle$). The intensity distribution is cigarshaped, with the long axis of the cigar pointing in $\langle 111 \rangle$ directions, and there are four such directions passing through each zone-centre. Orientational correlations hence are strong in planes perpendicular to these $\langle 111 \rangle$ directions. On the other hand, correlations between the planes are much more short-ranged. Using an Ornstein-Zernicke expression for the susceptibility $\chi(q) \sim 1/(1+\xi_{\perp}^{2} q_{\perp}^{2} + \xi_{\parallel}^{2} q_{\parallel}^{2})$ with $\vec{q}=\vec{Q}-\vec{Q}_{L}$ and $q_{\parallel}$, $q_{\perp}$ components parallel and perpendicular to the lines of high intensity, an experimental anisotropy $\xi_{\perp}/\xi_{\parallel} = 4$ results. The orientational fluctuations are very nearly two-dimensional in character, a phenomenon which is well known from antiferromagnets, for example.

On the basis of the James and Keenan model Hamiltonian (section 4.3.) the critical scattering may be calculated in MFA (32). We must evaluate the coherent scattering function integrated over energy.

$$S(\vec{Q})=N^{-1}\langle \sum_{ij} \int a(\vec{r}_i)\exp(i\vec{Q}(\vec{R}_i+\vec{r}_i))\,d\vec{r}_i \int a(\vec{r}_j)\exp(-i\vec{Q}(\vec{R}_j+\vec{r}_j)\,d\vec{r}_j \rangle \tag{5.1}$$

where N is the number of molecules. As has been done before, we introduce the concept of rotational form factors and relate to coordinate systems fixed in the individual molecules. Then the scattering function may be expressed in terms of a correlation function.

$$S(\vec{Q}) \sim B(\vec{Q}) \sum \langle \sum_{ij} U_{\tau\mu}^{(\ell)}(\omega_i) U_{\tau'\mu'}^{(\ell')}(\omega_j) \exp\left[i\vec{Q}(R_i-R_j)\right] \rangle \tag{5.2}$$

Here we have used, that orientational and translational motion are uncorrelated. $B(\vec{Q})$ is a $\vec{Q}$ dependent form factor. Since the James and Keenan model retains octupole octupole interaction only, $\ell = \ell' = 3$ and $\tau = \tau' = 1$.

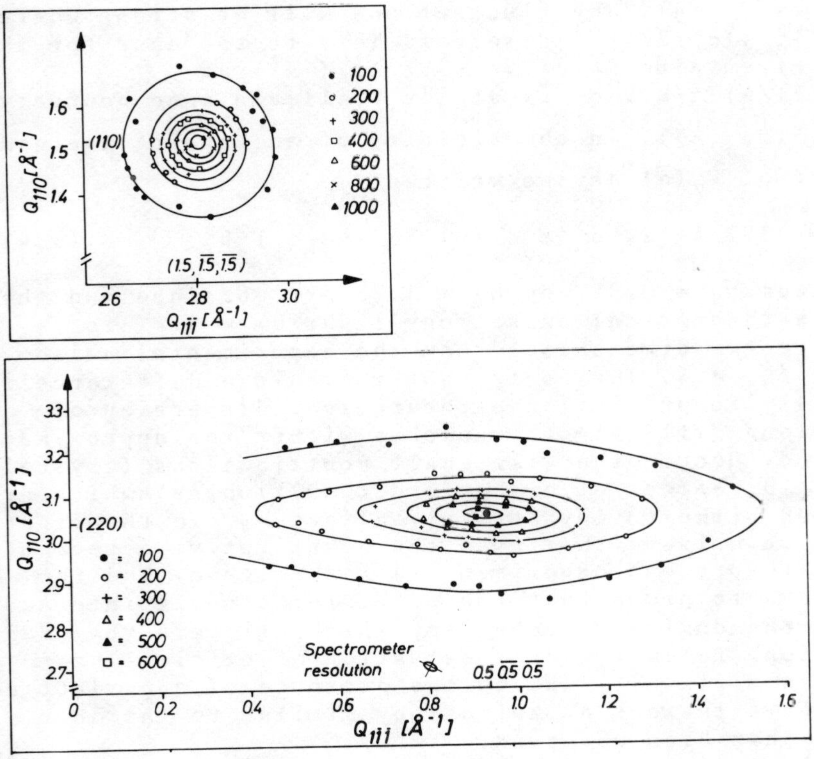

Fig. 9. Critical scattering in CD_4-I. Intensity contours in the [1, $\bar{1}$, 2] zone of scattering; (a) at (5, $\bar{3}$, $\bar{1}$)/2 the long axis of intensity is contained in the zone, (b) at [5, $\bar{1}$, $\bar{3}$] it is not contained and is 70.5 to the scattering zone (background: 50 counts).

In Ornstein-Zernicke approximation the following expression obtains (32):

$$S(\vec{Q}) = \sum_{\alpha\beta=1}^{7} F_{\alpha\beta}(\vec{Q})kT\sum_{\lambda=1}^{7} B^{-1}_{\alpha\beta}(\vec{q})\left[kT+(I_3^2/7R^7)\Gamma_\lambda(\vec{q})\right]^{-1}B_{\lambda\beta}(\vec{q}) \qquad (5.3)$$

with

$$F_{\alpha\beta}(\vec{Q}) = j_3(Q\rho)^2 b^{o2}_{31}K_{3\alpha}(\Omega_Q)K_{3\beta}(\Omega_Q)$$

Here the $\Gamma(\vec{q})$ are the eigenvalues of $C_{\alpha\beta}(\vec{q})$, which is the Fourier transform of the coupling matrix $C^{ij}_{\alpha\beta}$ (see Hamiltonian in section 4.3.). $B_{\lambda\beta}(\vec{q})$ is the matrix which diagonalizes $C_{\alpha\beta}(\vec{q})$. The fluctuations will be strong where $[kT + I_3^2\Gamma_\lambda(\vec{q})/7R^7]^{-1}$ diverges. This is the case for the lowest eigenvalue $\Gamma_7(q_L) = -357\ I_3^2/(4R^7)$ at $kT_c = (51/4)I_3^2/R$. q_L is at the Brillouin zone boundary ($q_L = \frac{2\pi}{a}(\frac{1}{2}, \frac{1}{2},\frac{1}{2})$). In the vicinity of an L-point the q-dependence of $\Gamma_7(\vec{q})$ may be written:

$$\Gamma_7(\vec{q}) = -357/4+12\sin^2(\alpha_o q_{\parallel}/\sqrt{12})b_{\parallel}^2+(\alpha_o q_{\perp})^2 b_{\perp}^2 \qquad (5.4)$$

The values $b_{\parallel} = 0.35$ and $b_{\perp} = 3.25$ are obtained and there fore the theoretical anisotropy is $b_{\perp}/b_{\parallel} = 9$. This is even more two-dimensional than the experimental anisotropy $\xi_{\perp}/\xi_{\parallel} = 4$. The fact, that there is a difference is not really surprising: the theoretical dispersion of $\Gamma_7(q)$ along $\langle 111\rangle$ almost vanishes within the approximation used. Hence otherwise small contributions (crystalline field, effect of higher-order multipoles) which are neglected hitherto may strongly affect $b_{\parallel}$. In the light of this we believe that even the quantitative agreement between theory and experiment is rather good. The result seems to prove that the octupole-octupole interaction is the dominant mechanism, which triggers the phase transition. Below T_c the observation of critical scattering is difficult, due to the presence of superlattice peaks. Nevertheless an anisotropy similar to the one above T_c has been observed.

5.4. Comparison between superlattice intensities and critical intensities

So far, we have paid attention to the static susceptibility, as observed at (531)/2 only. We may as well record intensities at various zone boundary points which become zone centres and compare with the superlattice intensities below T_c. Two major results are obtained: (i) there are no critical fluctuations visible at K,X-points. Especially (660)/2 has been carefully investigated and no indication of a temperature dependent peak was found. This is in agreement with the calculations (Eq. 5.3), which predict critical scattering peaking at L-points only. (ii) The form factor derived theoretically for $Q = Q_L$ agrees with the rotational form factors, which describe SL-intensities below T_c. In fact measured static susceptibilities $S(Q=Q_L)$ seem to scale very well with the corresponding SL-intensities.

6. FINAL REMARKS

We have discussed at length the temperature dependence of the order parameter and the Q-dependence of the orientational fluctuations in the molecular crystal CD_4. The order parameter has been identified with $\langle U^{(3)}_{1\mu}(\omega)\rangle$ on ordered sublattices, while fluctuations are governed by $\langle U^{(3)}_{1\mu}(\omega_i)U^{(3)}_{1\mu}(\omega_j)\rangle$. An anisotropy of the critical scattering has been detected, as well as a small critical exponent β and a difference in the temperature dependence of the SL-intensities. These experimental findings are rather well understood, theoretically. An approach has been made, which also seems to be promising in the context of other molecular crystals.

So far, we have dealt with static aspects of the I/II phase transition in CD_4 only. Certainly an extension of the analysis to the dynamical properties of CD_4 - in view of the phase transition - would be of great interest In detail these are: (i) single molecule reorientation (phase I, II), (ii) critical slowing down (phase I) and (iii) librational excitations (phase II). All these quantities have rather direct analogues in magnetic systems.

ACKNOWLEDGEMENT

The authors are very much indebted to Prof. H. Stiller and Dr. W. Stirling for many helpful discussions.

REFERENCES

1. K. Clusius, L. Popp and A. Frank; Physica 4, 1105 (1937)
2. J.H. Colwell, E.K. Gill and J.A. Morrison; J. Chem. Phys. 39, 635 (1963); J. Chem. Phys. 36, 2223 (1962)
3. J.S. Constantino, thesis, University of Princeton (1972)
4. A. Schallamach; Proc. Roy. Soc. A171, 569 (1939); Nature 143, 375 (1939)
5. S. Greer and L. Meyer; Z. Angew. Phys. 27, 198 (1969); J. Chem. Phys. 52, 468 (1970)
6. J. Herzeg and R.E. Stoner; J. Chem. Phys. 54, 2284 (1971)
7. D.N. Bolshutkin, V.M. Gasan, A.I. Prokhvatilov and A.I. Erenburg; Z. Strukt. Chem. 12, 313 (1971)
8. H. Grimm and W. Press; to be published
9. W. Press; J. Chem. Phys. 56, 2597 (1972)

10. W. Press, B. Dorner and G. Will; Phys. Letters 31A, 253 (1970)
11. W. Press, thesis, TH Aachen (1971)
12. M. Bloom and J.A. Morrison; "Orientational order and Disorder in the Solid Isotopic Methane", Surface and Defect Properties of Solids, Vol. 2 (The Chem. Society, London)
13. H.M. James and T.A. Keenan; J. Chem. Phys. 31, 12 (1959)
14. for example: K. Nishiyama and T. Yamamoto; J. Chem. Phys. 58, 1001 (1973)
15. S. Alexander and M. Lerner-Noar; Canad. J. Phys. 50, 1568 (1972)
16. Crystallographic Computing; Topic F. Edited by F.R. Ahmed, Munksgaard, Copenhagen (1970)
17. G.S. Pawley and B.T.M. Willis; Acta Cryst. A26, 260 (1970)
18. W. Press and A. Hüller; Acta Cryst. A29, 257 (1973)
19. W. Press; Acta Cryst. A29, 252 (1973)
20. for example: S.L. Altmann and A.P. Cracknell; Rev. Mod. Phys. 37, 19 (1965)
21. R.L. Mills, J.L. Yarnell and A.F. Schuch; Los Alamos preprint (1972)
22. H. Meyer, F. Weinhaus and B. Maraviglia; Phys. Rev. B6, 1112 (1972)
23. C.W. Garland and R. Renard; J. Chem. Phys. 44, 1120 (1966)
24. K.H. Michel; J. Chem. Phys. 58, 1143 (1973)
25. H. Kapulla and W. Gläser; in Neutron Inelastic Scattering IAEA-Symposium, Vienna (1972)
26. C. Chapados and A. Cabana; Chem. Phys. Letters 7, 191 (1970)
27. A.B. Harris; J. Appl. Phys. 42, 1574 (1971)
28. W. Press and A. Hüller; Phys. Rev. Letters 30, 1207 (1973)
29. D.C. Mattis and T.D. Schultz; Phys. Rev. 129, 175 (1963)
30. M.D. Rechtin, S.C. Moss and B.L. Averbach; Phys. Rev. Letters 24, 1485 (1970)
31. J. Sivardière and M. Blume; Phys. Rev. B5, 1126 (1972)
32. A. Hüller and W. Press; Phys. Rev. Letters 29, 266 (1972)

THE JAHN-TELLER EFFECT AS A MECHANISM FOR STRUCTURAL PHASE TRANSITIONS*

H. Thomas
Institut für Theoretische Physik
der Universität Franfurt

ABSTRACT

The mechanism of phase transitions resulting from the interplay between local Jahn-Teller distortions and their interactions is investigated. We discuss the properties of a single Jahn-Teller complex and review the types of interactions between such complexes. This leads to a classification of the different types of Jahn-Teller induced phase transitions. Some results obtained for specific models are reported.

1. INTRODUCTION

Jahn-Teller-induced phase transitions occur by the following mechanism: The interaction between the electronic state of a Jahn-Teller ion and the ionic configuration of its ligands gives rise to a distortion of the ionic complex, and to particular dynamics of coupled electronic and ionic motion. The interactions between different ionic complexes produce correlations which lead to long-range order and dynamical slowing down at the phase transition. The mechanism is studied here for the case of Jahn-Teller ions with orbital ground state doublets (non-Kramers doublets) in octahedral symmetry.

Transitions of this type have been observed in a number of spinel and perovskite type crystals containing d-state ions like Cu^{2+}, Ni^{+}; Mn^{3+}, Cr^{2+}; Ni^{3+},

Pt^{3+}, as well as in more complicated structures (1). Several theoretical treatments of such transitions have been given (2-7). The phase transitions occurring in crystals containing pseudo-Jahn-Teller rare earth ions which have been discussed recently (8,9), on the other hand, are outside the scope of the present paper.

The collective behaviour of the crystal depends strongly on the properties of its constituents. In Section 2, we discuss therefore the various types of behaviour of a single Jahn-Teller complex. The emphasis is put on a qualitative understanding of typical cases; for more details we refer to review papers (10,11) and to references to original papers contained therein. Section 3 contains a review of mechanism of interaction between different Jahn-Teller complexes. Based on the properties of the single complex and the strength of the interactions, we give in Section 4 a classification of the different types of Jahn-Teller-induced transitions. In Section 5 we report some results obtained for specific models (6,7).

2. SINGLE JAHN-TELLER COMPLEX

We consider a Jahn-Teller ion in octahedral symmetry with an electronic ground-state doublet (Fig.1)

$$\psi_1(\underset{\sim}{r}) = |2z^2-x^2-y^2\rangle; \quad \psi_2(\underset{\sim}{r}) = |\sqrt{3}\,(x^2-y^2)\rangle \, . \quad (2.1)$$

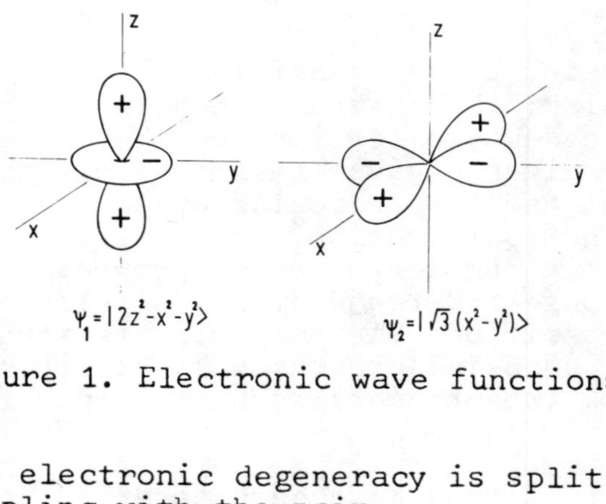

Figure 1. Electronic wave functions of an E-doublet.

The electronic degeneracy is split by the Jahn-Teller coupling with the pair

$$Q_3 = \rho \cos \Theta, \qquad Q_2 = \rho \sin \Theta \qquad (2.2)$$

of local normal coordinates (12) of the ionic octahedral complex (Fig.2), which have the same transformation properties as (ψ_1,ψ_2). Crystal symmetry induces C_{3v}-symmetry in the (Q_3,Q_2)-plane, i.e. elongations of the octahedron along the x and y directions are represented by vectors at angles $+2\pi/3$ and $-2\pi/3$ with respect to the Q_3-direction, respectively.

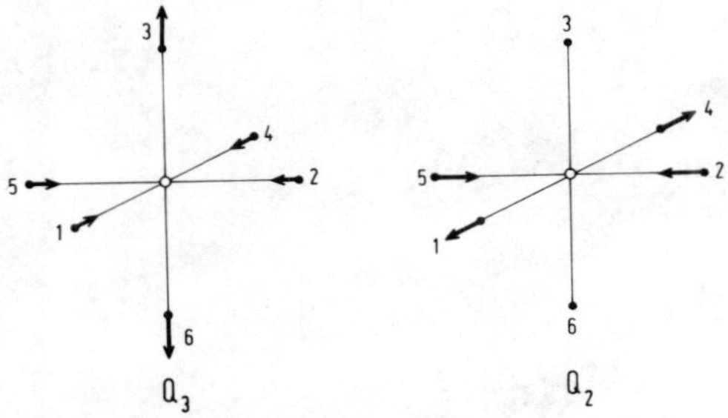

Figure 2. The normal coordinates Q_3, Q_2 of an ionic octahedral complex.

It is assumed that the behaviour of the system can be well described by a model which takes only electronic states from the space spanned by ψ_1,ψ_2, and only the ionic coordinates Q_3,Q_2 into account. This assumption is justified if all other electronic states have excitation energies much higher than the energies considered here, and if there exists only weak coupling to other ionic coordinates. The vibronic state of the complex is then described by a wave function

$$\Psi(\underset{\sim}{Q},\underset{\sim}{r}) = \phi_1(\underset{\sim}{Q})\psi_1(\underset{\sim}{r})+\phi_2(\underset{\sim}{Q}_2)\psi_2(\underset{\sim}{r}). \tag{2.3}$$

The Hamiltonian describing the coupled ionic and electronic motion is with respect to the electronic basis (ψ_1,ψ_2) represented by

$$H = \tfrac{1}{2}(P_3{}^2+P_2{}^2)+V_1(Q_3,Q_2)+\sigma_3 V_3(Q_3,Q_2)-\sigma_1 V_2(Q_3,Q_2), \tag{2.4}$$

where P_3,P_2 are the momenta conjugate to Q_3,Q_2, and σ_1, σ_3 are the Pauli matrices

$$\sigma_1 = \begin{pmatrix} 0 & 1 \\ 1 & 0 \end{pmatrix}; \qquad \sigma_3 = \begin{pmatrix} 1 & 0 \\ 0 & -1 \end{pmatrix}. \tag{2.5}$$

Crystal symmetry requires that V_1 is invariant under

C_{3v}, and that V_3, V_2 transform as Q_3, Q_2. The Taylor series expansion of the V_n with respect to Q_3, Q_2 thus starts with the terms

$$V_1(Q_3,Q_2) = \frac{1}{2}\,\Omega^2(Q_3{}^2+Q_2{}^2) + B_3(Q_3{}^3-3Q_3Q_2{}^2) \qquad (2.6a)$$

$$= \frac{1}{2}\,\Omega^2\rho^2 + B_3\rho^3 \cos 3\Theta$$

$$V_3(Q_3,Q_2) = AQ_3 + A_3(Q_3{}^2-Q_2{}^2) \qquad (2.6b)$$

$$= A\rho \cos \Theta + A_3\rho^2 \cos 2\Theta$$

$$V_2(Q_3,Q_2) = AQ_2 - 2A_3Q_3Q_2 \qquad (2.6c)$$

$$= A\rho \sin \Theta - A_3\rho^2 \sin 2\Theta$$

In the absence of V_3 and V_2, one has an ordinary anharmonic Einstein oscillator with harmonic frequency Ω and third order anharmonicity B_3, with all eigenstates electronically doubly degenerate. The coupling between ionic and electronic motion is described by the Jahn-Teller terms V_3 and V_2. For fixed ionic configuration (Q_3,Q_2), the linear Jahn-Teller coupling terms contribute an energy $\pm A\rho$, and one obtains together with the harmonic part of V_1 the well-known "Mexican-hat" energy surface (Fig.3)

$$E = \pm A\rho + \frac{1}{2}\,\Omega^2\rho^2 \qquad (2.7)$$

belonging to the electronic states

$$\varphi_1(\underset{\sim}{r},\underset{\sim}{Q}) = \psi_1(\underset{\sim}{r}) \sin\Theta/2 + \psi_2(\underset{\sim}{r}) \cos\Theta/2$$
$$\varphi_2(\underset{\sim}{r},\underset{\sim}{Q}) = \psi_1(\underset{\sim}{r}) \cos\Theta/2 - \psi_2(\underset{\sim}{r}) \sin\Theta/2 \,. \qquad (2.8)$$

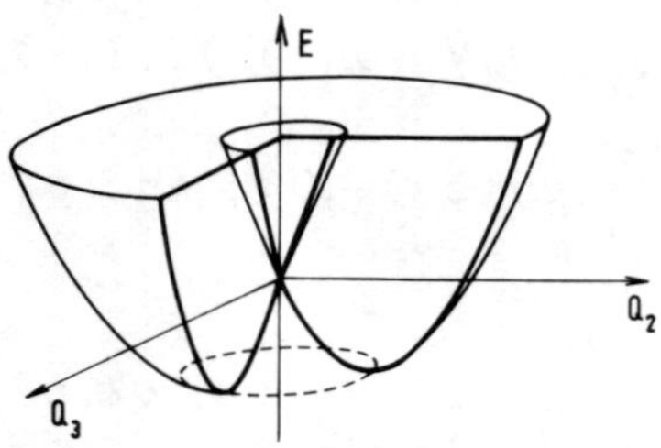

Figure 3. Energy surface for linear Jahn-Teller coupling ("Mexican hat").

The lower sheet of the energy surface assumes its minimum

$$E = -E_o \equiv -\frac{1}{2} A\rho_o = -\frac{1}{2} \Omega^2 \rho^2 \tag{2.9}$$

in the circular groove with radius

$$\rho = \rho_o \equiv A/\Omega^2 \ . \tag{2.10}$$

The destabilization energy E_o favoring a distorted configuration provides the energetic mechanism for the Jahn-Teller induced phase transition.

The dynamics of the complex depends strongly on the ratio of the Jahn-Teller distortion ρ_o to the width $\delta = \sqrt{\hbar/\Omega}$ of the harmonic oscillator wave function, or equivalently on the ratio of the Jahn-Teller stabilization energy E_o to the vibrational zero point energy $\hbar\Omega/2$. The limiting cases are the weak coupling limit

$$\rho_o \ll \delta \ , \qquad \text{i.e.} \qquad E_o \ll A\delta \ll h\Omega \tag{2.11}$$

and the strong coupling limit

$$\rho_o \gg \delta \ , \qquad \text{i.e.} \qquad E_o \gg A\delta \gg h\Omega \ . \tag{2.12}$$

The intermediate case must be treated numerically (14).

The states of the two-dimensional isotropic Einstein oscillator in the absence of the Jahn-Teller coupling are described by three quantum numbers: A radial quantum number $n=0,1,2...$, an azimuthal quantum number $\ell = 0, \pm 1, \pm 2...$, and a pseudospin quantum number $\zeta = \pm 1/2$ characterizing the electronic state. The energy levels are of course degenerate with respect to ζ. In the harmonic case, they are given by

$$\varepsilon_o(n,\ell) = [2n + |\ell| + 1] \ \hbar\Omega \ . \tag{2.13}$$

In the weak-coupling limit, one obtains a small perturbation of these levels. The pseudospin degeneracy is lifted, and the quantum numbers ℓ and ζ combine to

$$m = \ell + \zeta \tag{2.14}$$

(Fig.4). The wave functions have the form

$$\Psi(\underset{\sim}{Q},\underset{\sim}{r}) = [g(\rho)\varphi_1(\underset{\sim}{r},\Theta) + h(\rho)\varphi_2(\underset{\sim}{r},\Theta)] \ e^{im\Theta} \tag{2.15}$$

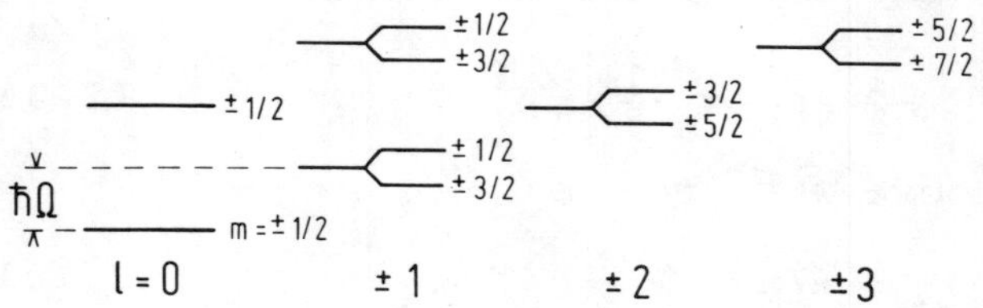

Figure 4. Splitting of the $\ell \neq 0$ levels of the two-dimensional harmonic oscillator by a weak Jahn-Teller coupling. Each of the new levels is doubly degenerate; they can be labelled by quantum numbers $m = \ell \pm 1/2$ giving the angular dependence $\exp(im\Theta)$ of the vibrational amplitude.

with vibrational amplitudes $g(\rho)$ and $h(\rho)$ which are concentrated near the origin in (Q_3,Q_2)-space.

In the strong coupling limit, the low-lying vibronic states are in good approximation described by vibronic wave functions of the form

$$\Psi(\underset{\sim}{Q},\underset{\sim}{r}) = \chi(\underset{\sim}{Q})\varphi_1(\underset{\sim}{r},\Theta) \;, \tag{2.16}$$

where the vibrational amplitude $\chi(Q)$ factors into a product of a radial harmonic oscillator function localized at ρ_o and an azimuthal rotator function,

$$\chi(\underset{\sim}{Q}) = \chi_n(\rho - \rho_o)\, e^{im\Theta} \;. \tag{2.17}$$

The energy eigenvalues are given correspondingly as the sum of a harmonic oscillator term and a rotator term,

$$\varepsilon(n,m) = n\hbar\Omega + \frac{\hbar^2}{2\rho_o^{\,2}}\, m^2 + \text{const.} \tag{2.18}$$

However, the Jahn-Teller coupling gives rise to an important peculiarity of the rotator states: For an ordinary rotator, where the distortion is due to destabilizing terms in V_1, uniqueness of the wave function requires the azimuthal quantum number m to be integer, $m=0,\pm1,\pm2\ldots$, and one obtains a singlet ground state (Fig.5a). But in the Jahn-Teller case, the electronic functions φ_1,φ_2 change sign for $\Theta\to\Theta+2\pi$. In order to make the total wave function (2.16) unique, $\chi(\underset{\sim}{Q})$ has thus to transform as a double representation

$$\chi(\Theta + 2\pi) = -\chi(\Theta) \ . \qquad (2.19)$$

Therefore, m takes on the half-integer values $m=\pm 1/2$, $\pm 3/2$...(see also Eq.(2.14)), and one obtains a doublet ground state (Fig.5b). This is to be required because the Jahn-Teller interaction is fully symmetric and cannot lift the degeneracy. Since

$$\hbar^2/\rho_o{}^2 = \hbar\Omega[\delta^2/\rho_o{}^2] \ , \qquad (2.20)$$

the excitation energies of the rotator levels are very small compared to the oscillator excitation energy.

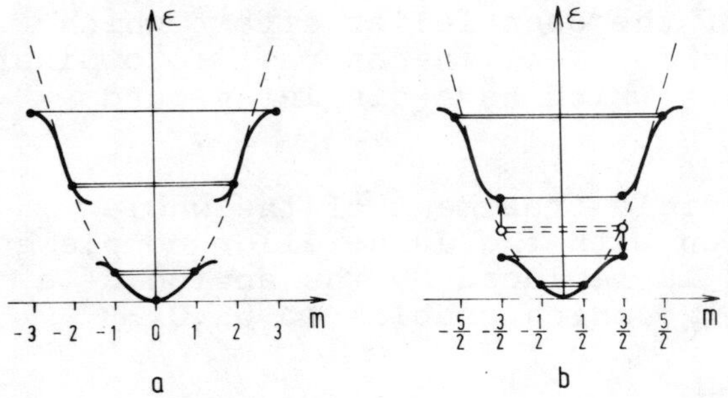

Figure 5. "Energy bands" of the anisotropic rotator.
a) ordinary periodic boundary condition
b) double-representation boundary condition.

The A_3 and B_3-terms give rise to an anisotropy of the energy surface: The "Mexican hat" is transformed into a "tricorn". Instead of a circular groove, there occur minima at $\Theta=0,\pm 2\pi/3$ or at $\Theta=\pm\pi/3,\pi$, corresponding to elongations or contractions of the octahedron along one of the three cubic axes, respectively, depending on the signs of the constants A_3 and B_3. This periodic potential leads to a separation of the rotator levels into "energy bands" each containing a doublet and a singlet (Fig.5). As the anisotropy increases, the vibrational amplitude becomes more and more localized also in the azimuthal direction, and the doublet and singlet within each band become nearly degenerate (flat band limit). The bands are separated from each other by an azimuthal oscillator energy

$$\Delta = \hbar\sqrt{\rho_o V_3} \qquad (2.21)$$

where V_3 is an effective third-order anharmonicity constant. For $A_3 \ll A$,

$$V_3 = 9(B_3 + A_3/\rho_o). \qquad (2.22)$$

This strong anisotropy limit represents the so-called static Jahn-Teller effect with the azimuthal oscillator localized in one of the three valleys corresponding to a static Jahn-Teller distortion along one of the three axes. It should be noted, however, that also in the dynamic case of non-vanishing doublet-singlet splitting

$$E^{singlet} - E^{doublet} = (3/2)\Omega_T, \qquad (2.23)$$

the average static distortion does not vanish. This is a direct consequence of the Jahn-Teller effect which leads to a ground-state doublet, in contrast to ordinary lattice dynamical models which have singlet ground states.

In the molecular-field treatment of the whole crystal, the interaction with the Jahn-Teller complexes at other lattice sites is replaced by the action of a molecular field $\underset{\sim}{F}=(F_3,F_2)$ which couples to Q_3,Q_2 according to

$$H^{field} = -(Q_3F_3 + Q_2F_2) . \qquad (2.24)$$

One has, therefore, to study the effect of such a field on the level structure. As an example, we consider a field in the Q_3-direction. The dependence of the three lowest levels on F for the case $\Omega_T \ll \Delta$ is given in Fig.6. The linear "Stark" effect of the ground state at small F demonstrates the existence of a static distortion mentioned above.

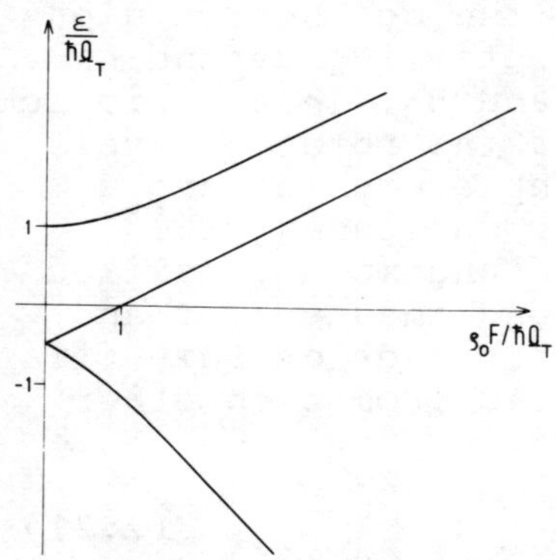

Figure 6. Dependence of the three lowest levels of a Jahn-Teller complex on an external field in the Q_3-direction (Ref. 6).

3. INTERACTIONS

Two Jahn-Teller complexes at lattice sites ℓ,ℓ' can interact in three different ways: The electronic charge distributions interact with each other (electron-electron coupling), the ionic configurations interact in the usual lattice-dynamical way (ion-ion-coupling), and the electronic charge distribution on one lattice site interacts with the ionic configuration at the other lattice site (electron-ion coupling). The three types of interactions can be represented in the following forms:

electron-electron coupling

$$H^{e-e} = -\frac{1}{2} \sum_{\ell\ell'}{}' \underset{\sim}{\sigma}_\ell \cdot \underset{\approx}{w}_{\ell\ell'} \cdot \underset{\sim}{\sigma}_{\ell'} \tag{3.1}$$

ion-ion coupling

$$H^{i-i} = -\frac{1}{2} \sum_{\ell\ell'}{}' \underset{\sim}{Q}_\ell \cdot \underset{\approx}{v}_{\ell\ell'} \cdot \underset{\sim}{Q}_{\ell'} \tag{3.2}$$

electron-ion coupling

$$H^{e-i} = - \sum_{\ell\ell'}{}' \underset{\sim}{\sigma}_\ell \cdot \underset{\approx}{u}_{\ell\ell'} \cdot \underset{\sim}{Q}_{\ell'} \, . \tag{3.3}$$

Since the electronic wave functions possess a quadrupole moment, the leading contribution to H^{e-e} will be the quadrupole-quadrupole coupling. The quadrupole operator is represented by

$$\begin{aligned} M_{xx} &= e\langle x^2\rangle - \tfrac{1}{3}M = \tfrac{4}{21}M \left[\sigma_1 \sin\tfrac{4\pi}{3} + \sigma_3 \cos\tfrac{4\pi}{3}\right] \\ M_{yy} &= e\langle y^2\rangle - \tfrac{1}{3}M = \tfrac{4}{21}M \left[\sigma_1 \sin\tfrac{2\pi}{3} + \sigma_3 \cos\tfrac{2\pi}{3}\right] \\ M_{zz} &= e\langle z^2\rangle - \tfrac{1}{3}M = \tfrac{4}{21} M\sigma_3 \\ & M_{xy} = M_{yz} = M_{zx} = 0 \, . \end{aligned} \tag{3.4}$$

The coupling constants $\underset{\approx}{w}_{\ell\ell'}$ can be obtained by substituting the expressions (3.4) into the quadrupole-quadrupole interaction energy

$$W = \frac{1}{8} \sum_{\substack{\ell\ell' \\ \alpha\beta\gamma\delta}}{}' M_{\ell,\alpha\beta} \frac{\partial^4}{\partial X_\alpha \partial X_\beta \partial X_\gamma \partial X_\delta} \left(\frac{1}{R_{\ell\ell'}}\right) M_{\ell',\gamma\delta} \, . \tag{3.5}$$

However, this interaction is always quite small. An estimate gives

$$w \sim \frac{e^2 \langle r^2 \rangle^2}{R^5} \lesssim 10K \; . \qquad (3.6)$$

Only in the weak-coupling case, where the Q are small, could the electron-electron coupling possibly compete with the ion-ion coupling.

The coupling constants $\underset{\approx}{u}_{\ell\ell'}$ can be obtained from the forces on the ligands in the quadrupole field, if the effective charges on the ligands are known.

We assume that the ion-ion coupling is the predominant interaction mechanism(15). For nearest-neighbor interaction of octahedral complexes in cubic arrangement, the interaction tensor $\underset{\approx}{v}_{\ell\ell'}$ can be written in terms of two parameters $v_{\parallel}$ and $v_{\perp}$ in the form

$$\underset{\approx}{v}_{\ell\ell'} = \begin{pmatrix} v_{\parallel}\cos^2\alpha + v_{\perp}\sin^2\alpha & (v_{\parallel} - v_{\perp})\sin\alpha\cos\alpha \\ (v_{\parallel} - v_{\perp})\sin\alpha\cos\alpha & v_{\parallel}\sin^2\alpha + v_{\perp}\cos^2\alpha \end{pmatrix} \qquad (3.7)$$

where

$$\begin{aligned} \alpha &= 0 && \text{for} \quad R_{\ell\ell'} \parallel z \\ &2\pi/3 && \qquad\quad R_{\ell\ell'} \parallel x \\ &4\pi/3 && \qquad\quad R_{\ell\ell'} \parallel y \; . \end{aligned}$$

It has the Fourier transform

$$\begin{aligned} v_{33}(q) &= (\tfrac{1}{4}v_{\parallel} + \tfrac{3}{4}v_{\perp})(\cos q_x + \cos q_y) + v_{\parallel}\cos q_z \\ v_{32}(q) &= v_{23}(q) = -\tfrac{1}{4}\sqrt{3}\,(v_{\parallel} - v_{\perp})(\cos q_x - \cos q_y) \qquad (3.8) \\ v_{22}(q) &= (\tfrac{3}{4}v_{\parallel} + \tfrac{1}{4}v_{\perp})(\cos q_x + \cos q_y) + v_{\perp}\cos q_z \; . \end{aligned}$$

So far, we have neglected the coupling to all other degrees of freedom. It is, of course, straightforward to add such coupling terms. The strong tetragonal distortion observed in the low-temperature phase of Jahn-Teller crystals shows that the coupling to the elastic displacement field is especially important. One obtains a coupling between the coordinates Q_3 and Q_2 and the strain coordinates

$$\varepsilon_3 = 2\varepsilon_{zz} - \varepsilon_{xx} - \varepsilon_{yy}; \qquad \varepsilon_2 = \sqrt{3}\,(\varepsilon_{xx} - \varepsilon_{yy}) \qquad (3.9)$$

with the same symmetry of the form

$$H^{i\text{-strain}} = g\,(\varepsilon_3 Q_3 + \varepsilon_2 Q_2)\ . \qquad (3.10)$$

This coupling gives an indirect contribution to the ion-ion coupling, which can be described as interaction between two elastic dipoles (16).

4. CLASSIFICATION OF JAHN-TELLER-INDUCED PHASE TRANSITIONS

We shall call a structural phase transition induced by the Jahn-Teller effect, if in the absence of the Jahn-Teller coupling the cubic phase would be stable down to T=0. This is the case if the restoring force of the local Einstein oscillator, Ω^2, is larger than the destabilizing force due to the interactions,

$$\Omega^2 > v, \qquad (4.1)$$

where v is the largest eigenvalue of the Fourier transform of the interaction tensor. Only in this case is the distortion caused by the Jahn-Teller effect alone. In the opposite case, the cubic phase would already be unstable in the lattice-dynamical sense, and one would have an ordinary displacive phase transition, possibly assisted by the Jahn-Teller effect.

Because the vibronic ground state of a Jahn-Teller complex is a doublet, the interaction energy has to compete with entropy, but not with a zero point splitting of the doublet. Therefore, a Jahn-Teller crystal always shows a phase transition, no matter how small the interaction, with a transition temperature of the order of the mean interaction energy,

$$kT_c \sim v\,\langle\rho^2\rangle\ . \qquad (4.2)$$

This is related to the existence of a spontaneous distortion of the single Jahn-Teller complex mentioned in Section 2.

Based on the relation of the interaction energy to the various characteristic energies of the single Jahn-Teller complex, we shall now obtain a classification of the different types of Jahn-Teller induced transitions. A similar classification scheme was already given by Englman and Halperin (5).

In the weak coupling case, Eq.(2.11), it follows from Eqs.(4.1) and (4.2) that

$$kT_c < \hbar\Omega; \quad <\rho^2> \sim \delta^2 . \qquad (4.3)$$

Thus, the higher states of the oscillator are not excited at $T \sim T_c$, and one can obtain a description of the phase transition in terms of the lowest doublet alone. The order parameter will be mainly of electronic nature, the ordering consisting essentially in an alignment of the electronic wave functions. The ionic distortion will be comparatively small.

In the strong coupling case, Eq.(2.12), the vibrational amplitude is concentrated at $\rho=\rho_o$, and one obtains from Eqs.(4.1) and (4.2)

$$kT_c < E_o . \qquad (4.4)$$

Now, a number of different cases can occur, according to the relation of the interaction energy $v\rho_o^2$ to the following energies of the single Jahn-Teller complex,

- the lowest doublet-singlet splitting $\hbar\Omega_T$, i.e. the tunnel splitting of the three lowest states localized in the three valleys,
- the excitation energy Δ to the next higher doublet-singlet set,
- the excitation energy $\hbar\Omega$ of a radial oscillation.

If $v\rho_o^2<\hbar\Omega_T$, one has Ising-like behaviour in the lowest doublet. The vibrational amplitude in the doublet states is not localized in one valley, but is tunnelling between the valleys. This case is somewhat similar to the weak coupling case, but now the order parameter is mainly the ionic displacement.

If $v\rho_o^2 \gtrsim \hbar\Omega_T$, the singlet state is also excited: Three-state regime. There occur interesting dynamic effects of the tunnel excitations.

If $\hbar\Omega_T << v\rho_o^2 < \Delta$, one is still in the three-state regime, but the interactions have become strong enough to completely localize the vibrational amplitude in one of the three valleys. The transition can therefore be treated as an order-disorder transition with three equivalent states.

If $\Delta < v\rho_o^2 < \hbar\Omega$, the whole set of azimuthal levels takes part in the transition, but radial oscillations are not yet excited. The detailed behaviour will depend on whether the anisotropy is weak (rotator regime) or strong (azimuthal oscillator regime).

If $h\Omega < v\rho_o^2$, also the radial excitations have to be taken into account (radial oscillator regime).

These considerations show that it is difficult to find a situation in which the mechanism of the Kanamori theory (2,3) could occur. Whereas in all the above cases, the local Jahn-Teller complexes remain distorted also above the transition, and only the long range order disappears at T_c, in the Kanamori theory the local Jahn-Teller distortion becomes temperature dependent and vanishes at T_c. Such behaviour would seem to require excitation of states belonging to the upper sheet of the energy surface (i.e. to the opposite pseudo-spin), but according to Eq.(4.4) such an excitation cannot occur in a genuine Jahn-Teller induced transition. - Moreover, the theory yields an unrealistic result for the transition temperature, in contrast to Eq.(4.2). In our notation it takes the form

$$kT_c = A/(\Omega^2 + \rho_o^2 v) \tag{4.5}$$

which shows that T_c stays finite for non-interacting complexes and decreases with increasing interaction strength.

5. RESULTS FOR SPECIFIC MODELS

For ferrodistortive ordering (largest eigenvalue of the interaction tensor at q=0), the static and dynamic behaviour has been studied in the three-state regime (6). The transition is found to be of first order, as expected from the Landau criteria because of the existence of a third order invariant. The dynamic-mode frequencies are obtained in RPA as a function of temperature (Fig.7). At high temperatures, where correlations between different complexes are small, their values can be read off the level scheme of Fig.6 at F=0; at low temperatures, where fluctuations are frozen out, their values can be read off the same level scheme at the T=0 value of the molecular field $F^{mol}=\rho_o v$. Because of the feedback-mechanism of the interactions, there occurs a dynamical slowing-down of all modes both above and below T_c. But at the stability limits of the two phases, all mode frequencies remain finite, i.e. none of the dynamic

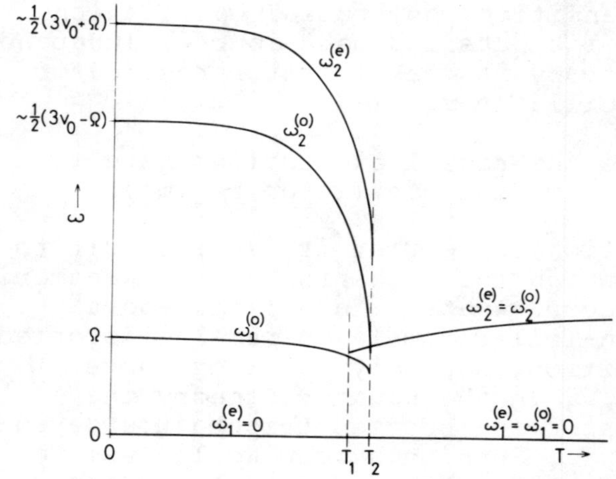

Figure 7. Temperature dependence of the dynamic mode frequencies for ferrodistortive ordering (Ref. 6).

modes is the soft mode connected with the stability limits. The soft mode will be one of the relaxation- or diffusion-type modes which have zero frequency in RPA.

For antiferrodistortive ordering (largest eigenvalue of the interaction tensor at $q=q_R$), the static behaviour has been studied in the three-state order-disorder regime (7). Here, a second-order transition is found. The cubic phase becomes unstable at T_c with respect to a doubly degenerate mode, and the order parameter is determined in this twodimensional space by higher terms. At T=0, two types of ordering have the same energy (Fig.8): The "Q_2-ordering" in which the octahedra in one sublattice are distorted along z and the octahedra in the other sublattice are distorted along x, and the "Q_3-ordering" in which the octahedra in one sublattice are distorted along z, but the other sublattice contains a statistical mixture of octahedra distorted along x and along y. For $T\neq 0$, the Q_3-ordering has an advantage over the Q_2-ordering because of the entropy contribution, and is the stable phase up to T_c (Fig.9).

The coupling to the elastic strain field is found to lead to an interesting effect. It favors the Q_2-ordering energetically, and competes thus with the entropy gain favoring the Q_3-ordering. As a result of

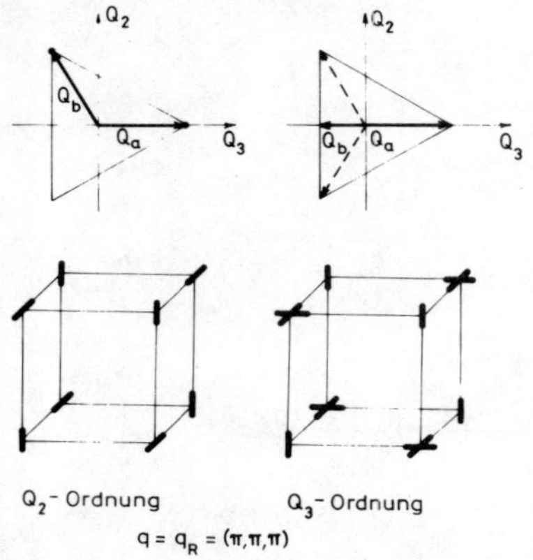

Figure 8. Two types of ordering for the $q=q_R$ antiferrodistortive case (Ref. 7). The bars give the directions of distortion of the octahedral complexes.

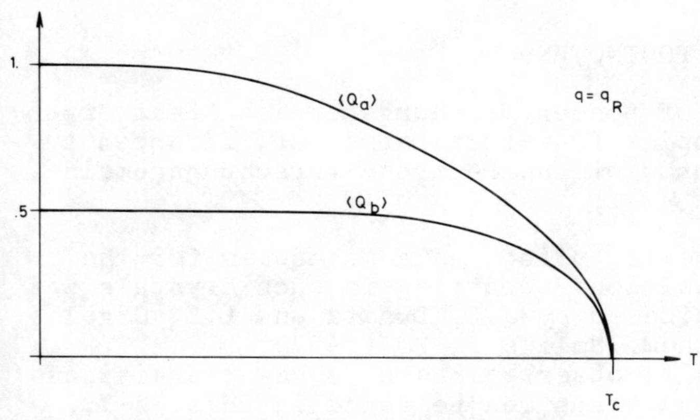

Figure 9. Temperature dependence of the sublattice magnetizations for the $q=q_R$ antiferrodistortive case (Ref. 7).

this competition, there occurs for a range of coupling parameters at some low temperature a first-order transition from Q_2-ordering to Q_3-ordering, at a higher temperature another first-order transition back to Q_2-ordering, and finally at T_c the second order transition to the cubic phase (Fig.10).

To our knowledge, such behaviour has not been reported until now, but we hope that this prediction will

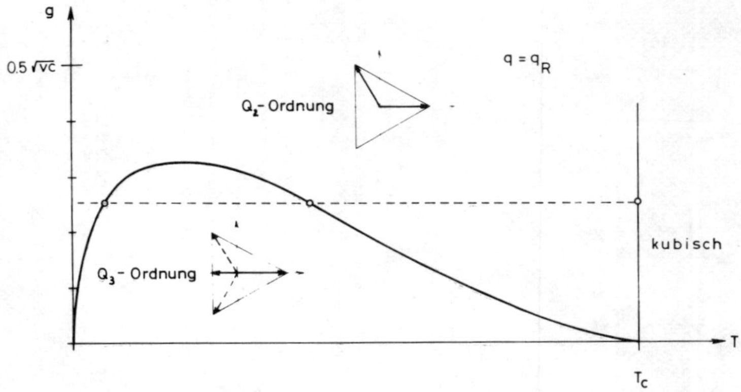

Figure 10. Phase transitions in the presence of coupling to elastic strain for the $q=q_R$ antiferrodistortive case (Ref. 7).

stimulate a search for these effects.

REFERENCES AND FOOTNOTES

* A project of Sonderforschungsbereich "Festkörperspektroskopie" Frankfurt-Darmstadt, financed by special funds of the Deutsche Forschungsgemeinschaft.

1. The Jahn-Teller effect as a mechanism for the phase transitions occurring in such crystals was first considered by J.D. Dunitz and L.E. Orgel, J. Phys. Chem. Solids 3, 20 (1957). References to observations of phase transitions in various systems can be found in Refs. 2-7.

2. P.J. Wojtowicz, Phys. Rev. 116, 32 (1959).

3. J. Kanamori, J. Appl. Phys. 31, 14S (1960).

4. E. Pytte, Phys. Rev. B3, 3503 (1971).

5. R. Englman, B. Halperin, Phys. Rev. B2, 75 (1970). B. Halperin, R. Englman, Phys. Rev. B3, 1698(1971).

6. H. Thomas, K.A. Müller, Phys. Rev. Letters 28, 820 (1972).

7. K.-H. Höck, H. Kaufmann, G. Schröder, Verhandl.

der Deutschen Physikalischen Gesellschaft 3, 489 (1973) and to be published.

8. E. Pytte, K.W.H. Stevens, Phys. Rev. Letters 27, 862 (1971).

9. R.J. Elliott et al., Proc. Roy. Soc. (London) A238, 217 (1972).

10. M.D. Sturge, Solid State Physics 20, 91 (1967).

11. F.S. Ham, Electron Paramagnetic Resonance, Ed. by S. Geschwind, Plenum Press, New York-London 1972.

12. The concept of local normal coordinates is discussed in Ref. 13.

13. H. Thomas, Structural Phase Transitions and Soft Modes, Ed. by E.J. Samuelsen, E. Andersen and J. Feder, Universitetsforlaget Oslo 1971, p. 15.

14. C.W. Struck, F. Herzfeld, J. Chem. Phys. 44, 464 (1965).

15. The interaction between two octahedral complexes mediated by the displacement field of the crystal has been discussed by P. Novák, J. Phys. Chem. Solids 30, 2357 (1969), 31, 125 (1970).

16. P. Novák, Czech. J. Phys. B20, 196 (1970).

SOFT MODES AT FIRST AND SECOND ORDER PHASE TRANSITIONS*

H. Thomas

Institut für Theoretische Physik
der Universität Frankfurt

ABSTRACT

The influence of the coupling between lattice modes and constants of motion of the system on the dynamics at first and second order phase transitions is investigated. It is shown that it gives rise to non-ergodic behaviour and zero-frequency anomalies. The thermodistortive coupling of a lattice mode to the heat diffusion mode is studied in detail, and the conditions for the occurrence of a soft heat diffusion mode are discussed.

1. GENERAL REMARKS ON FIRST-ORDER PHASE TRANSITIONS

We consider a physical system at given temperature T and external field F. Its equilibrium behaviour is described by the thermodynamic potential G(F,T) which is a continuous function of F and T. The system undergoes a first order phase transition at the point (F_c, T_c) if the gradient of G is discontinuous at this point. Such points form a coexistence line in the (F,T)-plane separating two phases, and one may pass from one phase to the other along any path crossing the coexistence line. As specific examples consider the line H=0, $0 \leq T \leq T_c$ in a ferromagnet separating the spin-up from the spin-down phase, the vapor-pressure curve in a gas-liquid system, or the line separating the cubic from the trigonal phase of $SrTiO_3$ in the (uniaxial stress, temperature)-plane (1).

One usually finds regions of metastability of the two phases extending out into the thermodynamically unstable regions to some intrinsic stability limit (spinodal line): The transition to the other phase does not occur at the thermodynamic transition point but at the stability limit. One therefore expects to find critical behaviour and soft modes only when approaching the stability limit (2). It appears as if the two pieces of the thermodynamic potential G(F,T) can be analytically continued across the transition. In order to comment on this picture, let us discuss the various modes of instability:

a) Homogeneous nucleation (3): A phase can become unstable by the formation and growth of a critical droplet of the other phase. This happens in principle for arbitrarily small supersaturation, so that the true stability limit coincides with the thermodynamic transition point. However, close to the thermodynamic transition point the times involved are so phantastically long, that this mode of instability can practically be observed only at high degrees of supersaturation. Moreover, it does not give rise to a singularity of any of the derivatives of the thermodynamic potential, but merely to a point of vanishing convergence radius of the Taylor series (4). Nevertheless it must be born in mind that the true thermodynamic potential cannot be continued across the thermodynamic transition point. In the following, we shall disregard this homogeneous nucleation mode. This requires in principle a change in the definition of the thermodynamic potential, which amounts to the omission of the large but extremely rare fluctuations consisting in the formation of a critical droplet (2). Only then can a thermodynamic potential be defined in the metastable region. This omission occurs automatically in all the pertinent approximation procedures of statistical mechanics like mean field approximation, random phase approximation etc, which therefore yield expressions for the thermodynamic potential also in the metastable region.

b) Inhomogeneous nucleation: The other phase can also be nucleated at defects or at the surface. For a given defect one expects a definite stability limit at which a dynamic local mode becomes unstable. For surface nucleation one expects an instability of a surface mode.

c) Intrinsic bulk instability: We are mostly interested in the intrinsic stability limit of the metastable phase, where one expects one of the bulk dynamic modes of the system to become unstable.

The behaviour of a system at the stability limit of a first order transition may differ from the behaviour at a second order transition in two important aspects: At a second order transition, the instability always occurs in the order-parameter susceptibility; at a first order transition, it may occur in some other component of the susceptibility tensor. In the uniaxial ferromagnet, for instance, one expects a changeover from one case to the other at a certain temperature (5). - Further, at a second order transition, one expects that the instability occurs always at zero frequency, with a dynamic mode becoming soft, so that it can condense into the static order parameter. At a first order transition, on the other hand, a mode could become undamped at a finite frequency, in which case the instability would show up not in the static, but in the dynamic susceptibility. However, in all cases studied so far, one always finds a soft mode. This is in contrast to phase-transition like phenomena in non-equilibrium systems, where undamping of finite-frequency modes can be found.

2. SOFT HEAT CONDUCTION MODE AT FIRST ORDER DISPLACIVE PHASE TRANSITIONS

We consider a lattice-dynamical model describing a crystal which undergoes a first-order displace phase transition, and assume that the instability occurs in the order-parameter susceptibility (Fig.1). For instance, the single-mode model discussed in Ref. 6 leads to a first order transition of this type for suitably chosen anharmonic single-ion potential. If the dynamic susceptibility $\chi(q,\omega)$ of such a model is calculated in random phase approximation (RPA), it is found that the frequency of the order parameter oscillation goes to zero at the stability limit T_1 of the high-temperature phase as expected, but stays finite at the stability limit T_2 of the low-temperature phase.

This behaviour is due to the fact that no entropy transport is taken into account in RPA. As a consequence, the q=0, ω=0 limit of χ^{RPA} corresponds to the adiabatic susceptibility χ_S, independent of the order in which the limit is taken. The stability limit, on the other hand, is determined by the divergence of the isothermal susceptibility χ_T. Therefore, whenever the adiabatic susceptibility stays finite at the stability

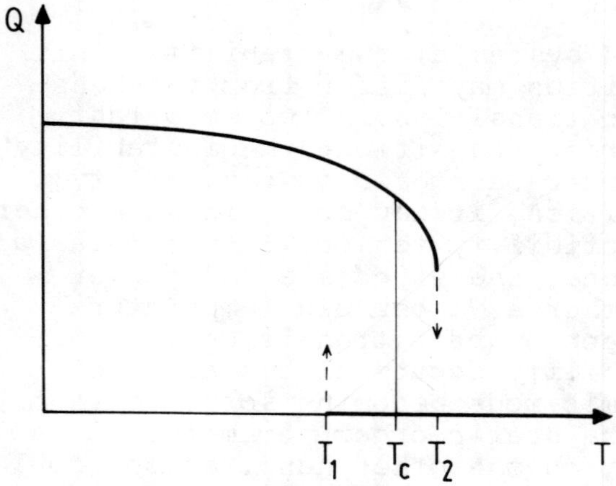

Figure 1. Variation of the order parameter with temperature at a first-order phase transition. T_C=thermodynamic phase boundary, T_1, T_2= stability limit of the high and low temperature phase, respectively.

limit, no soft mode will occur in RPA.

In order to study the dynamic nature of the instability, it is thus necessary to take entropy transport into account by coupling the order parameter to the entropy (thermodistortive coupling). This has been treated in Ref. 7 for the single-mode model. We present here a simple discussion of the general mechanism:

In the absence of thermodistortive coupling, the response δQ of the order parameter to an external field δF^{ext} is given by the bare susceptibility $\chi_o(q,\omega)$ (8),

$$\delta Q_{q,\omega} = \chi_o(q,\omega)\ \delta F^{ext}_{q,\omega} \tag{2.1}$$

(upper part of Fig.2), and the response δS of the entropy to an external power source δP^{ext} is described for small q and ω by the pair of equations

$$T\delta S_{q,\omega} = c_Q\ \delta T_{q,\omega} \tag{2.2}$$

$$\delta T_{q,\omega} = (1/\kappa q^2)\ [\delta P^{ext}_{q,\omega} + i\omega T\ \delta S_{q,\omega}] \tag{2.3}$$

(lower part of Fig.2). Here c_Q and κ are the bare

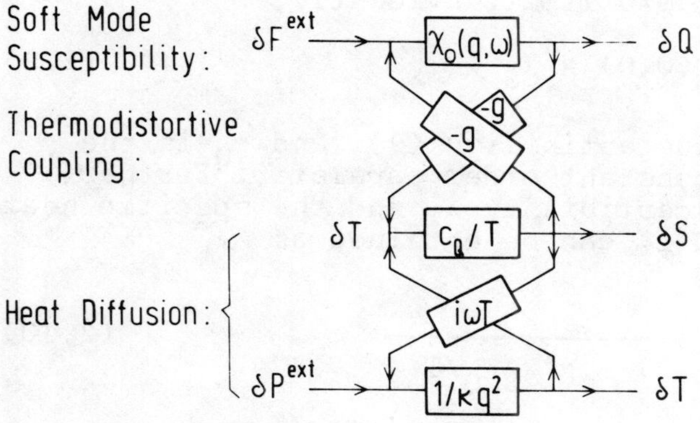

Figure 2. Block diagram showing the thermodistortive coupling between lattice mode and heat diffusion mode.

specific heat and the heat conductivity, respectively. Elimination of δT from Eqs.(2.2) and (2.3) yields

$$T\delta S_{q,\omega} = \rho_o(q,\omega)\ \delta P^{ext}_{q,\omega} \qquad (2.4)$$

with an entropy response function

$$\rho_o(q,\omega) = 1/(D_o q^2 - i\omega) \qquad (2.5)$$

the structure of which is determined by a heat diffusion mode with the bare heat diffusion coefficient

$$D_o = \kappa/c_Q\ . \qquad (2.6)$$

With thermodistortive coupling, a change of the lattice coordinate gives rise to an entropy change and vice versa, such that Eqs.(2.1) and (2.2) have to be replaced by

$$\delta Q_{q,\omega} = \chi_o(q,\omega)\ [\delta F^{ext}_{q,\omega} - g\delta S_{q,\omega}] \qquad (2.7)$$

and

$$T\delta S_{q,\omega} = c_Q[\delta T_{q,\omega} - g\delta Q_{q,\omega}]\ . \qquad (2.8)$$

(center part of Fig.2).
These equations show the significance of the coeffi-

cients in the q=0, ω=0 limit. Evidently,

$$\chi_o(0,0) = \chi_S \tag{2.9}$$

is the adiabatic susceptibility (9), and c_Q is the specific heat at constant order parameter. Further, the isothermal susceptibility χ_T and the specific heat for constant field c_F can be obtained as

$$\chi_T = \frac{\chi_S}{1 - g^2\chi_S c_Q/T} \tag{2.10}$$

and

$$c_F = \frac{c_Q}{1 - g^2\chi_S c_Q/T} , \tag{2.11}$$

respectively. Thus, $\chi_T \geq \chi_S$ and $c_F \geq c_Q$, as required by thermodynamic stability. The thermodistortive coupling constant g determines the adiabatic heating by an external field,

$$(\partial T/\partial F^{ext})_S = g\chi_S , \tag{2.12}$$

and the temperature dependence of the order parameter at constant field,

$$(\partial Q/\partial T)_F = -g\chi_T c_Q/T . \tag{2.13}$$

These quantities can be seen to satisfy the thermodynamic relations

$$\chi_T - \chi_S = (T/c_F)\,(\partial Q/\partial T)_F^{\,2} \tag{2.14}$$

$$c_F - c_Q = (T/\chi_T)\,(\partial Q/\partial T)_F^{\,2} \tag{2.15}$$

$$\chi_T/\chi_S = c_F/c_Q . \tag{2.16}$$

Equation (2.13) shows that $(\partial Q/\partial T)_F \to -\infty$ at the stability limit.

Solution of Eqs.(2.7), (2.8) and (2.3) for Q and S in terms of δF^{ext} and δP^{ext} yields

$$\delta Q_{q,\omega} = \chi(q,\omega)\,[\delta F^{ext}_{q,\omega} + \frac{g}{T}\,\rho_o(q,\omega)\,\delta P^{ext}_{q,\omega}] \tag{2.17}$$

$$T\delta S_{q,\omega} = \rho(q,\omega)\left[g c_Q D_o q^2 \chi_o(q,\omega)\,\delta F^{ext}_{q,\omega} + \delta P^{ext}_{q,\omega}\right] \qquad (2.18)$$

where the susceptibility

$$\chi(q,\omega) = \chi_o(q,\omega)\,\frac{D_o q^2 - i\omega}{D(q,\omega)q^2 - i\omega}\ , \qquad (2.19)$$

and the entropy response function,

$$\rho(q,\omega) = \frac{1}{D(q,\omega)q^2 - i\omega} \qquad (2.20)$$

are given in terms of a wavenumber and frequency dependent heat diffusion coefficient

$$D(q,\omega) = D_o\left[1 - g^2(c_Q/T)\chi_o(q,\omega)\right]\ . \qquad (2.21)$$

In the $q\to0$, $\omega\to0$ limit, one obtains

$$D = D_o\,\chi_S/\chi_T = \kappa/c_F\ . \qquad (2.22)$$

This equation shows that the heat diffusivity is strongly influenced by the coupling to the order parameter mode. Conversely, the renormalized heat diffusion mode determines the dynamics of the order parameter at small q and ω, as can be seen from Eq. (2.19). This heat diffusion dynamics leads to the correct limits

$$\lim_{q\to0}\ \lim_{\omega\to0}\ \chi(q,\omega) = \chi_T \qquad (2.23)$$

$$\lim_{\omega\to0}\ \lim_{q\to0}\ \chi(q,\omega) = \chi_S\ . \qquad (2.24)$$

It also gives rise to a Rayleigh peak

$$S(q,\omega) = 2kT\,(\chi_T - \chi_S)\,\frac{Dq^2}{\omega^2 + D^2q^4} \qquad (2.25)$$

in the dynamic structure factor.

In order for the thermodistortive coupling described by Eqs.(2.7) and (2.8) to exist, the lattice mode must be fully symmetric with respect to the symmetry group of the phase under consideration. For a lattice mode belonging to a representation different from the identity representation, thermodistortive coupling is forbidden by symmetry, and

$$g = 0; \quad \chi_T = \chi_S; \quad \chi = \chi_o; \quad D = D_o . \tag{2.26}$$

For the critical behaviour at a phase transition, a distinction must be made between transitions at which the symmetry is broken, and transitions in which the high and low temperature phase have the same symmetry.

Symmetry-breaking transitions: If the instability of the low symmetry phase occurs with respect to the order parameter mode, the above symmetry condition is satisfied, because the order parameter belongs always to the identity representation of the low symmetry phase, no matter whether the transition occurs at $q=0$ (ferrodistortive) or $q \neq 0$ (antiferrodistortive). Landau theory gives for a first-order transition

$$\chi_S \text{ finite}, \quad \chi_T \propto (T_2 - T)^{-1/2} \tag{2.27}$$

and we obtain a soft heat diffusion mode with

$$D/D_o \propto (T_2 - T)^{1/2} . \tag{2.28}$$

For a second-order transition, Eq.(2.13) shows that in scaling theory the coupling constant goes to zero as

$$g \propto (T_c - T)^{\alpha+\beta+\gamma-1} \tag{2.29}$$

whence according to Eqs.(2.14) to (2.16)

$$\chi_T/\chi_S \to \begin{cases} 1 \\ \text{const} > 1 \end{cases} \quad \text{for } \alpha+2\beta+\gamma \begin{cases} > 2 \\ = 2 . \end{cases} \tag{2.30}$$

Thus, the heat diffusion mode will not become soft at a second-order transition, in spite of the fact that it is coupled to the order parameter, and does give rise to a Rayleigh peak in the dynamic structure factor. The instability is already contained in $\chi_o(q,\omega)$; RPA predicts an ordinary lattice-dynamical soft mode for this case.

For an antiferromagnetic system below the Néel point, such behaviour has been discussed by Heller (10).

Here, the magnetic mode described by $\chi_o(q,\omega)$ is assumed to be a relaxation mode with temperature-dependent relaxation time $\tau_S(T)$ which is represented by a broad base of width $1/\tau_S$ in the dynamic structure factor. Superimposed on this base is the heat-diffusion peak of width Dq^2. As the temperature approaches the Néel point, it is expected that the width of the heat diffusion peak stays constant, while the width $1/\tau_S$ of the base narrows and goes to zero at T_N.

In the high-symmetry phase at a symmetry-breaking transition, on the other hand, thermodistortive coupling is forbidden by symmetry, and we obtain the behaviour described by Eq.(2.26), independent of the order of the transition. Therefore, no Rayleigh peak can appear in the dynamic structure factor. The central peak (11) found in some systems in the high-temperature phase must be produced by a different mechanism. - The same behaviour is found in the low-symmetry phase for the case that the instability occurs with respect to a lattice mode δQ of a symmetry different from the order-parameter mode.

Symmetry-conserving transitions: If the symmetry is not broken at the transition, the order parameter will be non-zero in both phases, and the transition will in general be of first order. One then obtains a soft heat diffusion mode as in Eq.(2.28) at the stability limits of both phases.

A second-order transition can only occur at a critical point of a first-order transition line in the (F,T)-plane,

$$F = F_c(T) . \qquad (2.31)$$

The temperature dependence of the order parameter for $T < T_c$ at constant field is related to that along the coexistence line by

$$-(\partial Q/\partial T)_F = -(\partial Q/\partial T)_{coex} + \chi_T(\partial F_c/\partial T) . \qquad (2.32)$$

The divergence of the second term, $(T_c-T)^{-\gamma}$, is stronger than that of the first term, $(T_c-T)^{\beta-1}$. Therefore, the coupling constant goes to zero only as

$$g \propto (T_c - T)^{\alpha} \qquad (2.33)$$

and one obtains

$$D/D_o \propto (T_c - T)^{\gamma-\alpha} . \qquad (2.34)$$

For $T>T_c$, a corresponding relation holds along the "critical isochore" $Q(F,T)=0$. - Now, according to dynamic scaling theory, the heat conductivity κ diverges as (12)

$$\kappa \propto (T - T_c)^{-\nu} . \qquad (2.35)$$

We therefore obtain at the critical point of a symmetry-conserving transition a soft heat diffusion mode with

$$D \propto (T - T_c)^{\gamma-\nu} \approx (T - T_c)^{\nu} . \qquad (2.36)$$

For second order transitions, experimental results are consistent with the described mechanism (see Refs. (10) and (12)). For first order transitions, however, no experimental results are available. Measurements of the dynamic structure factor in the metastable regions appear highly desirable in order to test predictions like Eq.(2.28).

3. COUPLING TO OTHER LOW-FREQUENCY MODES.

The above discussion shows that interesting dynamic effects result from the coupling of a lattice mode with a low-frequency mode. We consider now in general the possibilities for coupling of this type. Good candidates for such low-frequency modes are hydrodynamic modes resulting from conservation laws. In order for the coupling to exist, the lattice mode and the conserved quantity must belong to the same representation of the symmetry group of the phase under consideration.

In the above case of thermodistortive coupling, the low-frequency character of heat diffusion is a consequence of energy conservation, and the coupling exists only for modes belonging to the identity representation. The conservation law leads to a "zero-frequency anomaly" (13) in the susceptibility for the uniform (q=0) mode,

$$\lim_{\omega\to 0} \chi(0,\omega) \neq \chi_T , \qquad (3.1)$$

which results in the discontinuous behaviour of $\chi(q,\omega)$ at $q=0$, $\omega=0$ expressed in Eqs.(2.23) and (2.24).

Consider now a system in which there exists a general constant of motion A. We assume that A is spatially uniform, i.e. it belongs to $q=0$, because at $q\neq 0$ there always occur transport processes in real physical systems. But at $q=0$, A may transform as any of the representations of the symmetry group of the phase. A lattice mode which couples to A is a non-ergodic variable in the sense that it leads to a zero-frequency anomaly in the susceptibility $\chi(0,\omega)$ of linear response theory: The "isolated" susceptibility $\chi^{is}=\chi(0,\omega\to 0)$ differs from the isothermal susceptibility in a manner completely analogous to Eq.(3.1). The connection between constants of motion, ergodicity, and linear response theory is discussed in detail by Suzuki (14). In general, the isolated susceptibility is bounded by the adiabatic susceptibility (15) such that

$$\chi^{is} \leq \chi_S \leq \chi_T \ . \qquad (3.2)$$

In the low-symmetry phase, the order parameter itself is always non-ergodic, because of the thermodistortive coupling discussed above. Thus, the ergodicity assumption made in Ref. 16 always fails for the low-symmetry phase.

Of particular interest is the case that the constant of motion A belongs to a non-identity representation, such that $\chi_S=\chi_T$ by symmetry for lattice modes which couple to A. If only a single constant of motion has to be taken into account, one obtains from Ref. 14

$$\chi^{is} \equiv \chi_A = \chi_T - \beta\langle\Delta Q\cdot A\rangle^2/\langle A^2\rangle \ . \qquad (3.3)$$

For $q\neq 0$, one has to study the transport of the quantity A. One will in general find a hydrodynamic mode, either oscillatory as for sound waves, or diffusive as in the case of heat diffusion. Coupling of the lattice mode to this hydrodynamic mode will lead to the discontinuous behaviour at $q=\omega=0$,

$$\lim_{q\to 0}\ \lim_{\omega\to 0}\ \chi(q,\omega) = \chi_T \qquad (3.4)$$

$$\lim_{\omega\to 0}\ \lim_{q\to 0}\ \chi(q,\omega) = \chi_A \qquad (3.5)$$

in analogy to Eqs.(2.23) and (2.24). Thus, the dynamics at small q and ω will be determined by the hydrodynamics of the quantity A. Now, in real macroscopic systems there exist probably no general constants of motion except the trivial ones like total energy and total momentum. The only true hydrodynamic modes will then be heat diffusion and hydrodynamic sound waves. Coupling to heat diffusion has been treated in Section 2. Coupling to transverse sound waves occurs in crystals which have an optic mode transforming as an elastic strain. In these crystals, piezodistortive coupling gives rise to a zero-frequency anomaly which is reflected in the difference between susceptibilities at constant strain ε and at constant stress σ:

$$\chi^{is} = \chi_\varepsilon \neq \chi_\sigma \, . \tag{3.6}$$

Thus, when a crystal becomes unstable at a first-order transition against a mode which couples to elastic strain, there will appear a soft sound wave at the stability limit. For the case of piezoelectric crystals, this behaviour is well known (17,18).

Usually, physical systems are described by idealized models which contain only those degrees of freedom which play a dominant role in the effects to be studied. Such a simplified model may have additional constants of motion which are not conserved in the true physical system. Examples are the total energy of the phonons in a single band in a model neglecting interband coupling; the total energy in the spin system in models neglecting spin-lattice interaction; the total spin component in isotropic and uniaxial Heisenberg models; etc. Moreover, additional quantities may become constants of motion in the statistical-mechanical approximation schemes like RPA which one employs to treat the dynamics of the model. We want to argue that such approximate constants of motion are also good candidates for low-frequency modes which may couple to lattice modes of the appropriate symmetry: If the models or the approximation schemes are not completely unreasonable, then the quantities corresponding to the approximate constants of motion will vary only slowly with time in the real system. Although there does not exist an exact hydrodynamic theory in this case, one may construct a pseudo-hydrodynamic theory in which the slow motion of these quantities is phenomenologically taken into account (19). - If the central mode observed in the high temperature phase of $SrTiO_3$ (11) is due to a coupling to a low-frequency mode, there should exist an approximate constant of motion transforming as a Γ_{25}-mode at $q=q_R$.

REFERENCES AND FOOTNOTES

* A project of Sonderforschungsbereich "Festkörperspektroskopie" Frankfurt-Darmstadt, financed by special funds of the Deutsche Forschungsgemeinschaft.

1. K.A. Müller, W. Berlinger and J.C. Slonczewski, Phys. Rev. Letters 25, 734 (1970).

2. J. Feder, A Study of Hysteresis in the Phase Transition of Superconductors and Antiferromagnets. Universitetsforlaget Oslo 1970.

3. J. Feder, K.C. Russel, J. Lothe and G.M. Pound, Advances in Physics 15, 111 (1966).

4. A.F. Andreev, Soviet Phys. - JETP 18, 1415 (1964). (J. Exp. Theoret. Phys. (U.S.S.R.) 45, 2064 (1963)).

5. H. Thomas, Phys. Rev. 187, 630 (1969).

6. H. Thomas, Structural Phase Transitions and Soft Modes, Edited by E.J. Samuelsen, E. Andersen and J. Feder: Universitetsforlaget Oslo 1971, p. 15.

7. E. Pytte and H. Thomas, Solid State Communications 11, 161 (1972).

8. It should be noted that the bare susceptibility $\chi_o(q,\omega)$ does contain local entropy exchange of the lattice mode considered with the rest of the crystal, which is also neglected in RPA. Thus, when the RPA susceptibility is to be used as an approximation for $\chi_o(q,\omega)$, it has to be corrected for this entropy exchange.

9. Note that this relation holds only if the lattice mode δQ does not couple to any further constant of motion of the system (cf. the discussion in Section 3 below).

10. P. Heller, Intern. J. Magnetism 1, 53 (1970).

11. T. Riste, E. J. Samuelsen and K. Otnes, Structural Phase Transitions and Soft Modes, Edited by E.J.

Samuelsen, E. Andersen and J. Feder: Universitetsforlaget Oslo 1971, p. 395.
See also the contributions discussing the central peak in these Proceedings.

12. See the review by H.Z. Cummins and H.J. Swinney in these Proceedings, and references contained therein.

13. P.C. Kwok and T.D. Schultz, J. Phys. C (Solid state Physics) 2, 1196 (1969).

14. M. Suzuki, Physica 51, 277 (1971).

15. R.M. Wilcox, Phys. Rev. 174, 624 (1968).

16. T. Schneider, G. Srinivasan and C.P. Enz, Phys. Rev. A5, 1528 (1972).

17. E.M. Brody and H.Z. Cummins, Phys. Rev. Letters, 21, 1263 (1968).

18. V. Dvorak, Czech. J. Phys. B20, 1 (1970).

19. See for example Ref. 5 and F. Schwabl, Z. Physik 154, 57 (1972).

A QUALITATIVE PICTURE OF THE TRICRITICAL POINT*

M. Blume

Brookhaven National Laboratory
Upton, New York 11973 U.S.A.; and
State University of New York at Stony Brook
Stony Brook, New York 11790 U.S.A.

I. INTRODUCTION

The liquid-gas phase transition, when viewed in the pressure-temperature plane, consists of a line of first-order transitions which end in a critical point, beyond which a continuous transition between the two phases is possible. Such a phase diagram is not possible for all types of transitions, however, as was indicated by Landau.[1] In particular, if the two phases differ from one another by symmetry, as in the paramagnetic-ferromagnetic transition in the absence of an external field, or in most structural phase changes, it is always possible in principle to say in which of the two phases the system is. A line of first-order transitions could not, for these systems, end in a critical point, but it is possible for the line to change into a line of <u>second</u>-order. The critical phenomena at the point where the changeover from first- to second-order occurs are quantitatively different from those at an ordinary critical point or second-order transition. Theoretical examples of such changeovers were found in a number of models of magnetic systems, and some of the properties of these points were worked out.[2-4] In addition, experimental examples of this changeover, such as the behavior of the Ising antiferromagnet Dysprosium Aluminum Garnet in an external magnetic field[5] and phase separation in mixtures[6] of He^3 and He^4 were found.

*Work performed under the auspices of the U. S. Atomic Energy Commission.

A new period of activity, both experimental and theoretical, in the study of critical phenomena in the vicinity of this "change-over" point, and, indeed, a new name for this point, began with the observation by Griffiths[7] that, when viewed in a space of suitable thermodynamic variables, this point was the intersection of three lines of critical points. Accordingly, Griffiths proposed the name "tricritical" point, a name which was rapidly adopted. Since his paper there have been many studies of the properties of the tricritical point. The purpose of this lecture is to provide an introductory view of tricritical phenomena and an indication of some of the theoretical and experimental work in progress.

II. QUALITATIVE DISCUSSION OF A MODEL

To illustrate the concepts involved in the description of the tricritical point we will give a qualitative treatment of a model which displays a tricritical point. We consider a lattice with spin-one ions at each site. These ions interact with a nearest-neighbor Ising-type interaction. In addition we allow a crystal-field interaction and an interaction with an external magnetic field. The full Hamiltonian is

$$\mathcal{H} = -J \sum_{<ij>} S_i S_j + \Delta \sum_i S_i^2 - H \sum_i S_i \tag{1}$$

Here $S_i = \pm 1$ or 0. The term proportional to Δ gives single spin energy levels, for $\Delta > 0$, with an $S_i = 0$ singlet level lowest and $S_i = \pm 1$ degenerate and at an energy Δ above the singlet. Systems of this type have been studied, with application to magnetism in mind, in the molecular field approximation,[3,4] and, more recently, by the techniques of series expansion analysis[8,9] and Monte Carlo calculation.[10] The main conclusion of all of these analyses has been that the paramagnetic-ferromagnetic phase transition is, for small enough Δ, a second-order transition. As Δ is increased a critical value Δ_c is found at which the transition becomes first-order. As Δ is increased still further, the discontinuity in the order parameter at the transition increases and the transition temperature decreases. Finally, Δ becomes sufficiently large so that the $S_i = 0$ level lies sufficiently low that the exchange interaction, which acts only on the $S_i = \pm 1$ levels, is ineffective, and the system does not order. There are of course quantitative differences between the mean field results and the Monte Carlo and series expansion calculations, but on the whole the latter have shown that the mean-field results are qualitatively correct, and we will therefore rely on the latter for the qualitative discussion here.

The results of the mean-field calculation of the phase-diagram in the T-Δ plane for external magnetic field H = 0 are shown in figure 1.

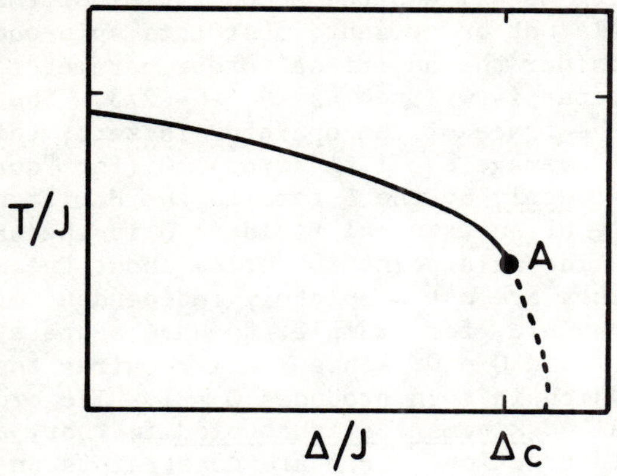

Figure 1. Phase diagram in the T-Δ plane for the mean field solution of Eq. (1). The solid lines represent second-order transitions; dashed lines represent those of first-order. The tricritical point is denoted by A (from Ref. 11).

Here the solid line represents the second-order transition between the ferromagnetic and paramagnetic phases (below and above the line, respectively). At the point A, the tricritical point, the transition changes from second- to first-order, and for Δ/J = 0.5, the transition disappears altogether, and the system is non-magnetic even at T = 0. To understand the nature of the tricritical point more completely we must consider the effects of an external field on the system. Before doing this, a few points about the Hamiltonian should be noted. There are two order parameters which are represented by terms in the Hamiltonian to be considered in this problem. These are, first, the ordinary magnetic order $M = \langle S_i \rangle$. The vanishing or nonvanishing of M, in the absence of an external field H, characterizes the system as paramagnetic or ferromagnetic, and the transition lines in figure 1 represent the points at which M changes from zero to a non-zero value. The change is continuous at the solid second-order line, and discontinues at the dashed first-order line. Two terms in the Hamiltonian (1) affect this order parameter <u>directly</u>. These are of course the exchange interaction

$-J\sum_{<ij>} S_i S_j$, which is responsible for the occurrence of the phase transitions, and the external field term $-H\sum_i S_i$. The presence of an external field will produce, through this term, a non-zero value of M at any temperature. If only these terms were present we would have an ordinary Ising model, and the complexities of the first-order transition would not be present. For this spin-one system, however, we must consider the additional order parameter $Q = \langle S_i^2 \rangle$. (Sometimes the parameter is written as $\langle S_i^2 \rangle - 2/3$. The 2/3 is subtracted so that the trace of the operator is zero; this makes its high temperature average equal to zero.) Q (for "quadrupolar" order) is affected directly by the Δ term in the Hamiltonian; indeed Δ plays the role of an external field to Q in the same way that H does with M. The main point to notice about the order parameters is that they are not completely independent of one another. It is not possible, for example, to have a state of the system in which M = 1 and Q = 0, since M = 1 requires that all the spins have $S_i = +1$ which in turn produces Q = 1. The order parameters can thus be called kinematically coupled in that, even in the absence of any interaction, there are constraints on the values which can be taken on by the two together. This kinematical coupling produces an indirect effect by the term $\Delta\sum_i S_i^2$ on the order parameter M. In fact, the presence of terms in the Hamiltonian which affect two kinematically coupled order parameters produces a competition between them which is ultimately responsible for the appearance of tricritical points, triple points, first-order transitions, etc. We can understand, on this basis, the change in the ground state from ferromagnetic to non-magnetic (for H = 0) with increasing Δ. The exchange term in (1) gives a lower energy for M = 1, while the crystal field term favors Q = 0. These two values are incompatible, however. The increase in Δ ultimately makes it more favorable energetically to have Q = 0 (which requires M = 0) than M = 1.

III. EFFECT OF A MAGNETIC FIELD

To see the full reasoning behind the name "tricritical" point we consider the effect of the magnetic field in (1) on the phase diagram in figure 1. Following Griffiths, we draw the phase diagram in a T-Δ-H space by adding an H-axis normal to the T and Δ axes in figure 1. The resulting phase diagram is shown in figure 2, and in the remainder of this section we give a qualitative discussion of the new features produced by the external magnetic field. At T = 0 it is clear that the external field is in competition with the field Δ. For H = 0 there is, as we have seen, a point, with increasing Δ, at which the ground state changes from M = 1, Q = 1 to M = 0, Q = 0.

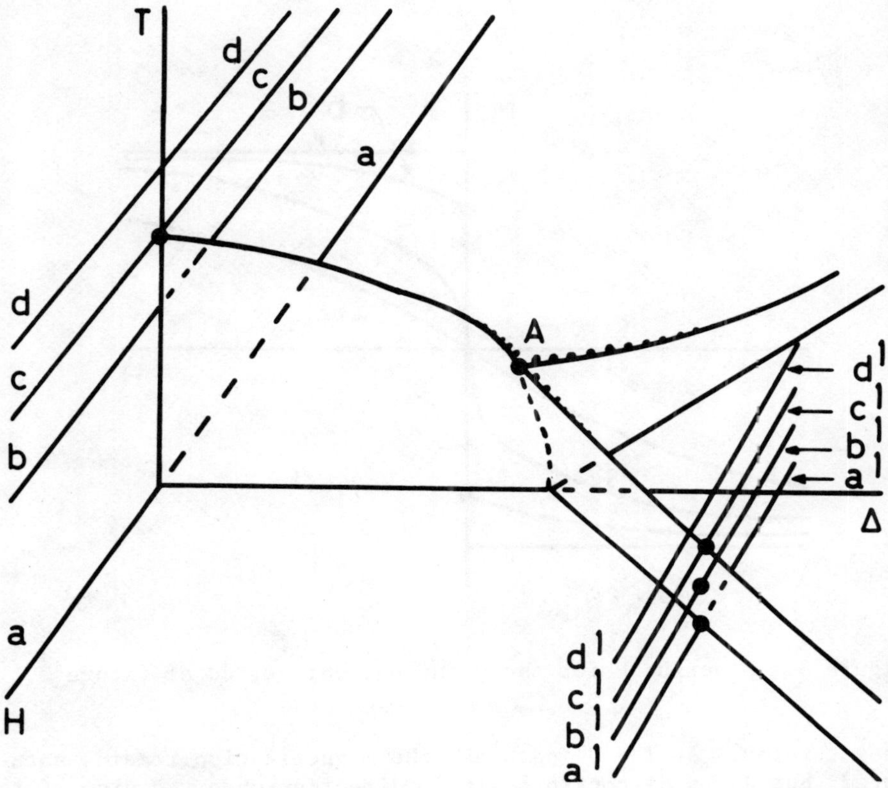

Figure 2. Phase diagram in T-Δ-H space, showing the tricritical point as the intersection of three lines of critical points.

The presence of an external field favors the M = 1 ground state so that, for non-zero H, this state will persist for higher values of Δ. This effect leads to the presence of the wings in figure 2. There are three surfaces in T-Δ-H space. The upper edges of each of these surfaces is a line of critical points, and these three lines meet at the tricritical point A, suggesting its name.

The meaning of the surfaces can be seen by plotting the order parameter M as a function of H for different paths which cross these surfaces. Whenever the path crosses one of the surfaces, there is a discontinuity in the order parameter, and when the path goes through one of the lines of critical points the order parameter has an infinite slope.

In figure 3 we show M versus H for the paths aa, bb, cc, dd in figure 2. For large positive values of H the magnetization

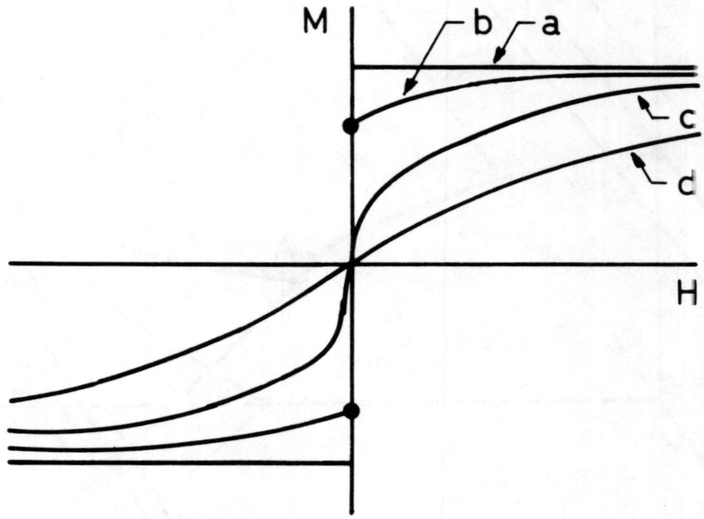

Figure 3. M versus H for the paths aa, bb, cc, dd of figure 2.

is saturated. At T = 0 (path aa) the magnetization remains saturated, but jumps discontinuously in direction when the sign of the field changes from positive to negative. At higher temperatures, the magnetization decreases with decreasing H before undergoing a discontinuous jump. At the Curie temperature (path cc) the slope of M versus H is infinite at zero field. Finally, for $T > T_c$ (path dd) M varies smoothly through H = 0.

The meaning of the wings becomes clear if we proceed similarly. In figure 4 we show M versus H for the four paths a'a', b'b', c'c', d'd' in figure 2. For T = 0 and H = 0, the ground state of the system is non-magnetic, with M = 0 and Q = 0. Hence the M-H curves vary smoothly through H = 0. Here too, for very large H, the magnetization is saturated, and the discontinuities take place along the wings. The critical temperature produces an M-H curve as in path c'c', with infinite slope occurring. The wings are symmetric in positive and negative H, as shown.

IV. DISCUSSION

The picture given above is applicable to many systems besides the theoretical model considered. That model was applied to the

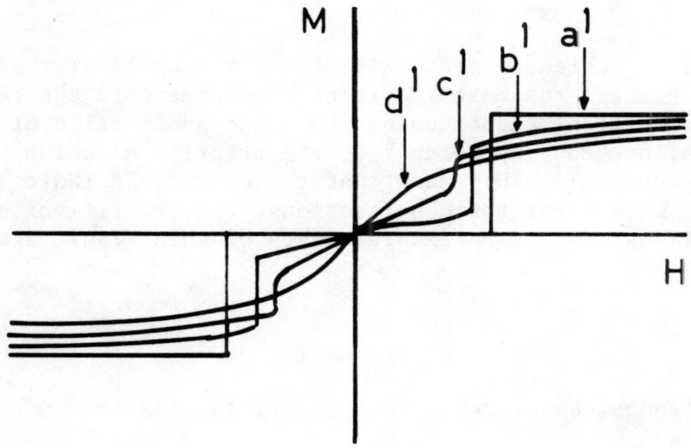

Figure 4. M versus H for the paths a'a', b'b', c'c', d'd' in figure 2.

description of tricritical phenomena in mixtures of He^3 and He^4.[11] The quantity Δ is then to be interpreted as the difference of the He^3 and He^4 chemical potentials, while the order parameter Q is 1 - x, where x is the concentration of He^3. The order parameter M is the condensate wave function $<\psi>$, and H is a physically non-accessible field η which, if applied to the system, would produce superfluidity. Similar phenomena can be found in antiferromagnets, where $\Delta \rightarrow H$, $Q \rightarrow M$, $M \rightarrow M_s$, and $H \rightarrow H_s$. Here H and M are the external field and magnetization, while M_s and H_s are the staggered magnetization and staggered field, respectively. Tricritical phenomena also occur in structural phase transitions, and a full description of experiments on NH_4Cl is given in the next paper.[12]

Theoretical developments have followed the path of the theories of ordinary critical points. Landau and mean field theories were of course the first to be "worked out", and they showed the change in critical indices at the tricritical point. Scaling theories have been also proposed[13-15] and the usual scaling laws are expected to hold at the tricritical point, with allowance for the changed indices. The indices are found to be:

	Mean Field	Tricritical (Mean Field)	3d Ising
α	0	1/2	"1/8"
β	1/2	1/4	"5/16"
γ	1	1	5/4
δ	3	5	5

Energy scaling, which predicts $\alpha + 2\beta + \gamma = 2$, holds for all three examples. Perhaps the most important development in the recent activity in critical phenomena has been the application of Wilson's renormalization group approach[16] to the tricritical point by Riedel and Wegner.[17] They show that the mean field indices should be correct, with logarithmic corrections, for tricritical points in three dimensions. Experimental tests of this result are clearly desirable.

REFERENCES

1. L. D. Landau, Phys. Zeit. Sowjet. 11, 26 (1937).
2. K. Motizuki, J. Phys. Soc. Japan 14, 759 (1959).
3. M. Blume, Phys. Rev. 141, 517 (1966); M. Blume and R. E. Watson, J. Appl. Phys. 38, 991 (1967).
4. H. W. Capel, Physica 32, 966 (1966).
5. D. P. Landau, B. E. Keen, B. Schneider, and W. P. Wolf, Phys. Rev. B3, 2310 (1971).
6. E. H. Graf, D. M. Lee, and J. D. Reppy, Phys. Rev. Letters 19, 417 (1967).
7. R. B. Griffiths, Phys. Rev. Letters 24, 715 (1970).
8. D. M. Saul and M. Wortis, Magnetism and Magnetic Materials - 1971, AIP Conference Proceedings No. 5 (AIP, New York, 1972).
9. J. Oitmaa, J. Phys. C 4, 2466 (1971).
10. D. P. Landau, Phys. Rev. Letters 28, 449 (1972).
11. M. Blume, V. J. Emery, and R. B. Griffiths, Phys. Rev. A 4, 1071 (1971).
12. W. Yelon, following paper.
13. E. K. Riedel, Phys. Rev. Letters 28, 675 (1972).

14. A. Hankey, H. E. Stanley, and T. S. Chang, Phys. Rev. Letters 29, 278 (1972).

15. R. B. Griffiths, "A Proposal for Notation at Tricritical Points" (to be published).

16. K. G. Wilson, Phys. Rev. B4, 3174 (1971).

17. E. K. Riedel and F. J. Wegner, Phys. Rev. Letters 29, 349 (1972).

TRICRITICAL STUDIES OF ND_4Cl

W. B. Yelon

Institut Laue-Langevin, Grenoble, France

THEORY

In the previous lecture (1), we have had a description of the origin of the tricritical point in terms of the ordering and non-ordering densities as well as the resulting phase diagram in the space of temperature and conjugate fields. I will discuss the experimental situation for one tricritical system, ND_4Cl, which we have extensively studied at Brookhaven (2). It is useful first to try to understand the relationship between ND_4Cl and those systems such as metamagnets, which Blume has described.

Tricritical points arise in systems which have two order parameters, which in the case of the metamagnets are M and M_s, the bulk and staggered (sublattice) magnetizations, respectively. Conjugate to these order parameters are the fields H, the direct field, and H_s, a physically inaccessible staggered field which has the same symmetry as J_A, the antiferromagnetic exchange. A plausible phase diagram for such a system, first given by Griffiths (3), is shown in Fig. 1. The second-order line in the H-T plane ($H_s = 0$) bifurcates at the TCP to form the so-called "wings", the boundaries of coexistence surfaces outside the H-T plane. Thus in the metamagnets one reaches the TCP by applying the field H. The system can then be studied as a function of both H and T.

The pressure-temperature diagram for the ammonium halides is shown in Fig. 2 (4). The region of these diagrams of particular interest surrounds the points labelled HCP (hypercritical point). This point which separates the first-order segment of the phase line from the second-order segment has been referred to

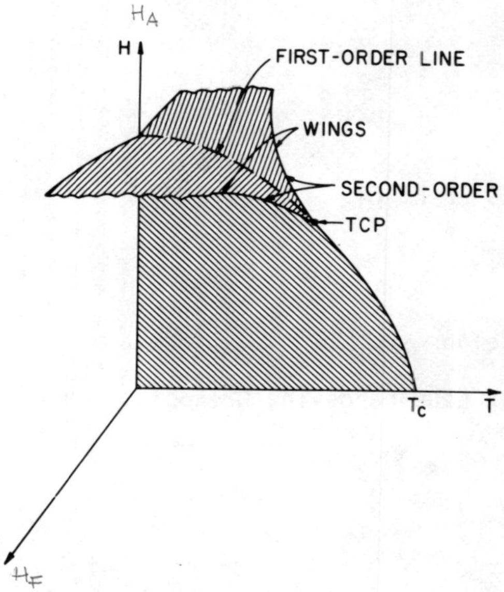

Figure 1. Postulated phase diagram for metamagnets in the field space H, $\zeta \equiv H_S$ (staggered field).

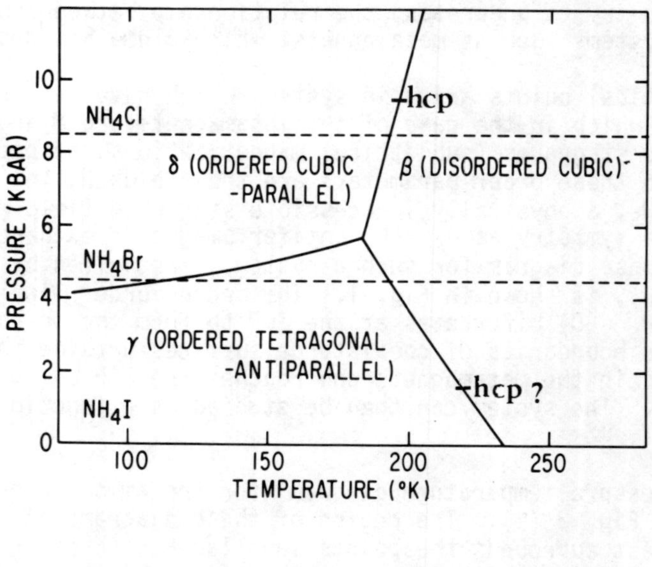

Figure 2. Generalized phase diagram for the ammonium halides (see ref. 4).

(and later we will again refer to it) as a tricritical point. Based on experimental evidence and on a simple model, we believe that this hypercritical point is in fact not a tricritical point in the simple sense, but a tetracritical point which is induced by a coupling of the ammonium tetrahedra to the underlying lattice.

The model that we have used to represent the ammonium halides is based on the assumption that the ammonium tetrahedra are rigid bodies residing in a cubic environment. Within a cubic unit cell formed by the halide ions there are two different tetrahedral positions as shown in Fig. 3. We assume that these sites are the only possible positions and hence we represent the tetrahedra by a Ising pseudo-spin 1/2 variable.

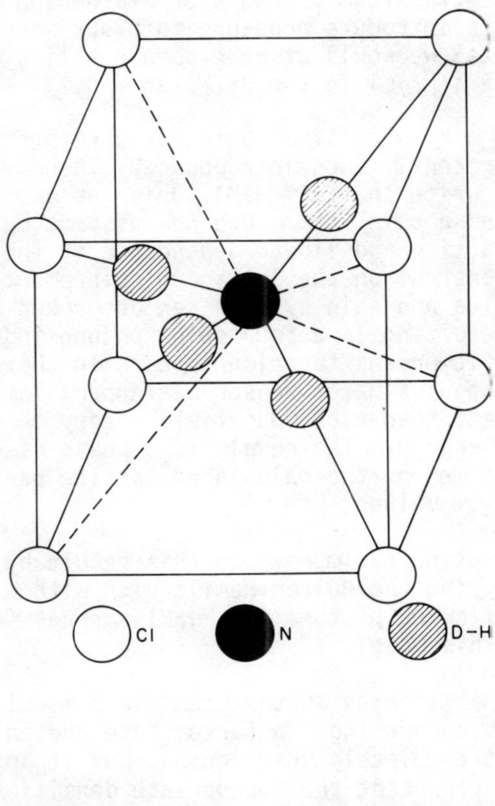

Figure 3. Arrangement of hydrogen atoms in NH_4Cl cell. Alternate tetrahedral sites are indicated by dotted lines

Our picture of ND_4X or NH_4X is based on the work of Nagamiya (5) and of Seymour (6) and more recently Hüller (7) who gives an electrostatic Hamiltonian governing the interactions of nearest neighbour tetrahedra and nearest neighbour tetrahedra and halide ions. The direct tetrahedral interaction is octopole-octopole and tends to produce a parallel alignment of nearest neighbour spins (J_F). The interaction between the tetrahedra and a halide ion is octopole-dipole, and can be approximated by purely inter-spin interactions which are :

a) n^2 antiferro, b) n^3 antiferro, c) n^4 ferro

This Hamiltonian can account for the ferro and antiferro states of the ammonium halides. It cannot, however, account for the hypercritical points, nor for any first-order behaviour observed in a transition from a disordered state to an ordered state, a result which follows directly from the work of Stephenson and Betts (8). Thus if we were to introduce pressure into our model via the exchange constants, we would at best obtain a line of continuous transitions for $ND_4(H_4)Cl$ in the P-T plane.

A model for a hypercritical point in a compressible Ising ferromagnet has recently been introduced by Theodorakopoulos (9). His method is to write the full Hamiltonian as two terms : a lattice term and a spin term. One now assumes quasi-harmonic theory for the lattice and linear dependence of the nn ferromagnetic exchange constant on the lattice displacement. The coupling between the lattice and spin systems can be broken by a unitary transformation resulting in a four spin "phonon-induced" interaction. The static Greens function appears in this four spin term. Theodorakopoulos makes reasonable approximations for this Greens function and then uses mean field theory to obtain an approximate free energy. The result is a phase diagram similar to that observed - however, the calculated lattice parameter is only about 1/2 the ND_4Cl value.

We are attempting to improve on this result by using the same methods and employing the Hüller Hamiltonian with slight modifications. The following picture for ND_4Cl emerges from preliminary calculations on this model.

The Hüller Hamiltonian in the mean field model provides for two types of stable ordering : a ferro state and an antiferro state. It is not completely unambiguous, but it appears that these two states represent the appropriate densities for the system. This is supported strongly by a recent X-ray (10) study near the order-disorder transition in NH_4Cl which shows critical scattering corresponding to the "antiferro" state even though the system orders in the "ferro" state. Fluctuations in both order parameters are expected near a TCP (1).

We introduce for convenience of description two fields, H_F and H_A, which couple directly to the order parameters of the two states. Thus we will ultimately require four field variables, H_F, H_A, P and T, to describe ND_4Cl. If we fix the pressure at some value above the HC pressure, we obtain the same phase diagram as shown in Fig. 1 with H_A replacing H and H_F replacing H_s. That is, under these conditions the system is completely analogous to the metamagnets.

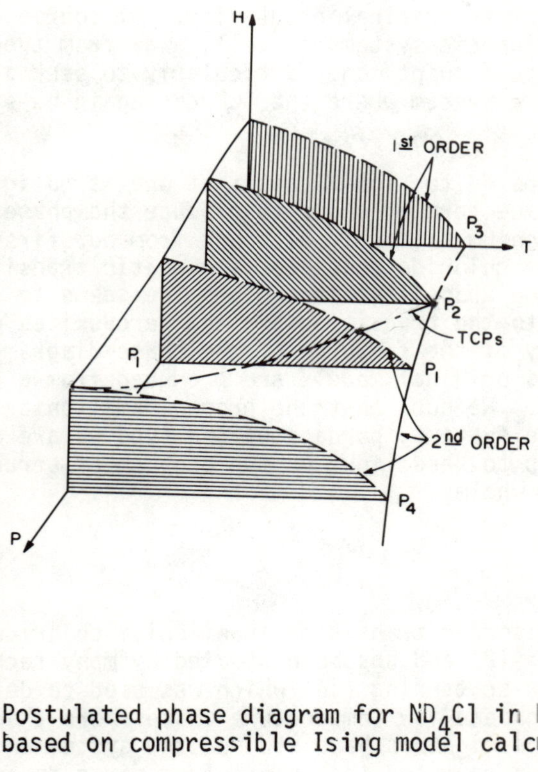

Figure 4. Postulated phase diagram for ND_4Cl in H-T-P space based on compressible Ising model calculations.

It is now convenient to look at the P-T-H_A phase diagram as shown in Fig. 4. Anywhere above P_2 we have segments of both first and second-order transitions with a TCP between, and the H_F-H_A-T diagram is as in Fig. 1. In this space, we have surfaces of first and second-order transitions with a line of TCP's between terminating at P_2 and $H_A = 0$. At P_2, however, the wings in Fig. 1 begin at $H_A = 0$ and when one includes the positive and negative H_A

directions it is clear that these are four second-order lines all meeting at one point. The tricritical point has thus become a tetracritical point.

Although it is not completely clear at this time, it is probable that the critical behaviour at this point is the same as at any TCP on the line. One encounters a similar situation in the H-T plane for an antiferromagnet where the symmetric joining of two second-order lines at H = 0 (for the ± H directions) does not affect the critical behaviour, at least in practice. An important consequence of finding a TCP at $H_A = 0$ is that the tricritical properties can be studied without the necessity of making demagnetizing corrections, such as are necessary for the metamagnets. In this respect, the ammonium halides are an ideal system for studies of tricritical properties. Of course, it is desirable to study magnetic systems as well, away from the ambiguity of the tetracritical point, and particularly to seek a pressure sensitive magnetic system where the TCP can again be studied at H = 0.

It is perhaps quite significant that one is obliged to introduce a spin-lattice term in order to produce the phase structure seen for the ammonium halides. Although from our first view the transition in the chloride resembles a magnetic transition in most respects, the transition in the bromide seems to show many features common to the transitions in the perovskites et al. Furthermore, many of the features of the phase-diagrams not produced by Hüller's original model, are produced from a phonon-based model (11). We hope that the proper Hamiltonian for these relatively simple systems, perhaps of the type we are now examining may also help to shed light on the problem of structural transitions as a whole.

EXPERIMENTAL

The order-disorder transition in ammonium chloride has been known since 1922 (12) and has been studied by many techniques (13) including neutron scattering (14) which was used to determine the nature of the high and low temperature phases more than 20 years ago. Several investigators have studied the system under pressure (15) and most have reported the change from first to second-order behaviour, but there has been wide disagreement about the location of the TCP. In part the disagreements may be due to sample differences, but more likely they reflect the difficulty of accurately locating the TCP, a point we will return to.

The only study in the vicinity of the TCP of sufficient precision to report critical exponents, other than the neutron work I will report on, was performed by Garland and Weiner (16) who have used capacitance techniques to measure the length of a good single crystal of NH_4Cl as a function of pressure and temperature. They localize the TCP to about ± 30 bars (at $P \approx 1500$ bars) and report also the values for the critical exponents α and α' (determined from the compressibility) as between about 0.4 and 0.9. The uncertainties were considerable but values of $\alpha \approx 0.5$ at the TCP are not inconsistent with present models.

Neutron scattering is particularly useful for a study of this system, since it can, at least in principle, provide all of the susceptibility exponents and it may be the only direct method for measuring the order parameter, since this is a structural transition involving the motion of the hydrogen atoms. The work which we have performed at Brookhaven during the past year covers three basic problems.

1. Is the transition in fact "Ising-like"?
2. Where is the TCP?
3. What are the tricritical exponents?

To the extent that the system is described by a pseudo-spin 1/2, one expects that the Ising model will hold. The pseudo-spin model may, however, not be adequate under various circumstances, for example, if the tetrahedra are in fact not well localized on their sites due to rotational freedom; if the tetrahedra do not transform as a unit, but instead via the movement of individual members; if the rotational freedom in the ordered state is significantly different than in the disordered state, or if the transformation is coupled to the lattice via a phonon. This problem has been studied by various investigators and the conclusions have generally been that the Ising model is adequate, but we sought to confirm this to the extent possible by structure studies of ND_4Cl at standard pressure as a function of temperature. The data collected are fitted to the following structure factor expression

$$F_T = \psi F_{LT} + (1 - \psi) F_{HT}$$

where F_{LT} and F_{HT} are the structure factors for the low and high temperature phases, which differ only in that the low temperature phase has all tetrahedra parallel-ordered while the other phase has them oriented at random. ψ, the order parameter is defined as

$$\psi = \frac{|N(111) - N(\overline{111})|}{N(111) + N(\overline{111})}$$

where $N(111)$ and $N(\overline{111})$ are the number of the tetrahedra in the (111) and ($\overline{111}$) orientations respectively, in a given domain. Although the model does not include free rotations of the ammonium tetrahedra, a substantial rotation would produce large thermal

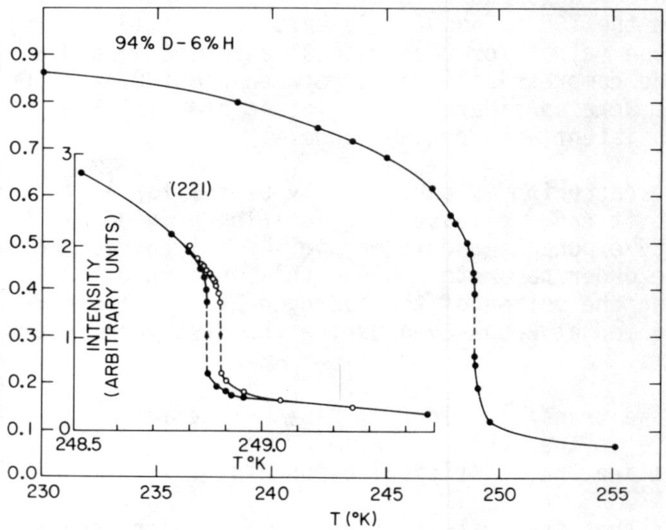

Figure 5. The order parameter S at P = 0 determined from the structure refinements and (insert) the intensity of the (221) reflection close to T_c.

ellipsoids for the deuterium atoms and a significant shortening of the N-D bond, neither of which are observed.

The results of these measurements, some of which are shown in Fig. 5, show that at P = 0, ND_4Cl has a first-order transition with ≈ 0.035°K hysteresis. Furthermore, the thermal parameters are continuous through T_c and the rotation of the tetrahedra is small (≲ 10°). These results are consistent with the Ising description. We found at the same time that the order parameter ψ can be accurately determined by studying only the intensity (and lineshape) of the (221) reflection, and we began to investigate the other problems utilizing this reflection.

In many respects, locating the TCP was the most difficult problem we encountered. Attempts to localize the TCP by observation of hysteresis or the discontinuity in ψ were accurate to only about 70 bars since critical scattering is quite significant and tends to obscure small changes in ψ.

By looking at the critical scattering at large scattering angles (where the Bragg scattering does not contribute) we were able to define the phase line quite accurately, and locate the TCP to within ± 30 bars. This is possible because the susceptibility

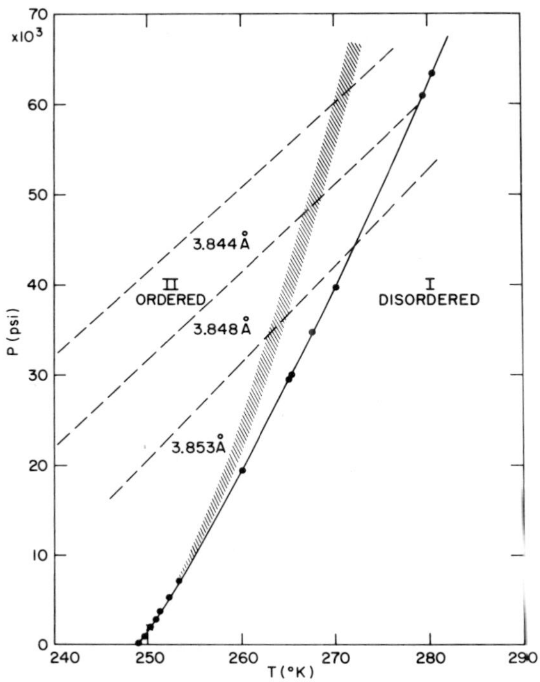

Figure 6. The phase line for ND_4Cl (solid line). The shaded region represents the crossover from second-order to tricritical behaviour.

(and hence the critical scattering) diverges along the phase line in the second-order region but not in the first-order region. Hence the intensity of the critical scattering drops rapidly (even for large scattering angles) as one moves into the first-order region. However, this fact can be taken advantage of for a more accurate determination of the TCP.

It is clear that if the susceptibility (critical scattering) could be measured along the phase line one could see the change from divergent to non-divergent behaviour. This is not practical experimentally since the phase line is not sufficiently well known, and the pressure and temperature stabilities not sufficiently good to preclude the possibility of crossing into the ordered region and measuring Bragg scattering as well as the desired critical scattering. If, however, we measure in the disordered region,

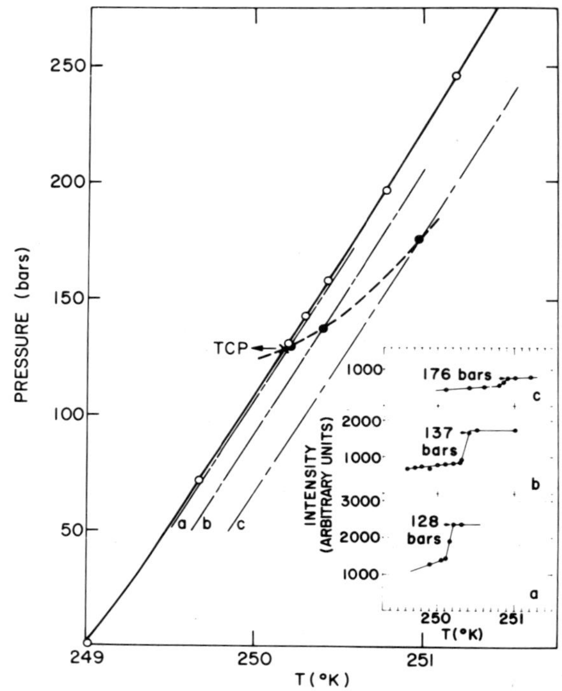

Figure 7. The phase line near the TCP showing the scans made to localize the TCP and (insert) the peak intensity of the critical scattering corresponding to the scans.

along lines roughly parallel to the phase line, we can expect to see the susceptibility fall from a high, non singular, value near the second-order portion of the phase line to a much lower value below the TCP. Therefore, we studied the peak intensity at the (221) Bragg position along the lines indicated in Fig. 7. As expected, the susceptibility falls quite rapidly both as one moves away from the phase line, and also as we move "past" the TCP (in some geometric sense). However, the sharpness of the drop along the lines was quite unexpected and has not yet been fully explained. It can be understood in a qualitative way, however, if it is realized that constant susceptibility contours are parallel to the phase-line in the second-order region but must curve into the phase line in the first-order region. If, in fact, the susceptibility falls quickly in the first-order region, then these lines will curve in steeply in the region of the TCP perhaps producing the structure seen experimentally. In this way we have localized the TCP at 1850 ± 30 psi and 250.17 ± 0.01°K in ND_4Cl (~6% H). Nevertheless the extremely sharp drop in susceptibility and other features in this region remain not well understood. Further experiments, with better sensitivity are planned to study in detail the susceptibility both in the ordered and disordered region in the vicinity of the TCP.

Another point of interest in this vicinity of the P-T diagram is the joining of the first and second-order portions of the phase line. It appears that the two portions join with continuous slope, but the first-order line does not coincide with the analytic continuation of the second-order line determined by fitting the second-order portion by either quadratic or cubic equations. At P = 0, the actual transition occurs about .1°K below the fitted temperatures. Further experiments are also planned on a system such as $N(D_{0.5}H_{0.5})_4Cl$ where the first-order region is larger in order to study in greater detail the behaviour of the phase line, as well as the discontinuity of the order parameter and the susceptibility at the phase line as one moves away from the TCP, to see if the predicted scaling relations are valid.

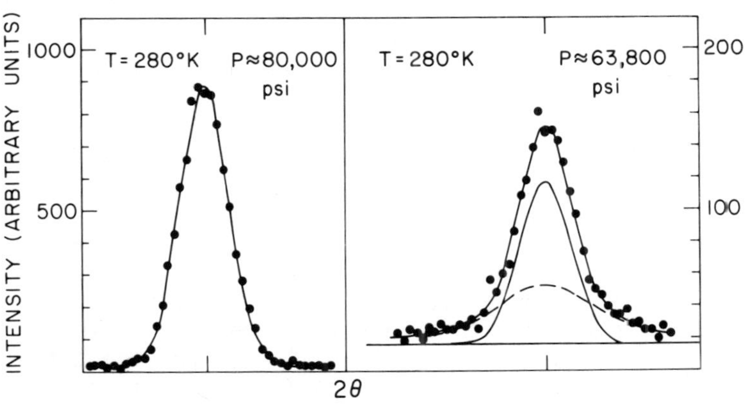

Figure 8. Results of fitting the measured peak (221) both far and near to the phase line. The peak at 80,000 psi is fitted by a Gaussian only, while that at 63,800 is a Gaussian plus a Lorentzian folded with the resolution function.

After localizing the TCP, the measurement of ψ and β was relatively straightforward. However, the possible existence of "crossover" impelled us to seek a method of separation of the Bragg and critical scattering which would permit us to analyse reliably the data quite close to the phase line. This was accomplished by measuring far into the wings of the Bragg peak and then fitting the data to a Gaussian with the resolution width plus a Lorentzian (Ornstein-Zernike form) folded with the resolution function (Fig. 8). This approach would not be practical if one used the direct folding of the three-dimensional resolution function, as has been used in the past for studies of the critical scattering (17) because of the long computing times involved (typically 1 hr/point) and the large number of points collected (more than 200 have already been analysed). However, we were fortunate to have had shown to us a Laplace transform (18) which reduces the folding to a one-dimensional problem speeding the calculations by a factor of about 100. This approach was checked with various test functions and appears to be more accurate than the direct method and is suitable for two-axis spectrometer problems when the Ornstein-Zernike or the Fisher-Burford (19) first or second approximants are used to describe the critical scattering. Even though we were only interested in the Bragg portion, we found that the best results were obtained only when this method was used as opposed to approximating the critical scattering by a simpler form such as a second Gaussian, or even a one-dimensional Lorentzian folded with a Gaussian resolution, and our results were generally reliable to a

reduced distance $g = \dfrac{T_c - T}{T_c}$ or $\dfrac{P - P_c}{T_c}\dfrac{\partial T_c}{\partial P}$ of 2×10^{-4}.

Determinations of β were made both as a function of temperature and of pressure near the TCP and at several other places along the phase line. The results of these measurements show both the presence of the TCP and the effects of the second-order line. Measurements as a function of temperature at P = 2275 psi produce a markedly low $2\beta = 0.36 \pm 0.01$ (Fig. 9). The fitted power law seems to be valid over quite a wide range perhaps extending even past 0.1 in reduced temperature. I hasten to point out that in this and all the other cases to be mentioned, the transition temperature or pressure were parameters of the fit, but in all cases the resulting values were in excellent agreement with the values deduced from the peak in the critical scattering.

At 250.17°K, the measurements as a function of pressure show the same general behaviour as those just discussed (Fig. 10). However, the fitted value for $2\beta = 0.28 \pm 0.02$ is significantly less than the value for the constant pressure run near the TCP.

Far from the TCP, along the second-order line one hopes to see the Ising value $\beta = .3125$. At the $T_c = 280°K$ ($P \approx 60{,}000$ psi)

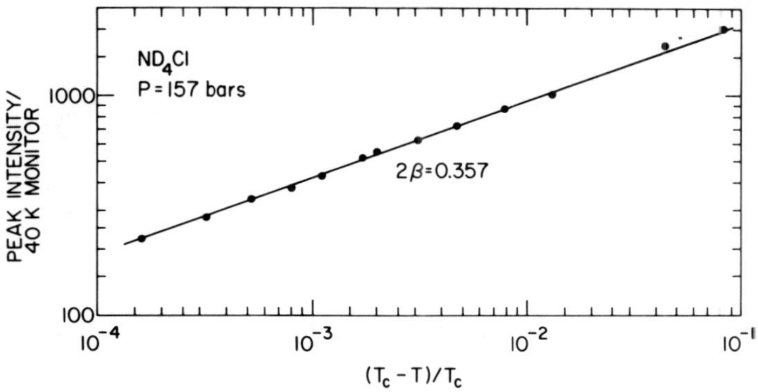

Figure 9. Logarithmic fit of I vs T at 157 bars (2275 psi).

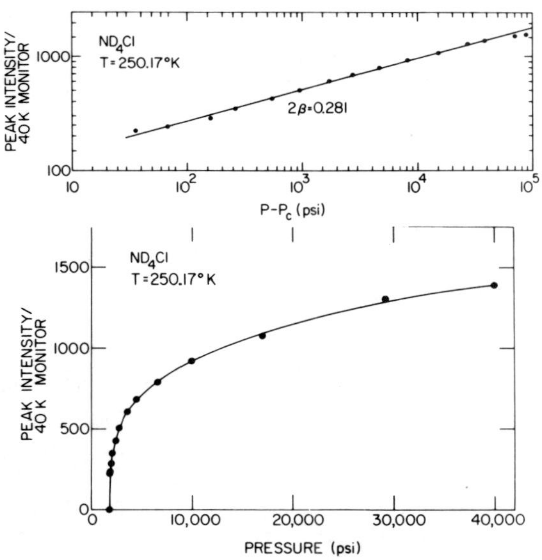

Figure 10. Linear and logarithmic fits of I vs P at 250.17°K.

the highest value for which we could measure both as a function of temperature and of pressure, this is not completely achieved. As a function of pressure (Fig. 11), one measures $2\beta = 0.50 \pm 0.04$. The greater uncertainties reflect a larger contribution from critical scattering at the same reduced temperature as well as a smaller value for the order parameter close to T_c.

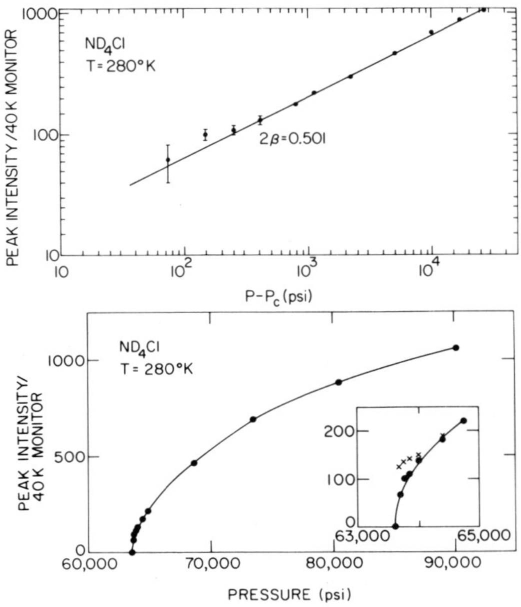

Figure 11. Linear and logarithmic sets of I vs P at 280°K with insert showing data before (crosses) and after correction for critical scattering.

The constant pressure measurements at 60,000 psi (Fig. 12) show an additional interesting feature. All of the data cannot be fitted to a single power law; instead, the data seem to be divided into two regions of higher and lower exponents. For $g \leq 0.02$, $2\beta = 0.61 \pm 0.04$ which agrees well with Ising value for a second-order transition. However, it is clear that a lower value is needed to describe the outer region $0.02 \leq g \leq 0.1$. This change is thought to show the "crossover" from the region in which the singularity associated with the second-order line is dominant to a region in which the tricritical point dominates the behaviour. If current models are correct, the width of the inner region should shrink down as one approaches the TCP and this is observed for a region of T_cs above 30,000 psi. Below this, only an average exponent

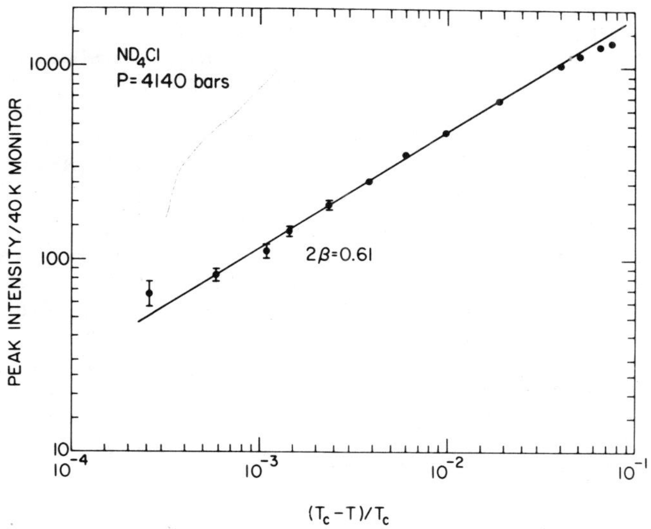

Figure 12. Logarithmic fit of I vs T at 4140 bars (60,000 psi).

can be defined, one cannot separate the data into two regions. However, enough observations of crossover were made to roughly define these two regions.

The data collected between the TCP and 280°K suggests that the values observed at 280°K are close to the limit one will measure along the second-order line, i.e. the 280°K results are near the true second-order values. Although the absolute values of the amplitude factors $A\left(\frac{\psi}{\psi_o} = A\,\frac{(T_c-T)^{\beta}}{T_c}\right)$ were not determined, the ratio for different temperature scans (where the proper reduced quantity is unambiguously defined) was well determined. In particular, the amplitude A for the scan at 60,000 psi is (1.25 ± 0.02) A_{TCP} the amplitude for the scan at 2275 psi.

I would also like to mention two scans made to the first-order line. At 249.68°K, the β determined is consistent with the tricritical value, but at P = 0, 2β = 0.31 ± 0.02 compared to 0.36 ± 0.01 at the TCP, a difference which must be considered significant. In both cases the transition point was a fitting parameter and deviated from the experimental value in the expected direction.

CRITICAL SCATTERING

For reasons of resolution, sample quality, size and time, a thorough study of the critical scattering was not undertaken, although work will be resuming on this aspect shortly. Nevertheless considerable information was obtained while performing the separation of Bragg and diffuse scattering contributions and one careful scan was made near the TCP. The general features are of sufficient interest to bear commenting upon here.

First, in the ordered phase the critical scattering near the TCP is quite small compared to that at 280°K. The corrections for critical scattering become negligible for $|g| > 0.002$ at 2275 compared with $|g| > 0.006$ at 60,000 psi. This behavior can easily be understood in a qualitative way. Near the TCP where the exponent β is small, the system orders very rapidly as the temperature is lowered (despite the somewhat smaller amplitude). As a result of this rapid long-range ordering the system is not free to fluctuate and critical scattering is small.

In the disordered phase, on the other hand, the critical scattering near the TCP is very much larger than in the second-order region, having perhaps three times as much intensity at the same reduced distance. Furthermore, at 1900 psi critical scattering is readily observed more than 20°K above T_c and is still weakly visible 30°K above.

The data at 1900 psi have been carefully analyzed by the same methods used in the ordered region. In this case there is a small residual Bragg peak which was fitted far above the transition and held constant near T_c.

The wave vector susceptibility, shown to have the form (19)

$$\chi(q)/\chi(0) = \left(\frac{1}{r_1}\right)^{2-\eta} (\kappa^2 + \phi^2 q^2)^{\eta/2} \Big/ (\kappa^2 + \psi q^2)$$

was originally analyzed using the Fisher-Burford first approximant

$$\chi(q)/\chi_0 = \frac{A}{(\kappa^2+q^2)^{1-\eta/2}}$$

where A, κ and η were all to be determined.

However, it was pointed out (20) that the first approximant is probably a worse model than the zeroth (O-Z form) approximant. A better approximation is likely to be

$$\chi(q)/\chi(0) = A\kappa^{\eta}/(\kappa^2 + q^2)$$

for $T > T_c$ and $q < \kappa/\phi$, which cannot be used to determine η unless A is held constant.

In the limit $T \approx T_c$ ($\kappa << \phi q$), the form

$$\chi(q)/\chi(0) = A\phi^{\eta}/q^{2-\eta}$$

should be valid, enabling one to evaluate η. Unfortunately the data were not of sufficient quality to permit this because of the relatively poor resolution.

In the region where the analysis was successful, A was found to be nearly constant, and κ can be fit to the power law

$$\kappa \propto (T-T_c)^{\nu} \quad \text{with} \quad \nu = 0.52 \pm 0.08$$

and the resulting value for γ is 1.05 ± 0.20. Because of the large uncertainty, this result is only to be considered suggestive.

DISCUSSION

There are two general results which I would like to discuss further, the very low value of β near the TCP and the inequality of the measured β's for pressure and temperature scans near the same transition point.

Several models for the tricritical point in an Ising system have recently been proposed. One of these, largely due to Riedel and Wegner (21), predicts that at the TCP, $\psi \propto (g \ln g)^{1/4}$. In the region of g studied in our experiment, 10^{-1} - 10^{-4}, this would be best represented in a simple power law by $g^{0.19}$. Another model (22) based on the Schofield (23) equation of state gives a value for β which is dependent upon some details of the model, particularly the geometry of the phase lines, but is on the order of 0.17.

Both these results are in general agreement with our measurements. The crossover behavior that has been seen is also predicted in these models. It is on the question of "smoothness" that we have to dwell further.

According to the hypothesis of Griffiths and Wheeler (24) we should measure the same β for all "non-asymptotically-parallel" approaches to the critical line and furthermore this value should remain the same along that part of the phase line where the nature of the underlying first-order transition remains the same. The results of our work at first sight appear to be in contradiction to this hypothesis. However the smoothness postulate deals with an asymptotic exponent which is the limit of an effective exponent obtained by fitting ln ψ vs lng over several decades. It is quite possible that in some or all of our data we have not reached that limiting value, in which case our results are understandable.

According to the theory of Griffiths and Wheeler, critical behavior along a second-order critical line can be described by two independent thermodynamic fields, one of which must be chosen in some sense parallel to the phase line. Each of these fields defines a direction of approach to the critical line and would be characterized by a different exponent. It is now natural to expect that some optimum direction of approach will yield only the asymptotic exponent while other directions will sense the parallel direction to greater or lesser extent and will thus reach the asymptotic limit at different reduced distances. It is in fact probable that the pressure axis is closer to the parallel direction of approach than the temperature axis since, as the lattice becomes less compressible at very high pressures, the phase line will become more and more parallel to the P axis. We expect then that the measurements of ψ at constant pressure will be characterized by exponents nearly representing the optimum approach direction, whereas measurements of ψ at constant T would feel the parallel field direction to a much greater extent and possibly reflect a rather different exponent.

This is consistent with the observation of β near the Ising value at 60,000 psi for the temperature scan, as well as the value for β for this type of scan at the TCP close to the existing theoretical models. On the other hand, we would expect that for scans parallel to the P axis, a simple power law should in fact not be adequate. Perhaps further experimental studies measuring to smaller reduced distances can show this to be in fact the case.

The variation of the "effective exponent" defined by a set of ψ vs g measurements along the critical line appears to show, in accord with current theory, two limiting types of critical behavior. As one moves from the second-order region (high pressures) to lower pressures, one leaves a region where the second-order results are correct, moves through some intermediate region and finally arrives in a region where the tricritical exponents are valid. The region of crossover from second order to tricritical behavior moves out in g as one moves away from the TCP and while the crossover lies within the experimental regime 10^{-1} - 10^{-4}, intermediate exponents are seen if the data are fitted over the entire interval.

These experiments were part of a major experimental program lasting several months on the HFBR. We are presently readying to begin an equally broad program at the ILL (Grenoble) to study in further detail some of the features here reported, and particularly to attempt to measure all of the susceptibility exponents both in the vicinity of the TCP and along the phase line. With the completion of those studies we hope to be able to make more definitive statements concerning the validity of current models and in particular the success or failure of scaling at the TCP.

ACKNOWLEDGMENTS

I am very grateful for the fruitful collaboration with D. E. Cox and W. B. Daniels and would like to thank also M. Blume, R. J. Birgeneau, G. Shirane and J. Skalyo Jr. for help on many aspects of this work. I particularly wish to thank P. J. Kortman for his assistance and continuing interest on the many theoretical questions involved in this work as well as his help with the manuscript.

REFERENCES

1. M. Blume, Proceedings of the NATO A.S.I. "Anharmonic Lattices, Structural Transitions and Melting" 1973 (This volume, previous lecture).

2. W. B. Yelon and D. E. Cox, Solid State Comm. 11, 1011 (1972). W. B. Yelon, D. E. Cox, P. J. Kortman and W. B. Daniels (to be published).

3. R. B. Griffiths, Phys. Rev. Letters 24, 715 (1970).

4. This generalized phase diagram is taken from Ref. 11, but was originally suggested by Stevenson. (R. Stevenson, J. Chem. Phys. 34, 1757 (1961)).

5. T. Nagamiya, Proc. Phys. Math. Soc. Japan 25, 572 (1943).

6. R. S. Seymour, Acta Cryst. A27, 348 (1971).

7. A. Hüller, Zeitschrift Physik 254, 456 (1972).

8. J. Stephenson and D. D. Betts, Phys. Rev. B2, 2702 (1970).

9. N. Theodorakopoulos, Solid State Comm. 12, 955 (1973).

10. M. Lambert (private communication).

11. Y. Yamada, M. Mori and Y. Noda, J. Phys. Soc. (Japan) 32, 1565 (1972).

12. F. Simon, Ann. Phys. 68, 241 (1922).

13. For a more complete list of references see Ref. 2.

14. H.A. Levy and S.W. Peterson, Phys. Rev. 86, 766 (1952).

15. N.J. Trappeniers and Th.J. Van der Molen, Physica 32, 1161 (1966).
N.J. Trappeniers and W. Mandema, Physica 32, 1170 (1966).
I.J. Fritz and H.Z. Cummins, Phys. Rev. Letters 28, 96 (1972).

16. C.W. Garland and B.B. Weiner, Phys. Rev. B3, 1634 (1971).
B.B. Weiner and C.W. Garland, J. Chem. Phys. 56, 155 (1972).

17. See for example A. Tucciarone, H.Y. Lau, L.M. Corliss, A. Delapalme and J.M. Hastings, Phys. Rev. B4, 3206 (1971).

18. M. Lax (private communication).

19. M.E. Fisher and R.J. Burford, Phys. Rev. 156, 587 (1967).

20. R.J. Birgeneau, J. Skalyo Jr. and G. Shirane, Phys. Rev. B3, 1736 (1971).

21. E.K. Riedel and F.J. Wegner, Phys. Rev. Letters 29, 349 (1972).
E.K. Riedel, Phys. Rev. Letters 28, 676 (1972).

22. P.J. Kortman, Phys. Rev. Letters 29, 1449 (1972).

23. P. Schofield, Phys. Rev. Letters 22, 206 (1969).

24. R.B. Griffiths and J.C. Wheeler, Phys. Rev. A2, 1047 (1970).

COMPUTER SIMULATION OF A DISCONTINUOUS PHASE TRANSITION IN THE TWO-DIMENSIONAL ONE-SPIN-FLIP ISING MODEL

T. Schneider and E. Stoll

IBM Zurich Research Laboratory
8803 Rüschlikon-ZH, Switzerland

ABSTRACT

It is the purpose of this work to shed some light on the physics of discontinuous phase transitions by means of a computer simulation. We consider the two-dimensional one-spin-flip Ising model. Calculations are presented of the equation of state, the decay rate of metastable states and of cluster distributions. The numerical data reveals the existence of long-lived metastable states in this model, and supports the prediction of the droplet model, according to which the phase boundary is a line of essential singularities. It does not support, however, the existence of a spinodal, as predicted by the mean-field approximation, or by Gaunt and Baker, using numerical series expansions. Finally, we present snapshots of the temporal development of spin configurations during a discontinuous phase transition, indicating how spin clusters form, coalesce and disintegrate, and how the nucleation process takes place. A brief discussion of the nucleation theory and its relationship to the motion picture is also given.

I. INTRODUCTION

The classical theory of the liquid-vapor phase transition represents the transition point as the point of intersection of two analytic functions (1), each corresponding to the Gibbs free energy of one of the phases. The analyticity guarantees that the system may go over, instead of making the appropriate phase transition, continuously into a one-phase state, called metastable state (Fig.1).

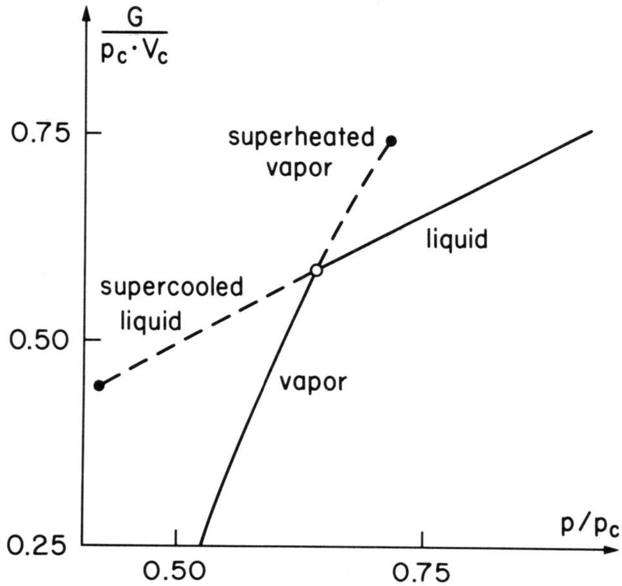

Figure 1. Gibbs free energy as a function of pressure at $T/T_c = 0.9$ according to the van der Waals equation of state.

This metastable phase can only be continued in a region in which the individual phase is thermodynamically stable. In a one-component system it is traditionally supposed that as such a metastable isotherm is followed, the compressibility will increase, until it becomes infinite at some point, which then locates the so-called spinodal (Figure 2). An infinity of the compressibility implies mechanical instability and further prolongation of the isotherm would necessarily result in an immediate collapse into separated phases.

This pattern is consistent with the van der Waals equation and is also supported, as the existence of metastable states is concerned, by experimental results. In fact nature provides many examples of metastable states; they include supercooled vapors and liquids, supersaturated solutions, superheated He^3(Reference (2)), ferromagnets in the part of the hysteresis loop where the magnetization and applied field are opposite in direction, and diamond.

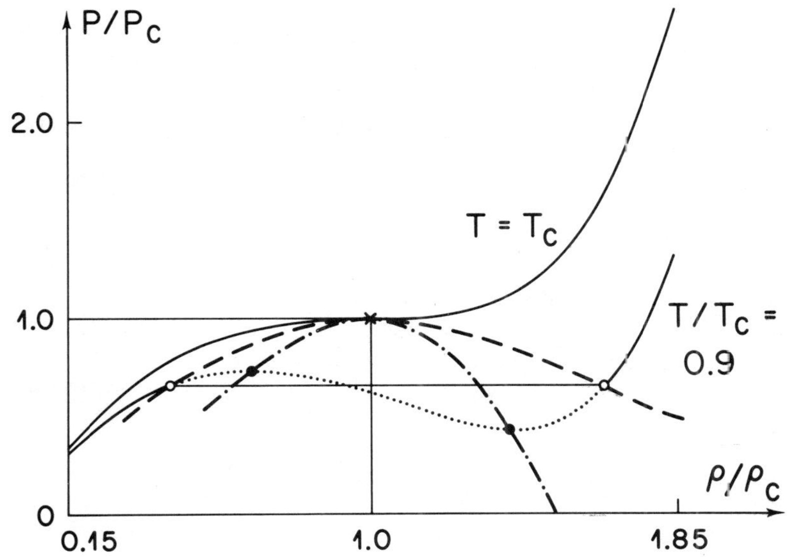

Figure 2. Van der Waals equation of state: —— isotherm; ---- phase boundary; -·-·- spinodal.

Nevertheless metastable states represent an extrapolation into regions of the phase diagram where the stable thermodynamic state is one of inhomogeneous two-phase equilibrium. As a consequence metastable states are not equilibrium states and have to evolve, either through external disturbances or spontaneous nucleation of the other phase, towards the new stable equilibrium (3).

In view of this it becomes evident that the significance of metastable states and in particular of the spinodal curve heavily depends on the lifetime for the decay of a metastable state (4). Moreover, Fisher (5) suggested, on the basis of the droplet model, that for short-range forces the transition point on the liquid-vapor phase boundary should correspond to an essential singularity of the Gibbs energy. The existence of such a singularity has a marked effect on the description of discontinuous phase transitions (5)-(7) since no real analytical continuation is possible and as a consequence there is no prescription for calculating metastable isotherms and spinodal curves. So far, however, a rigorous proof of the presence or absence of this singularity on the phase boundary in some Ising models is still missing. Nevertheless, Fisher's suggestion is supported by negative results of Lanford and

Ruelle (8), tending to prove that metastable states, as close analogues or analytic continuation of equilibrium states, cannot exist.

Nevertheless, the droplet model also offers an explanation for the existence of long-lived metastable states. These states are understood in terms of the extremely small rate of formation of nuclei of the missing phase, unless substantial supercooling has taken place. Clearly, the partition function without these nuclei configurations has no singularity and can be continued into the metastable region.

In contrast to the prediction of the droplet model Gaunt and Baker (9), using numerical series expansions, found no evidence for an essential singularity on the phase boundary in the Ising model and located by assuming their existence, a spinodal curve. A metastable continuation of the isotherms is also found in the linear model equation of state (10), (11). In this model, however, extrapolation is possible only until the susceptibility has a cusp (11). Similarly, Monte-Carlo simulations (12) established the existence of long-lived metastable states in finite square Ising systems with free boundaries.

Apart from these questions concerning static properties, the dynamics, including the nucleation process and decay of the metastable states also attracted considerable attention. The idea of nucleation was developed by Becker and Döring (13), who used quasi-thermodynamic arguments to find the smallest "liquid droplet" in a supersaturated vapor that will "grow" to become the new liquid phase. The probability of forming such a critical droplet is found to be extremely small (for not too great supercooling), which explains the metastability of supercooled vapor. The basic ideas of this approach have been used rather successfully by many authors for a variety of phenomena (14)-(17). Despite their success, however, these attempts are empirical in character and suffer from the lack of precise formulation of the problem (3).

It is the purpose of this work to shed some light on the physics of discontinuous phase transitions, which, so far, has either been empirical or conflicting. In this situation it is natural to study a model system, simple enough to obtain some fundamental knowledge of the nonequilibrium phenomena associated with discontinuous phase transitions. The type of model to be taken up in the following, which was initiated by Glauber, answers this need (18). This is the so-called two-dimensional one-spin-flip Ising model. The subject matter concentrates almost exlusively upon phenomena associated with our motion picture of the same title.

Section II is designed to provide a brief review of the Monte-Carlo method as a tool to investigate the kinetic Ising model. Section III concerns a discussion of the calculated equation of state, the decay rate of metastable states and the numerical evidence for the applicability of dynamic scaling to nonequilibrium phenomena. Section IV is devoted to the discussion of the calculated cluster distributions in the light of the droplet model. The concept of clusters of reversed spins distributed over the background of up-spins, leads then to section V, providing a short summary of the motion picture. A brief discussion of the nucleation theory and its relationship to the motion picture is presented in section VI.

II. THE KINETIC ISING MODEL AND THE MONTE-CARLO TECHNIQUE

To introduce dynamic processes the assembly of Ising spins with Hamiltonian

$$\mathcal{H} = - J \sum_{<i,j>} \mu_i \mu_j - \mu_B H \sum_i \mu_i \ , \qquad \mu_i = \pm 1 \tag{1}$$

is thought to be in contact with a heat bath which induces random flips of the spins from one state to another. $<ij>$ indicates a sum over the nearest-neighbor pairs of sites on the lattice. In the original Glauber model, considered here, only one spin is permitted to flip at a time. To formulate this model, we introduce the probability $P(\mu_1 \ldots, \mu_N)$ of finding the Ising spins in the configuration $\{\mu_1, \ldots, \mu_N\}$ and the transition probability $W_j(\mu_j \to - \mu_j)$ that the j-th spin flips from μ_j to $-\mu_j$. The time dependence of P is assumed to be governed by the master equations (18)-(20)

$$\frac{d}{dt} P(\mu_1,\ldots,\mu_N;t) = \sum_{j=1}^{N} W(-\mu_j \to \mu_j)\, P(\mu_1,\ldots,-\mu_j,\ldots \mu_N;t)$$

$$- \sum_{j=1}^{N} W(\mu_j \to - \mu_j)\, P(\mu_1,\ldots,\mu_j,\ldots,\mu_N;t) \ . \tag{2}$$

The first sum corresponds to the total number of ways that the system can flip into the state $\{\mu_1,\ldots,\mu_N\}$, whereas the second sum corresponds to the total number of ways that the system can flip out of the state $\{\mu_1,\ldots,\mu_N\}$. In equilibrium the left-hand side of equation (1) is equal to zero. A stronger condition is the principle of detailed balance which asserts that

$$W(-\mu_j \to \mu_j)\, P_0(\mu_1,\ldots,-\mu_j,\ldots,\mu_N) = W(\mu_j \to - \mu_j)$$

$$P_0(\mu_1,\ldots,\mu_j,\ldots \mu_N) \ . \tag{3}$$

$P_0(\mu_1,\ldots,\mu_N)$ denotes the probability of finding the Ising spins in the configuration $\{\mu_1,\ldots,\mu_N\}$ when the system is in equilibrium. Observing that

$$P_0(\mu_1,\ldots,\mu_N) \sim e^{-\beta\mathcal{H}}, \beta = \frac{1}{k_B T} \tag{4}$$

equation (2) leads to

$$\frac{W(\mu_j \rightarrow -\mu_j)}{W(-\mu_j \rightarrow \mu_j)} = \frac{e^{-\beta E_j \mu_j}}{e^{\beta E_j \mu_j}} = \frac{1 - \mu_j}{1 + \mu_j} \frac{\tanh \beta E_j}{\tanh \beta E_j} . \tag{5}$$

The local field E_j is defined by

$$E_j = \mu_B H + \sum_i J_{ij} \mu_j . \tag{6}$$

Following Suzuki and Kubo (20) one might choose for W_j a form consistent with equation (5),

$$W(\mu_j \rightarrow - \mu_j) = \frac{\alpha}{2} (1 - \mu_j \tanh \beta E_j) . \tag{7}$$

The parameter α is the inverse relaxation time of a free Ising spin interacting with the heat bath, and determines the time scale of the dynamic processes. As a consequence the time scale is determined only within a factor α . Using equations (2) and (3) and the definition of the expectation value,

$$\langle\mu_j\rangle = \sum_{\{\mu\}} \mu_j P(\mu_1,\ldots,\mu_N; t) , \tag{8}$$

where the sum is taken over all possible configurations, equations of motion (12)

$$\frac{1}{\alpha}\frac{d}{dt}\langle\mu_j\rangle = -[\langle\mu_j\rangle - \langle\tanh(\beta E_j)\rangle]$$

$$\frac{1}{\alpha}\frac{d}{dt}\langle\mu_j\mu_k\rangle = - 2\langle\mu_j\mu_k\rangle + \langle\mu_j \tanh \beta E_k\rangle + \langle\mu_k \tanh \beta E_j\rangle \tag{9}$$

can be derived. Thus the computational problem is reduced to that of solving a hierarchy of differential equations subject to certain initial conditions. This has been done only for the case of one-dimensional systems by Glauber (18). For systems with higher dimensionality one must make approximations, or one has to solve the problem numerically, in order to get explicit predictions. Since the approximate methods available are rather unsatisfactory, one is left with numerical techniques using digital computers. There are two such approaches, namely, computer simulation (12), (21)-(24) and the series-expansion techniques (25),(26). The latter method, however, is suited only to estimate the static and dynamic phenomena occurring at second-order phase transitions. For discontinuous transitions and non-equilibrium phenomena, such as metastability and nucleation, computer simulations seem to be the only technique which makes it possible to obtain approximations of the exact solutions.

The simulation technique which we adopted is the so-called Monte-Carlo method (27). Let us now sketch some of the essential features of this technique. The basic quantity is the transition probability equation (7). Instead of equation (7) we used the expression

$$W(\mu_j \to -\mu_j) = \begin{cases} \alpha' \exp(2\beta\mu_j E_j), & \text{if } 2\beta\mu_j E_j < 0 \\ \alpha', & \text{otherwise} \end{cases} \quad . \tag{10}$$

This choice is of course also consistent with the condition of detailed balance (Equation (3)).

The method of simulation is as follows: one starts with the input data, such as nearest-neighbor interaction J, external field H, temperature T, number of spins N and an initial Ising spin configuration (see figure 3). Then one chooses a spin by a random number and calculates the transition probability for flipping. If $W_j/\alpha' = 1$ the spin is flipped. The process is continued by choosing the next spin again by random number. If $W_j/\alpha' < 1$ the j-th spin is flipped only if W_j/α' equals or exceeds a random number between 0 and 1. However, if W_j/α' is smaller than this random number the spin is not flipped. Again the process is continued by choosing another spin by random number. In this manner a sequence of N spin configurations is generated. For further applications one saves every N-th configuration generated by these Monte-Carlo steps, where N is the number of spins. The whole procedure is then repeated M times. Obviously, there is a correspondence between the time lapse and the sequence of saved configurations. Thus to describe the evolution of the system we may use a parameter t, called the time which takes on the sequential values $t_k = k$; k denoting the k-th saved configuration. On this basis, we may then define time-dependent averages, such as the

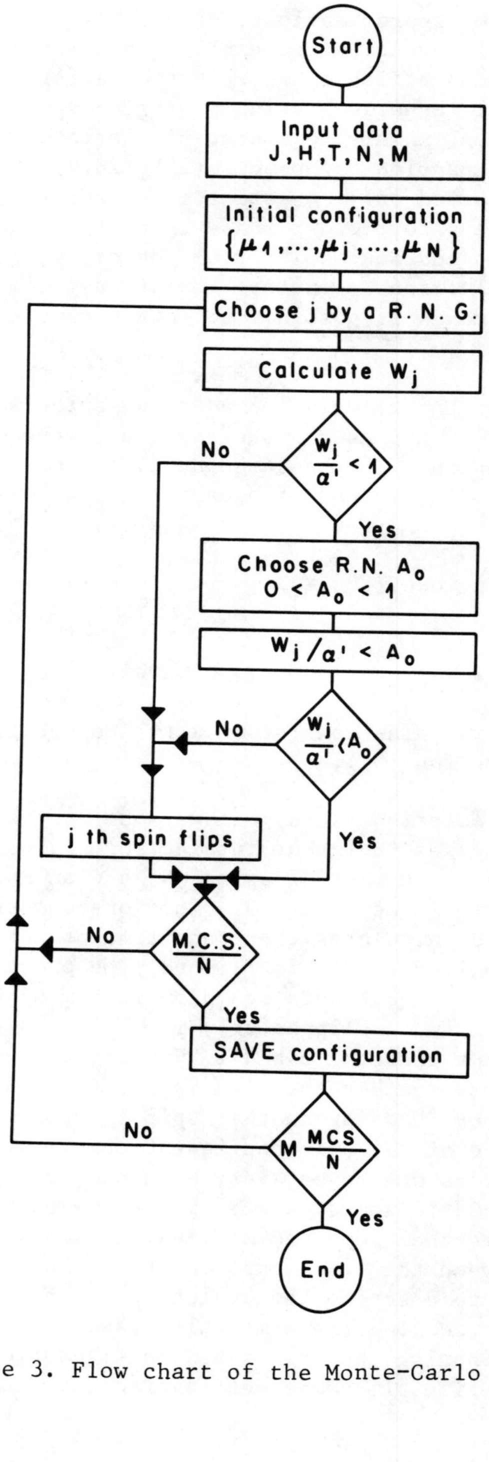

Figure 3. Flow chart of the Monte-Carlo process.

magnetization

$$\bar{\mu}(t) = \frac{1}{N} \sum_{i=1}^{N} \mu_i(t) = \frac{1}{N} \sum_{i=1}^{N} \mu_i(k) \quad , \tag{11}$$

where k again denotes the k-th configuration in the chain of saved configurations. A characteristic behavior of $\bar{\mu}(t)$ is shown in figure 4 for $T < T_c$.

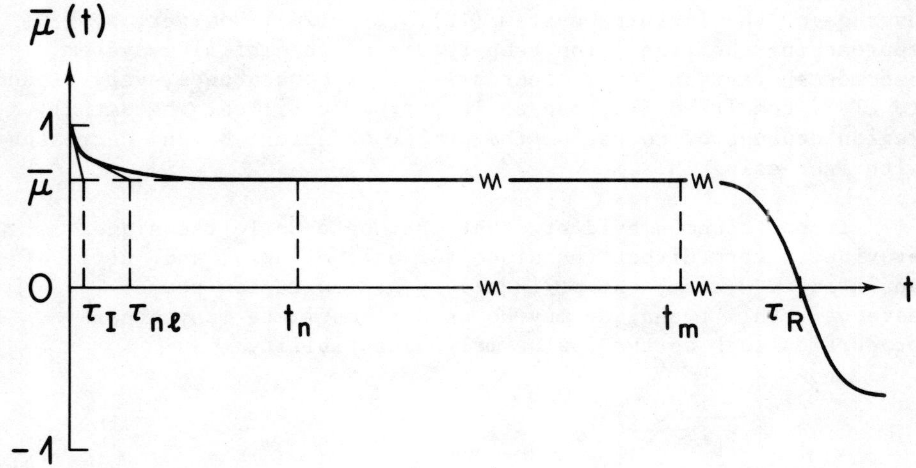

Figure 4. Schematic sketch of the time evolution of the average magnetization $\bar{\mu}(t)$. The initial state is the completely ordered ferromagnet. Three time intervals may be distinguished: 1) The initial decay is noncritical (τ_I); 2) the decay to "equilibrium" which becomes critical ($\tau_{n\ell}$); 3) this "equilibrium" state is a metastable state with lifetime τ_R. Monte-Carlo averages are taken in the interval $t_n \gg \tau_{n\ell}$ and $t_m \ll \tau_R$.

In the initial stage the system relaxes rapidly from the nonequilibrium initial configuration towards "equilibrium" (τ_I). In a second interval the relaxation becomes slower and the system develops towards a metastable state. This metastable state may be long-lived (τ_R) so that time averages of the type

$$\bar{\bar{\mu}} = \frac{1}{t_m - t_n} \int_{t_n}^{t_m} \bar{\mu}(t)\, dt \tag{12}$$

become meaningful quantities. However, at some time τ_R the system may also undergo a "discontinuous transition". This behavior expresses the fact that in any finite system at zero field the canonical ensemble average of μ vanishes $(\langle\mu\rangle = 0)$. Nevertheless, as long as

$$\tau_R >> t_m - t_n , \quad (T < T_c) \tag{13}$$

and the interval $t_m - t_n$ permits reasonable time averages, $\bar{\bar{\mu}}$ should provide a reliable estimate for the canonical ensemble average of the infinite system (12), (21)-(24). However, by approaching the transition temperature of the infinite system τ_R becomes shorter and $\tau_{n\ell}$ increases. As a consequence, very close to T_c , condition (13) can no longer be fulfilled. The actual region depends of course on the number of spins N and decreases with increasing N.

It now becomes evident, that the Monte-Carlo technique provides a very direct technique for estimating $\bar{\bar{\mu}}$ and $\bar{\mu}(t)$ of an infinite system, except in a very narrow region round T_c. Moreover, this technique may be used to estimate many other properties such as the isothermal susceptibility

$$\begin{aligned}\chi_{\delta\mu\delta\mu} &= \frac{1}{k_B TN} \langle (\sum_i \mu_i - \langle\mu_i\rangle)^2 \\ &= \frac{N}{k_B T} \frac{1}{t_m - t_n} \int_{t_n}^{t_m} [\overline{\mu(t)} - \bar{\bar{\mu}}]^2 \, dt ,\end{aligned} \tag{14}$$

and correlation functions, such as

$$\hat{\Phi}_{\delta\mu\delta\mu}(t) = \frac{N}{k_B T\chi_{\delta\mu\delta\mu}} \frac{1}{t_m - t - t_n} \int_{t_n}^{t_m - t} [\bar{\mu}(t') - \bar{\bar{\mu}}][\bar{\mu}(t+t') - \bar{\bar{\mu}}] \, dt' . \tag{15}$$

In figure 5 we show as an example the calculated temperature dependence of the isothermal susceptibility for a square 55 × 55 lattice (24). It should be emphasized that the susceptibility is a second derivative of the free energy, and therefore harder to calculate accurately than first derivatives, such as magnetization. In view of this, agreement between simulated and "exact" results is satisfactory and reveals again the possibility to estimate critical exponents of static quantities with the Monte-Carlo method (12), (22)-(24).

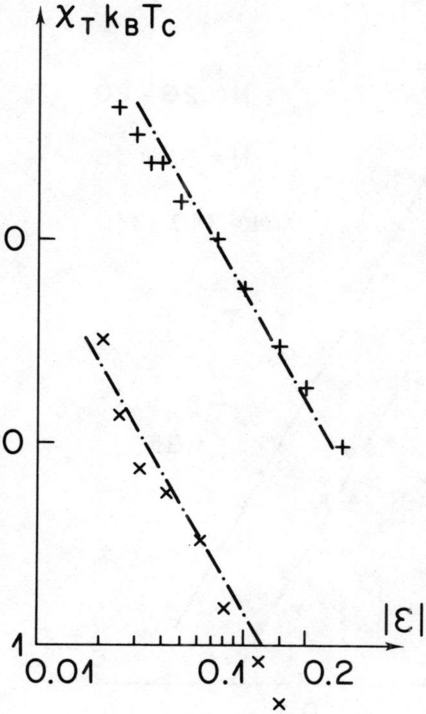

Figure 5. Computed temperature dependence of the isothermal susceptibility for a 55 × 55 square lattice with nearest-neighbor interactions subjected to periodic boundary conditions. +: $T > T_c$, X: $T < T_c$ (24). The broken lines correspond to the asymptotic behavior obtained from series expansions and exact solutions, respectively.

In figure 6 we show as an example the calculated temperature dependence of the relaxation time

$$\tau_{\delta\mu\delta\mu} = \int_0^\infty \hat{\Phi}_{\delta\mu\delta\mu}(t)\, dt \tag{16}$$

characterizing the critical slowing-down of the order-parameter fluctuations. This relaxation time is expected to diverge as

$$\tau_{\delta\mu\delta\mu} \sim \begin{cases} (-\varepsilon)^{-\Delta'_{\delta\mu\delta\mu}} & \text{, for } T_c - T \to 0^+ \\ (\varepsilon)^{-\Delta_{\delta\mu\delta\mu}} & \text{, for } T - T_c \to 0^+ \end{cases} , \tag{17}$$

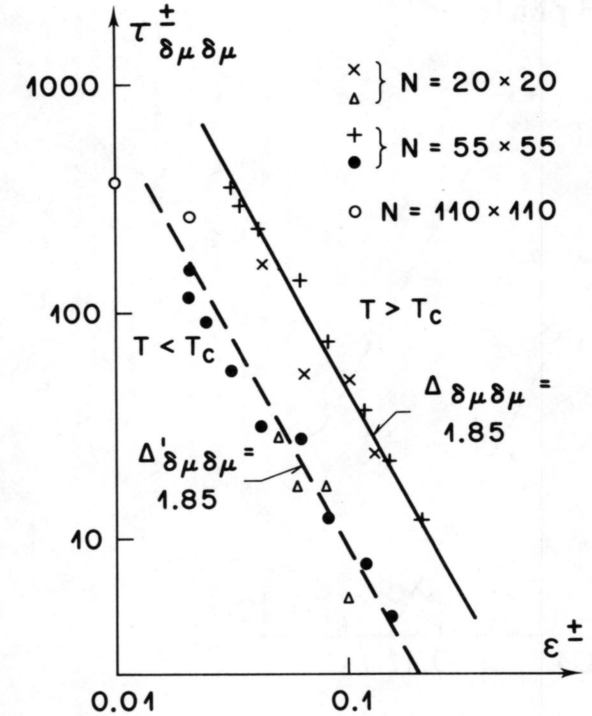

Figure 6. Calculated temperature dependence of the relaxation time $\tau_{\delta\mu\delta\mu}$ (Equation (16)) associated with the order-parameter correlation function (24). The slopes by the full and broken lines, respectively, correspond to $\Delta_{\delta\mu\delta\mu} = 1.85$.

where

$$\epsilon = \frac{T_c - T}{T_c}$$

The data shown in figure 6, taken from reference (24) leads to the estimate

$$\Delta'_{\delta\mu\delta\mu} \approx \Delta_{\delta\mu\delta\mu} \approx 1.85 \pm 0.10 \tag{18}$$

which is consistent with the exact inequalities (28)

$$\begin{aligned} \Delta'_{\delta\mu\delta\mu} &\geq \gamma' = 7/4 \\ \Delta_{\delta\mu\delta\mu} &\geq \gamma = 7/4 \end{aligned} \tag{19}$$

and the estimates as obtained from a high-temperature expansion method (25), (26) ($\Delta_{\delta\mu\delta\mu} \sim 2$) and an extension of Wilson's expansion methods ($\Delta_{\delta\mu\delta\mu} = 2$) (29).

III. THE EQUATION OF STATE AND DECAY RATE OF METASTABLE STATES

Of particular interest in the context of first-order transitions is the equation of state, that is, the mean magnetization as a function of field at constant temperature. A schematic sketch of the equation of state is shown in figure 7(a). At H = 0 an infinite system with interactions of finite range is expected to undergo an abrupt transformation from a spin-up to a spin-down state. It is important to note that the isothermal susceptibility

$$\mu_B \chi_T = \left.\frac{\partial \bar{\bar{\mu}}}{\partial H}\right|_T = \left.\frac{\partial \langle\mu\rangle}{\partial H}\right|_T \tag{20}$$

should remain finite as the transition point is approached (5). In finite systems, considered here, there will be a rounding effect in an exact treatment, as shown in figure 7(a). Nature, however, provides many examples of a "metastable" continuation of the isotherms. In fact, instead of making the appropriate transition, the system may go over continuously into a one-phase state, called the metastable state, which may have a very long lifetime. Even though the resulting metastable isotherm might be quite reproducible, the system is no longer in a state of true thermodynamic equilibrium. In fact, eventually, either through external disturbance or spontaneous fluctuations which nucleate the phase with lower free energy (Figure 7(b)), the system begins an irreversible process which leads to the new equilibrium phase. The irreversibility of this first-order transition is associated with a decrease in free energy.

Next we turn to some specific results of our computer simulation. In this simulation we considered the single-spin-flip Ising model consisting of a 220 × 220 square lattice subjected to periodic boundary conditions (p.b.c.). The use of p.b.c. is an advantage if one is interested in the simulation of an infinite system (12), (22)-(24).

Figure 8(a) shows some results of the equation of state at $T/T_c = 0.958$ and subjected to p.b.c.. A significant fact is the appearance of rather well-defined metastable states, such as spin-up states at negative field, which are metastable with respect to the corresponding spin-down states.

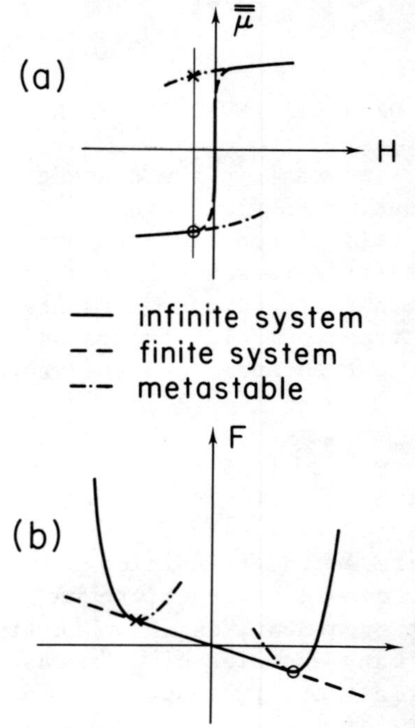

Figure 7(a) Schematic sketch of the equation of state, for a finite and infinite system, respectively, including metastable states. (b) Free energy per spin as a function of average spin $\bar{\mu}$.

The decay rate $1/\tau_R$ of the metastable states has been computed for intervals up to 10^4 Monte-Carlo steps per spin and is plotted in figure 8(b). An interesting feature of these results is the fact that the decay rate decreases with decreasing field in such a way, that for small negative fields the lifetime becomes so large that it exceeds reasonable computing times. As a consequence $\bar{\mu}(t)$ exhibits an extended flat region (Figure 4), so that time averages of the type equation (17) become meaningful quantities, even for $\bar{\bar{\mu}} > 0, H < 0$ and $\bar{\bar{\mu}} < 0, H > 0$, respectively. This characterization of metastable states is consistent with that given by Penrose and Lebowitz (3). These authors call a state metastable if

1) one thermodynamic phase is present;
2) a system that starts in this state is likely to take a long time to get out;

3) once the system has got out it is unlikely to return.

In fact we only made the meaning of "long time to get cut" more precise in the sense that time averages of the type equation (17) can be performed to characterize the metastable one-phase state.

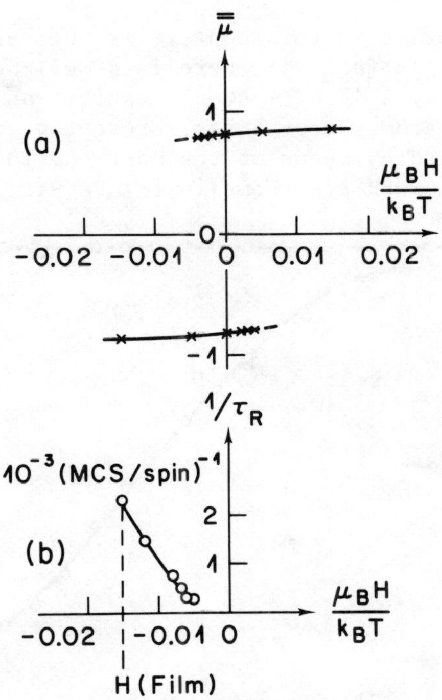

Figure 8(a) Computed equation of state for a 220 × 220 square lattice with nearest-neighbor interactions subjected to periodic boundary conditions, at $T/T_c = 0.958$. (b) Computed decay rate of the metastable states for the system defined above.

As an alternative measure of the lifetime of metastable states, one might also introduce the concept of the non-equilibrium relaxation times (30). For this purpose we assume that at $t = -\infty$ the system is in thermal equilibrium. At $t = 0$ a sudden change of the temperature ($T \rightarrow T + \Delta T$) and of the applied field ($H \rightarrow H + \Delta H$) is performed. The resulting relaxation to the new equilibrium or metastable state is described by the relaxation function

$$\Phi_{\mu}^{\Delta T \Delta H}(t) = \frac{\bar{\mu}(t) - \bar{\mu}(t_n)}{\bar{\mu}(0) - \mu(t_n)} \tag{21}$$

and the associated relaxation time is

$$\tau_{\mu}^{\Delta T \Delta H} = \int_{o}^{t_n} dt \, \Phi_{\mu}^{\Delta T \Delta H}(t) \quad . \tag{22}$$

Of special interest in this context is the case where $\Delta T = 0$ and $\Delta H < 0$ $(H = 0)$. In fact, if there is a well-defined metastable state at $\Delta H < 0$, $\Phi_{\mu}^{0,\Delta H}(t)$ should exhibit an extended flat part, so that $\tau_{\mu}^{0,\Delta H}$ becomes very large. Presently, these quantities have been studied by means of the Monte-Carlo method (30). Some results of that work are shown in figure 9.

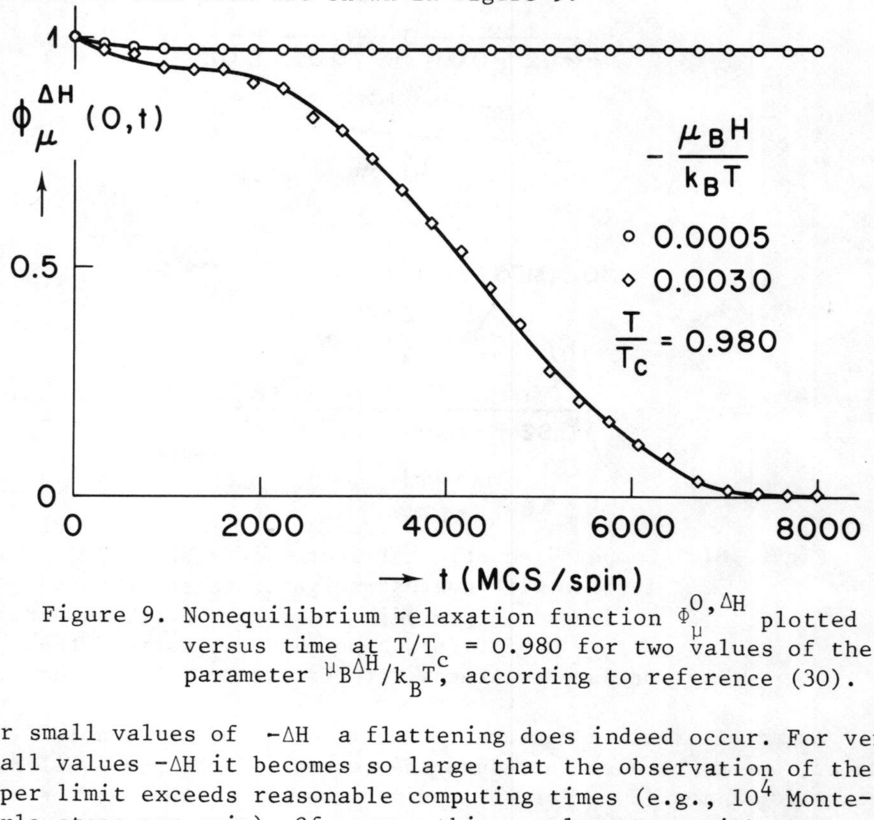

Figure 9. Nonequilibrium relaxation function $\Phi_{\mu}^{0,\Delta H}$ plotted versus time at $T/T_c = 0.980$ for two values of the parameter $\mu_B \Delta H / k_B T$, according to reference (30).

For small values of $-\Delta H$ a flattening does indeed occur. For very small values $-\Delta H$ it becomes so large that the observation of the upper limit exceeds reasonable computing times (e.g., 10^4 Monte-Carlo steps per spin). Of course this result agrees with those shown in figure 8(b), in the sense that the lifetime of the metastable states becomes extremely large for small negative fields.

Another interesting result of this Monte-Carlo investigation is the scaling property of the relaxation time $\tau_{\mu}^{0,\Delta H}$. Plotting

$$\frac{(-\varepsilon)^{-\Delta_{\delta\mu\delta\mu}}}{\tau_{\mu}^{0,\Delta H}} \quad \text{versus} \quad \frac{(\Delta H)\ \mu_B}{k_B T (-\varepsilon)^{\beta\delta}} \ , \tag{23}$$

where

$$-\ \varepsilon = \frac{T_c - T}{T_c} \ , \ \beta = \frac{1}{8} \ , \ \delta = 15 \ , \ \Delta_{\delta\mu\delta\mu} \approx 2 \ , \tag{24}$$

Binder, Stoll and Müller-Krumbhaar (30) found that the numerical data falls on a single curve as required by dynamic scaling, extended to nonequilibrium phenomena. These results are shown in figure 10. We observe that for small negative fields and $T \neq T_c$, τ_R becomes very large, indicating again the existence of well-defined metastable states.

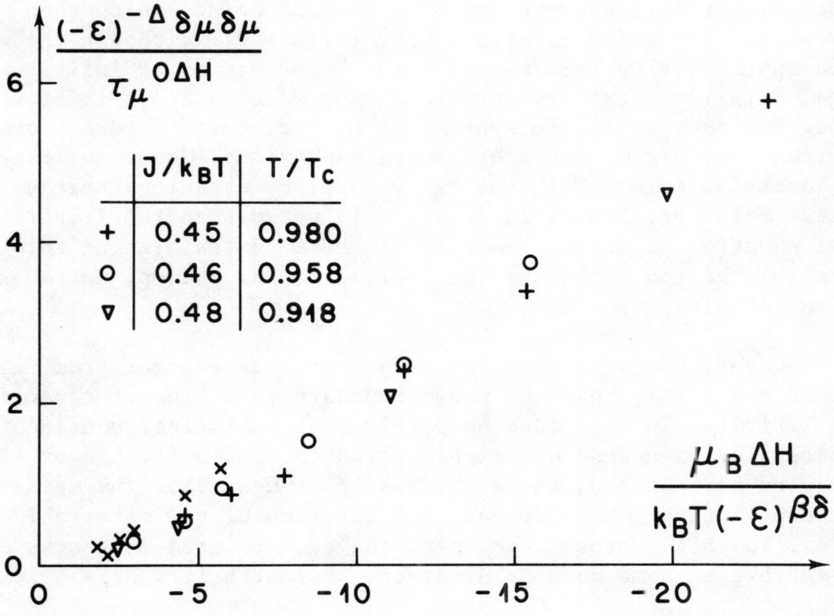

Figure 10. Scaled nonequilibrium relaxation time $\tau_{\mu}^{0,\Delta H}$ versus scaled field (Equation (23)), according to reference (30).

The numerical results so far discussed indicate, in accordance with an earlier investigation in finite systems with free boundaries (12), that in the two-dimensional one-spin-flip Ising model long-lived metastable states exist. In view of the fact that we considered only finite systems, one might ask whether the occurrence of long-lived metastable states is an artifact of finite systems. In systems with free boundaries we found (12) a pronounced dependence on the number of spins N, suggesting that the lifetime of the metastable states increases in the region N = 30 × 30 to N = 110 × 110 with increasing N (12). However, in systems subjected to periodic boundary conditions no N dependence was found in the interval N = 55 × 55 - N = 440 × 440 spins (30). Emphasizing that systems subjected to periodic boundary conditions are more suitable to simulate on infinite systems, we expect no N dependence of the lifetime of metastable states, provided that the system is reasonably large (e.g., $N \geq 55 \times 55$).

Next we discuss the question whether or not the continuation of the isotherm in the metastable region may exhibit some end point. The mean-field approximation predicts, for example, following a metastable isotherm, that the susceptibility should increase, until it becomes infinite at some point, which then locates the so-called spinodal curve (Figure 2). Although the mean-field approximation becomes exact for interactions of infinite range, this limit is very artificial and we should not trust the answer for more realistic systems with short-range interactions. Nevertheless, using numerical series expansions, Gaunt and Baker (9) located a spinodal in the Ising model by assuming their existence. Moreover, Schofield et al. (11) noted that the linear-model equation of state, leads to isotherms extrapolating into the metastable region until the susceptibility has a cusp, but does not become infinite.

However, there is also the conjecture, as emerged from the droplet model (5), that the phase boundary is a line of essential singularities. In this case no purely real analytical continuation is possible. Of course one might introduce, following Langer (31), a complex free energy, where the real part describes the metastable isotherm and the imaginary part the lifetime of the metastable states. In this approach, however, the extrapolated isotherms do not exhibit at some point a divergent susceptibility or a cusp.

Nevertheless, the common feature of these approaches is the possible existence of long-lived metastable states, as observed in nature and in the computer experiments. The distinguishing feature concerns the existence of an end point of the metastable isotherms and its properties.

In figure 11 we show the calculated field dependence of the

susceptibility

$$\chi = k_BT\ (<\mu^2> - <\mu>^2)N \tag{25}$$

in the metastable region. It is seen that χ increases with increasing negative field, as expected from the equation of state (see figure 9). We also included the positions where the linear-model equation of state predicts a cusp (11) and Gaunt and Baker (9) locate a divergent susceptibility, respectively. Although the divergence of the susceptibility is predicted to be visible only in a rather small region (9), it is clear that its existence implies an instability which would necessarily result in an immediate transition into the equilibrium state.

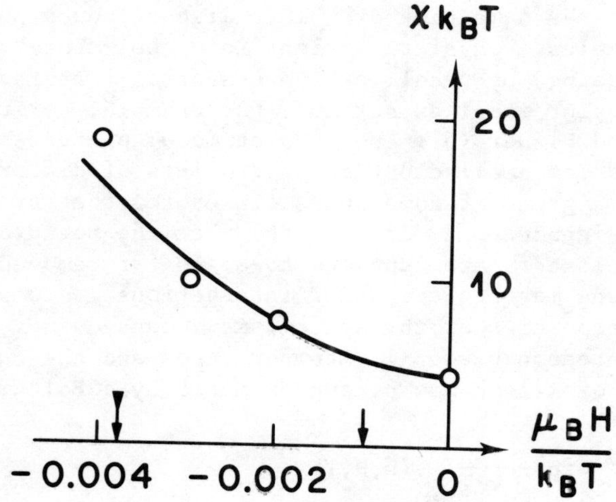

Figure 11. Field dependence of the susceptibility in the metastable region at $T/T_c = 0.958$.

Position where the linear-model equation of state predicts a cusp.

Position of the spinodal located by Gaunt and Baker (9).

Due to the observation of long-lived metastable states at considerably higher negative fields, our data does not support the existence of the spinodal located by Gaunt and Baker (9). Concerning the cusp in the susceptibility predicted by the linear-model equation of state, we note that the occurrence of this cusp may also represent the limit of applicability of the linear model (2).

The preceding discussion revealed the existence of long-lived metastable states in the one-spin-flip Ising model. The numerical data presented does not support the existence of a spinodal, as predicted by the mean-field approximation, or by Gaunt and Baker (9), using numerical series expansions. A further interesting result was the numerical evidence for the applicability of dynamic scaling to nonequilibrium phenomena (30).

IV. THE DROPLET MODEL

When we inquire into the nature of the metastable states and the associated nucleation of the stable states we are faced with the difficulty that no model systems are known which are exactly soluble and at the same time exhibit a discontinuous phase transition. Nevertheless, physical insight into the "cause" of nucleation can be gained by recalling the essential features of the cluster or droplet model as set out, for example, by Frenkel (13), Langer (31) and Fisher (5). The droplet model pictures this system as a rare "gas" of small clusters or droplets of different sizes, consisting of a group of down-spins linked together by at least one nearest-neighbor bond, distributed over the background of up-spins. These clusters are supposed to be so far removed from one another that one may neglect their interactions in computing the equilibrium properties of the system. As a consequence the droplet model is most reasonable at low temperatures and the number of clusters P(n) of size n may be approximated by a Boltzmann factor

$$P(n) = P_0 \exp - \frac{1}{k_B T} \Phi(n,H,T) \quad , \tag{26}$$

where $\Phi(n,H,T)$ is the free energy of a cluster of n down spins. Within this approximation the free energy per spin is given by

$$F(H,T) = - H\mu_B - \frac{1}{2}qJ - \sum_{n=1}^{\infty} P(n,H,T) \quad , \tag{27}$$

q denotes the number of nearest neighbors. $\frac{1}{2}qJ$ is the energy of the fully-aligned state. The crucial assumption of Frenkel's droplet model is that, at least for large enough n, $\Phi(n)$ appearing in equation (26) may be written as

$$\Phi(n) = 2Hn\mu_B + \sqrt{32}\left[\left\{\frac{n}{\pi}\right\}^{1/2} + 1/6\right]2J \quad . \tag{28}$$

Here we assumed a square lattice. The first and second terms describe the "volume energy" and "surface energy", respectively.

Substituting equations (26) and (28) into equation (27) we find for the square lattice (q = 4)

$$F(H,T) = - \mu_B H - 2J - \sum_{n=1}^{\infty} z^n \exp\left[- \frac{1}{k_B T} 2J\sqrt{32}\left(\left\{\frac{n}{\pi}\right\}^{1/2} + 1/6\right)\right] \tag{29}$$

$$z^n = \exp\left\{ - \frac{1}{k_B T} 2H\mu_B n\right\} = \left\{ \exp\left[- \frac{2\mu_B H n}{k_B T}\right]\right\}^n . \tag{30}$$

Note that each term of the sum consists of two factors depending on droplet size: the first, z^n , with exponent proportional to the "volume" of the droplet, and the second with exponent related to the surface. For $z < 1$, the first of these factors is the most important for large droplets. However, even for $z = 1$ $(H = 0)$ the number of very large droplets is still small because of the second factor, the surface term. For any $z > 1$ $(H < 0)$ we have a "catastrophe" in which the sum in equation (29) diverges due to contributions from extremely large droplets of down-spins, indicating that at $z = 1$ $(H = 0)$ the system "knows" it is due to undergo a discontinuous phase transition, to a state where the overwhelming majority of spins are reversed. Note that (5) according to equation (29), $F(H)$ has an essential singularity at the phase boundary $(H = 0)$. This has led to the conjecture that a similar singularity exists on the phase boundary of actual systems.

We now turn to discuss on a qualitative basis the relation between the singularity of $F(H,T)$ on the phase boundary and the condensation mechanism. The quantities $P(n,H,T)$ and $\Phi(n,H,T)$ as a function of n are illustrated in figure 12(a) for a positive and negative value of H for $T << T_c$. For positive H, $P(n)$ decreases rapidly as n increases. But for $H < 0$, $P(n)$ first decreases until n equals

$$n_c = \frac{8}{\pi} \frac{J^2}{(\mu_B H)^2} \tag{31}$$

at which point the "volume energy" $- |H| n\mu_B$ starts to decrease faster than the "surface energy" increases, so that $\Phi(n,H,T)$ is maximum (Figure 12(b)). As a consequence, clusters larger than n_c , the so-called supercritical clusters, will find it energetically more favorable to grow, thus nucleating the missing phase. There is, however, a "free-energy barrier" at $n = n_c$ (Figure 12(b)).

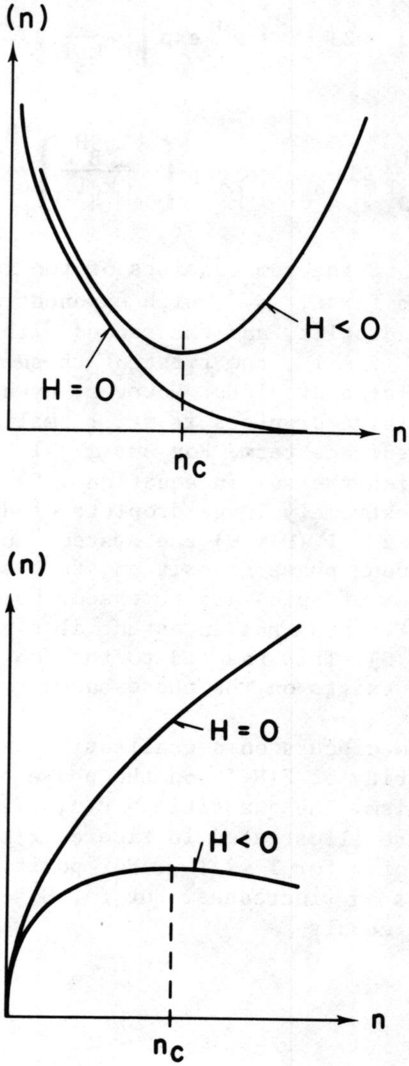

Figure 12. (a) Schematic plot of the cluster distribution P(n) according to the droplet model. (b) Schematic plot of the cluster-free energy of down-spin clusters.

For $n < n_c$ the clusters are in mutual equilibrium, continually associating and dissociating. Only occasional chance fluctuations which will be extremely infrequent when n_c is large will carry the system over the barrier. Nevertheless, even a single cluster with $n > n_c$ will tend to grow and thus constitute a nucleus of the missing phase. We can thus understand that long-lived metastable states are simply due to the fact that for small negative fields, the chance to overcome the free-energy barrier is extremely small. As a consequence, the rate of formation of nuclei of the missing phase is very slow. It should be noted that we consider only the phenomenon of homogeneous nucleation. Here it is assumed that the fluctuation carrying the system over the barrier occurs spontaneously as a thermodynamic fluctuation. The more common but fundamentally less interesting case is known as "inhomogeneous nucleation", and occurs when the disturbance leading to the phase transition is caused by a foreign object or an irregularity of the walls of the container.

We have so far essentially reproduced some of the essential features of the droplet or cluster model of condensation as set out, for example, by Frenkel (13) and Langer (31). This model has been extended by Fisher (5). Fisher started from the high-field expansion of the free energy which is a power expansion in $z = \exp\left\{-\frac{\mu_B H}{k_B T}\right\}$. Fisher's droplet model simulates this expansion by considering the model partition function as a gas of independent clusters of any possible size. Then the model free energy is given by equation (27), where

$$P(n) = q(n)\, z^n \quad , \tag{32}$$

$q(n)$ is the partition function of a cluster of n spins. If the dominant droplets are rather regular in shape, then $q(n)$ can be written as the product of two terms. The first term is a surface-energy contribution, due to the misalignment of the spins at the surface of a cluster. The second factor contained in $q(n)$ is an entropic quantity corresponding to the number of distinct configurations of spins forming a cluster of n spins. Some configurations are shown in figure 13. The product of these factors is assumed to be

$$q(n) = q_o\, e^{-\frac{W}{k_B T} a n^{\sigma}} \cdot n^{-\tau} \lambda^{a n^{\sigma}} \quad . \tag{33}$$

W is the surface energy per spin. σ is expected to have a value close to the geometrical one of spheres

$$\sigma \approx 1 - \frac{1}{d} \ , \tag{34}$$

where d is the dimensionality. τ , λ , a and q_o are model parameters. The second entropic term in equation (33) has been neglected by Frenkel (13) (Equation (29)). Identifying the critical temperature T_c as the temperature at which the microscopic surface free energy

$$k_B T \ln\lambda - W \tag{35}$$

vanishes, one finds (5) as $n \to \infty$,

$$q(n) = q_o \ n^{-\tau} \ \exp - [a \frac{T_c - T}{T_c} n^{\sigma}] \ . \tag{36}$$

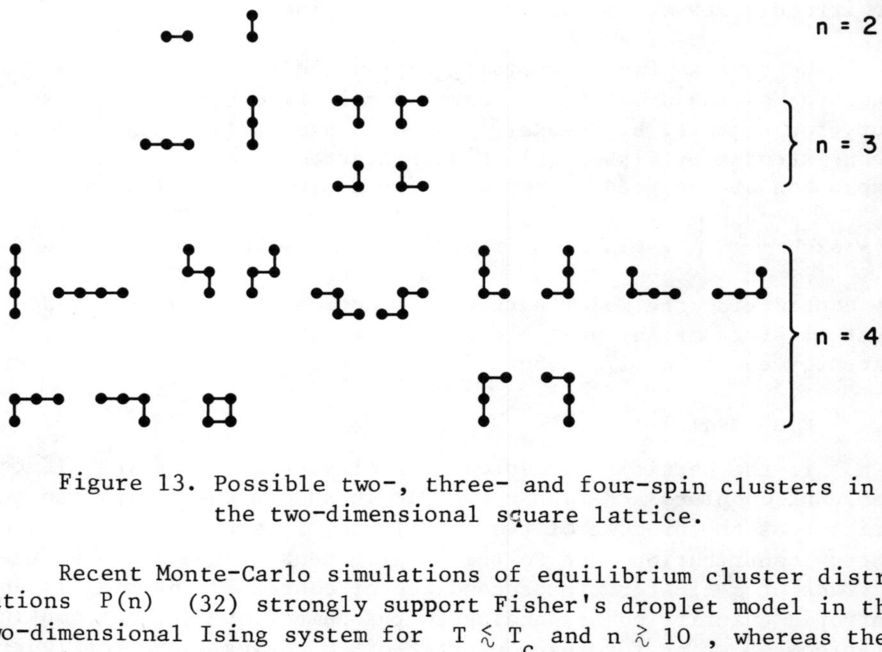

Figure 13. Possible two-, three- and four-spin clusters in the two-dimensional square lattice.

Recent Monte-Carlo simulations of equilibrium cluster distributions $P(n)$ (32) strongly support Fisher's droplet model in the two-dimensional Ising system for $T \lesssim T_c$ and $n \gtrsim 10$, whereas the comparison with Frenkel's formulation revealed considerable discrepancies (12), due to the neglect of the entropic factor. Nevertheless, this neglect does not affect the essential predictions. In fact Fisher's expression for $P(n)$ (Equations (32) and (36)) also leads to an essential singularity at the phase boundary, a critical cluster and the associated free-energy barrier. Fisher's droplet model leads to (5)

$$n_c \propto \left[a_\sigma \frac{J}{2|H|} (-\varepsilon)\right]^{\frac{1}{1-\sigma}} \quad . \tag{37}$$

The primary purpose of reviewing some of the essential features of the droplet model of condensation was to shed some light on the nature of condensation and metastable states. Moreover, the droplet model has the merit that it can be extended to provide a description of the nucleation process. Apart from the prediction that the phase boundary might be a line of essential singularities, it leads to the concept of critical clusters and a free-energy barrier. Even the existence of long-lived metastable states is understood in terms of the extremely small chance to overcome this barrier if the "supercooling" is not too great. This view accords with the description of Pippard (1) of metastable states: it is clear that to the purist the supercooled state is not acceptable as an equilibrium state. Nevertheless, if the supercooling is not too great, the contribution to the Gibbs function of the system by the small number of large droplets present after a reasonable time is quite immeasurably small, so that the equation of state apparently runs smoothly across the phase boundary. Under these circumstances we can afford for thermodynamic purposes to treat the supercooled system as if it were genuinely metastable.

Let us now turn to the numerically calculated cluster distributions. Defining a cluster as a group of reversed spins (-) linked together by nearest-neighbor interactions, we obtain in equilibrium the following expression for the positive magnetization

$$\langle\mu\rangle_+ = \bar{\bar{\mu}}_+ = 1 - 2 \sum_{n=1}^{N} nP(n) \quad . \tag{38}$$

Figure 14 shows some calculated cluster distributions in zero applied field (32) for different temperatures in comparison with Fisher's droplet model (Equations (32) and (36)). Of particular interest is that the number of large clusters increases with increasing temperature and accordingly, with decreasing order parameter. In fact, for $T_c - T \to 0^+$ the cluster distribution tends to the power law

$$P(n) \cdot n^\tau \sim 1 \ , \ \tau = 31/15 \quad , \tag{39}$$

τ denotes a critical exponent.

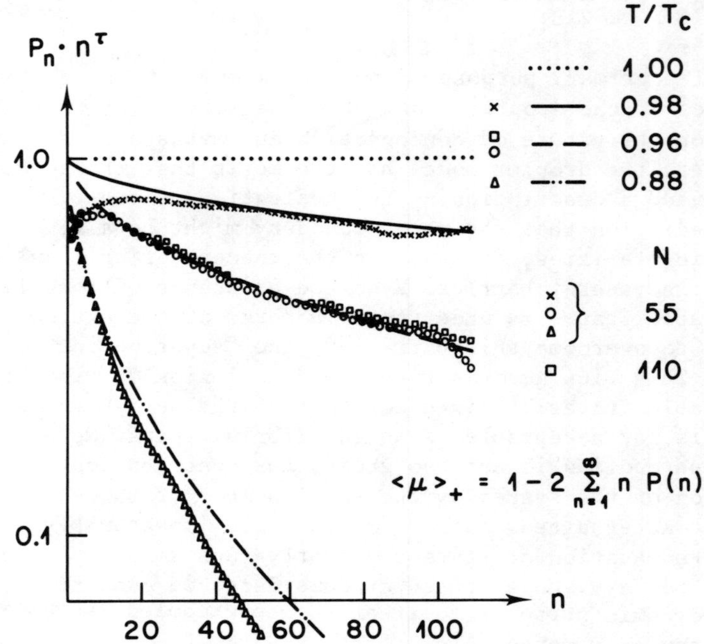

Figure 14. Log $[P(n)n^{\tau}]$ versus n at zero applied field for three different temperatures as taken from reference (32). The curves represent Fisher's droplet model, with $\tau = 31/15$ and $\sigma = 8/15$.

Next we consider the cluster distribution at constant temperature $T < T_c$ for different applied fields. According to the prediction of the droplet model, that the phase boundary is a line of essential singularities, one expects that the system undergoes a discontinuous transition at $H = 0$, provided one waits sufficiently long. In fact, for negative applied fields, $P(n)$ first decreases until n equals the critical size and then starts to increase (see figure 12). As a consequence, the sum

$$\sum_{n=1}^{\infty} n\, P(n) \tag{40}$$

appearing in the expression for the magnetization, is not expected to converge for $H < 0$. The numerical results shown in figure 15 reveal that the number of large clusters increases with decreasing field. This behavior explains the decrease in magnetization with decreasing field (Figure 8(a)). Moreover, this decrease may be interpreted as a precursor of the non-convergence of the series (Equation (40)), signaling the essential singularity on the phase boundary.

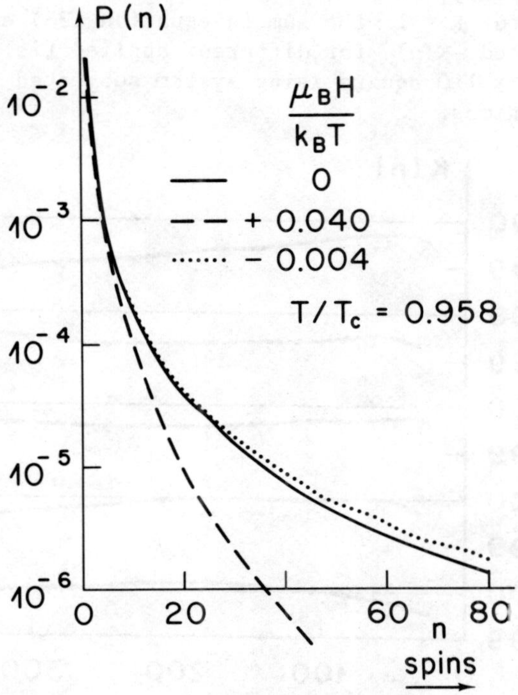

Figure 15. Calculated cluster distribution at $T/T_c = 0.958$ for three different fields with strength $\mu_B H/k_B T = 0, + 0.04, - 0.004$.

To investigate this possible non-convergence, we might consider the criterion of d'Alembert, stating that $\sum_{n=1}^{\infty} n\, P(n)$ converges if

$$G(n) = \frac{(n+1)}{n}\,\frac{P(n+1)}{P(n)} \leq \rho < 1 \tag{41}$$

for $n \geq m$, where m can be chosen arbitrarily large. Instead of equation (41), we consider for the sake of convenience, the convergence parameter

$$K(n) = \frac{(n+1)^2}{n^2}\,\frac{P(n+1)}{P(n)} \leq K \tag{42}$$

satisfying

$$K(n) \geq G(n) \quad . \tag{43}$$

Obviously, if $K < 1$, d'Alembert's criterion is satisfied. Moreover, for $K < 1$ the sum in equation (38) exists. In figure 16 we plotted $K(n)$ for different applied fields, as obtained from a 110 × 110 square Ising system subjected to periodic boundary conditions.

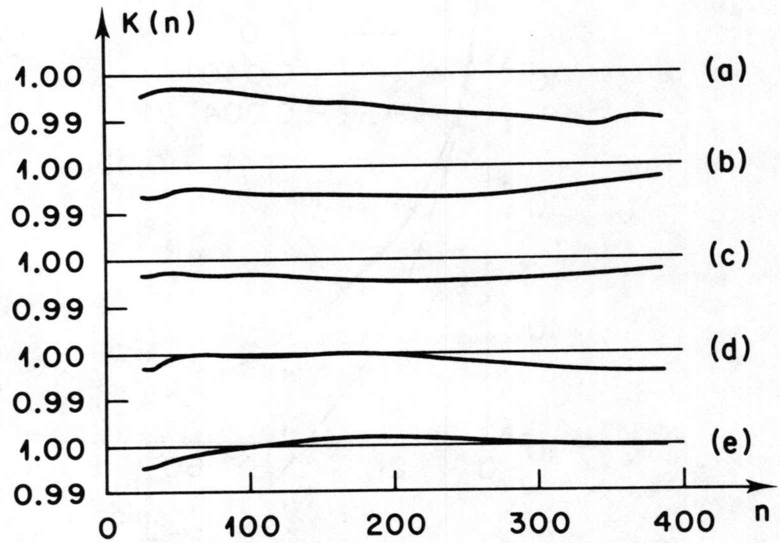

Figure 16. Convergence parameter K(n) versus n (Equation (42)) at $T/T_c = 0.980$ for different applied fields.

	$\frac{\mu_B H}{k_B T}$
(a)	0.001
(b)	0.0005
(c)	0
(d)	- 0.0005
(e)	- 0.001

It is seen that for positive fields, including zero applied field, K(n) satisfies

$$K(n) \leq K < 1 \tag{44}$$

for any n in the range n = 20 - 400. Consequently, d'Alembert's criterion is satisfied in this range and the existence of the order parameter $\bar{\bar{\mu}}$ (Equation (38)) is guaranteed. At negative fields, as figures 16(d) and 16(e) indicate, the convergence becomes worse with increasing field. This behavior is consistent with that of the corresponding nonequilibrium relaxation function (Figure 9). In fact, according to the results shown in figure 9 one expects between $\mu_B H/k_B T = -0.0005$ and -0.003 a changeover from long-lived to shorter-lived metastable states. Figure 16(e) indicates that this changeover takes place around $\mu_B H/k_B T \approx -0.001$. Here, the convergence of the series in equation (38) is no longer guaranteed. As a consequence, the value of $\bar{\bar{\mu}}$ of the corresponding metastable state strongly depends on the value of the upper limit of the summation (Equation (38)).

This analysis and the distinct behavior of the calculated K(n) with decreasing field and in particular at negative fields (Figure 16) clearly support the conjecture that the phase boundary is a line of singularities. Of course, our numerical results do not allow to distinguish between an essential singularity and a pole.

V. DESCRIPTION OF THE MOTION PICTURE

"A picture says more than thousand words"

In the spirit of this Chinese proverb we prepared a motion picture, demonstrating the temporal development of spin configurations during a discontinuous phase transition in the two-dimensional kinetic Ising model. The film shows in detail how spin clusters form, coalesce and disintegrate, and how the nucleation process takes place.

Observation of this film demonstrates the following features:

a) We start from the initial configuration $\bar{\mu}(t = 0) = +1$, at zero external field (Figure 17). Due to the interaction with the heat bath with temperature $T/T_c = 0.958$, the 220×220 square Ising system, subjected to periodic boundary conditions, relaxes very fast towards thermal equilibrium with mean magnetization $\bar{\bar{\mu}} = 0.81$.

b) In this equilibrium state, small clusters of down-spins are distributed over the background of up-spins. They reach a relatively small size, decay and appear again at a different location, without a pronounced tendency for further growth (Figure 18). As a consequence large clusters of the other phase have a rather small chance of occurring (Figure 19).

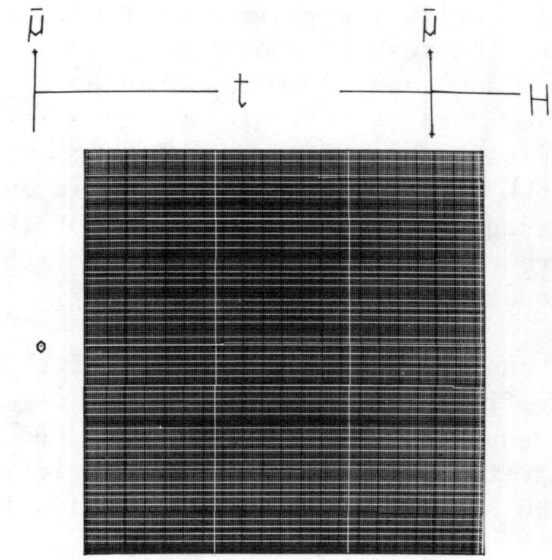

Figure 17. Initial configuration $\bar{\mu}(t=0) = +1$ at $H = 0$.

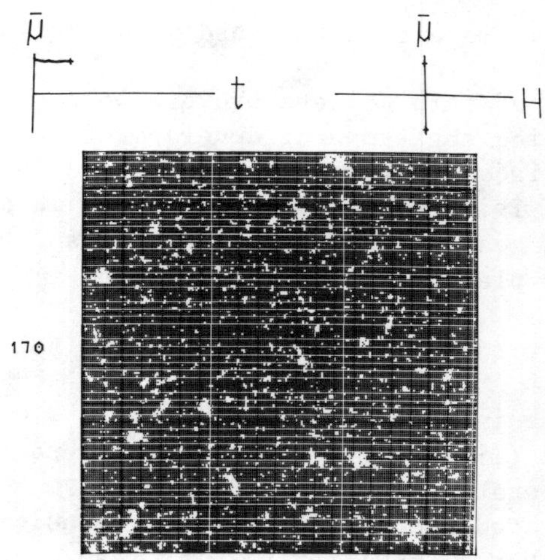

Figure 18. Snapshot of an equilibrium spin configuration at $H = 0$ and $T/T_c = 0.958$; small spin-down clusters are visible.

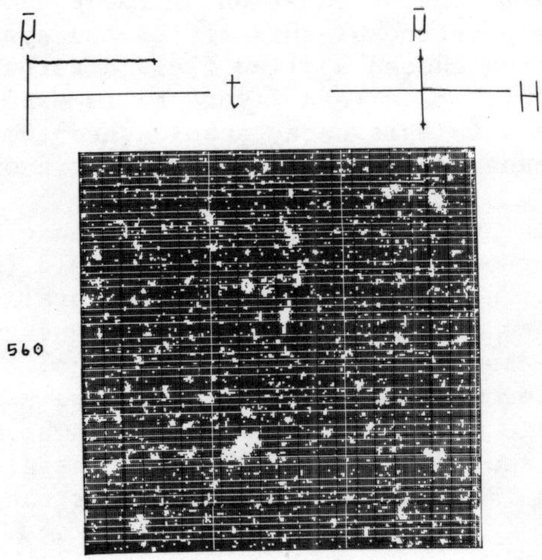

Figure 19. Snapshot of an equilibrium configuration in which, by chance, a relatively large spin-down cluster occurs.

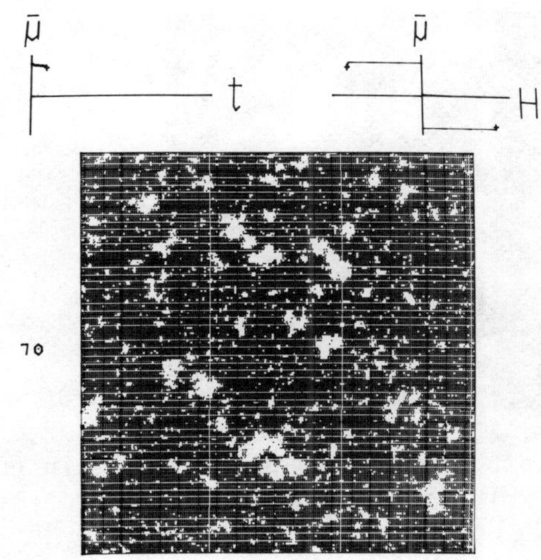

Figure 20. Snapshot of a nonequilibrium spin configuration in the presence of an applied field $\mu_B H/k_B T = -0.015$, which guarantees a reasonably short lifetime of the metastable state. As a consequence the spin-down clusters exhibit a tendency to grow after some delay. In fact the spin-down clusters are distinctly larger than in zero field (Figure 19).

c) Now we switch-on an external field of strength $\mu_B H/k_B T = -0.015$ antiparallel to the orientation of the predominant spin-up distribution. In the presence of this field, the system is metastable. However, in the chosen applied field a rather short lifetime of this metastable state (see figure 8) is expected. The processes associated with this decay should shed some light onto the essential phenomena connected with a discontinuous transition.

d) After some time delay, clusters of the new phase show a pronounced tendency to grow, due to the applied field (Figure 20). This growth brings with it a reduction of the total surface of the clusters and hence a lowering of the total energy carrying the system over the free-energy barrier. Later on, the tendency for further growth becomes more pronounced in the sense that certain clusters are able to reach a critical size which seems to prevent them from dissociation. At this stage one also finds amalgamation of clusters (Figure 21), leading to a strongly interacting cluster gas.

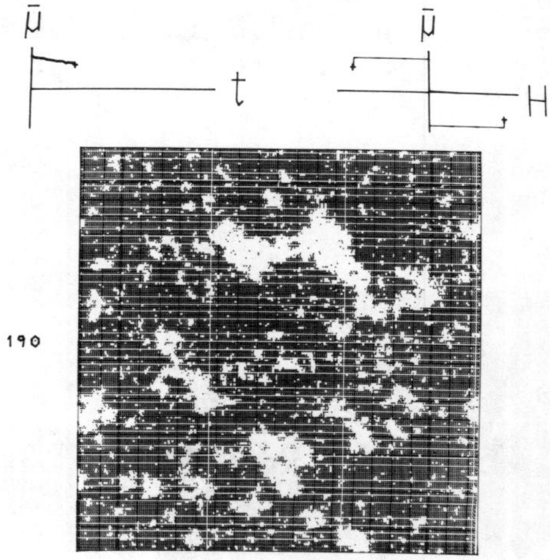

Figure 21. Snapshot of a nonequilibrium spin configuration in which the amalgamation of two clusters has just occurred.

e) The clusters which reach this critical size represent the embryos of the new phase. Such embryos, which grow due to the reversal of spins of the other phase at the border and by linking with surrounding small clusters, are often denoted as the "nuclei" of the new phase (Figure 22).

f) In the next stage, the nuclei continue to grow and associate so that a "macroscopic" cluster is formed, which in turn indicates that nucleation of the new phase has taken place (Figure 23). However, within this "macroscopic" cluster, small clusters of the old phase appear and disappear, indicating that local equilibrium has already been reached within this "macroscopic" cluster.

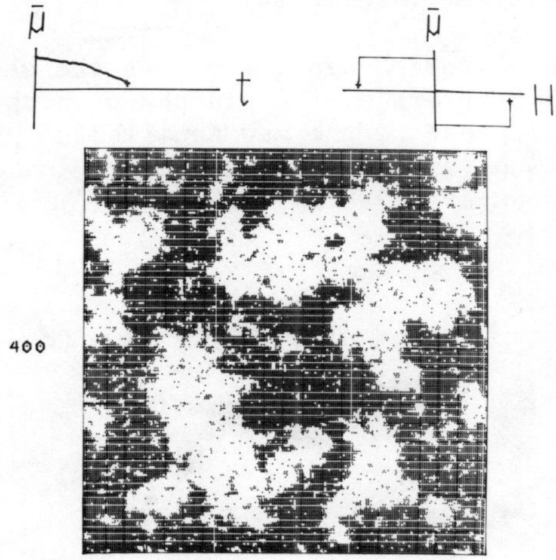

Figure 22. Snapshot of a nonequilibrium spin configuration in which several embryos of the new phase are clearly visible.

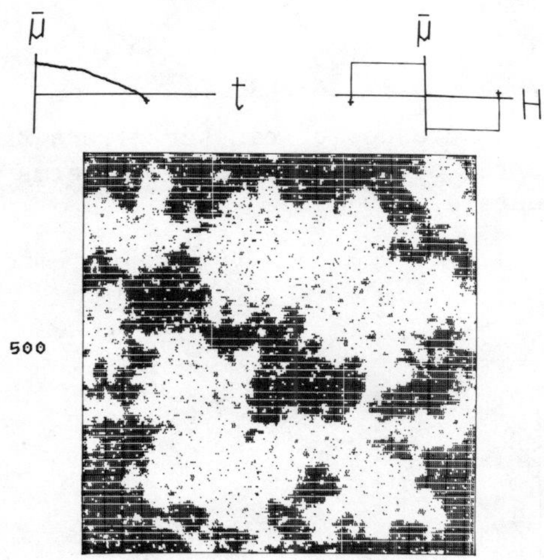

Figure 23. Snapshot of a nonequilibrium spin configuration where embryos link together forming a "macroscopic" cluster of the new spin-down phase. Within this "macroscopic" cluster, small clusters of the old phase are seen.

g) The remaining temporal development is mainly characterized by further growth of the "macroscopic" cluster and the origination and disintegration of small clusters of the initial phase within the "macroscopic" cluster (Figure 24).

h) Finally, the "macroscopic" cluster covers the whole system, with inclusions of small clusters of the old phase which appear and disappear. This stage represents the new equilibrium phase. In fact, the pattern consists of small clusters of up-spins distributed over the background of down-spins (Figure 25).

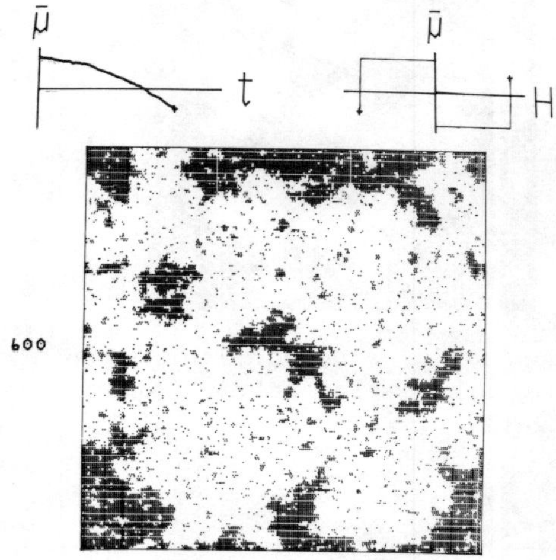

Figure 24. The "macroscopic" cluster covers the whole system and the old phase appears in terms of small dark clusters.

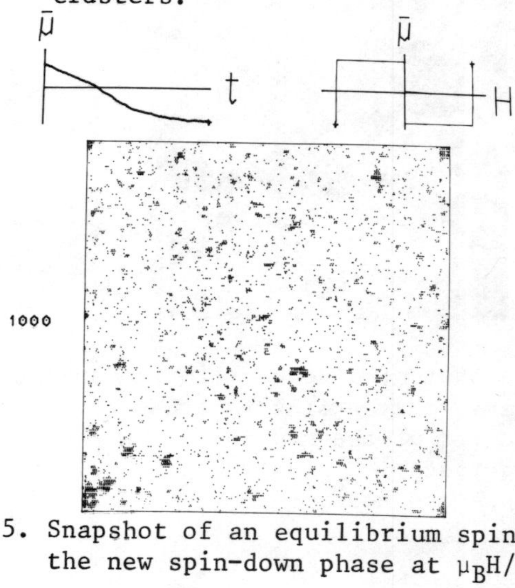

Figure 25. Snapshot of an equilibrium spin configuration of the new spin-down phase at $\mu_B H/k_B T = -0.015$.

Finally, let us point out that we computed this motion picture for two reasons. Firstly, we thought it worthwhile to demonstrate the evolution of an equilibrium state and of a discontinuous phase transition in a simple system. We feel that the film gives a good pictorial illustration of these processes. Secondly, however, we hope that the film will stimulate theoretical work in this area.

VI. NUCLEATION THEORY

The motion picture and the sequence of snapshots taken from it (Figures 17-25) clearly reveal that the processes associated with the dissociation, the formation, the growth and the interaction of clusters may be the key for a theory of discontinuous phase transitions. An attempt in this direction is the nucleation theory of Becker and Döring (13). The basic ideas of this approach have been extended and used with success by many authors (13)-(17).

Despite their success, these attempts suffer from the lack of a precise formulation of the problem (3). It is never entirely clear just what, if any exact quantity, one is trying to compute approximately. In view of the recent controversy about a factor of 10^{17} in the spontaneous nucleation rate of a supercooled vapor (33), these defects are not only aestetic. A more fundamental theory is therefore necessary.

In principle, all information needed is contained in the equations of motion for the correlation functions equation (9) or, in master equation (2). The solution of these equations, however, is known only for the one-dimensional problem (18), (19). For systems with higher dimensionality one must make approximations, or, what we have done, one must solve the problem numerically.

A closer contact with the conventional nucleation theory may be obtained by translating the master equation (2) into an equation of motion for the cluster distribution (30). As an example, we sketched in figure 26, the possible creation processes of an $n = 20$ cluster in a two-dimensional square lattice. The difference between the creation and annihilation processes, including the corresponding transition probabilities, determined the rate of change of $P(n = 20, t)$. Obviously, this constitutes an equation of motion. The hierarchy of these equations is exact as the master equation is. Their language, however, corresponds to that of nucleation theory. Clearly, an analytic solution of this hierarchy leads to unsurmountable difficulties and requires approximations. For $T << T_c$ and close to equilibrium one might neglect processes 2 and 3 shown in figure 26. Moreover, if the transition probability of the process is assumed to depend only on the cluster size, but not on the spin configuration, we again obtain

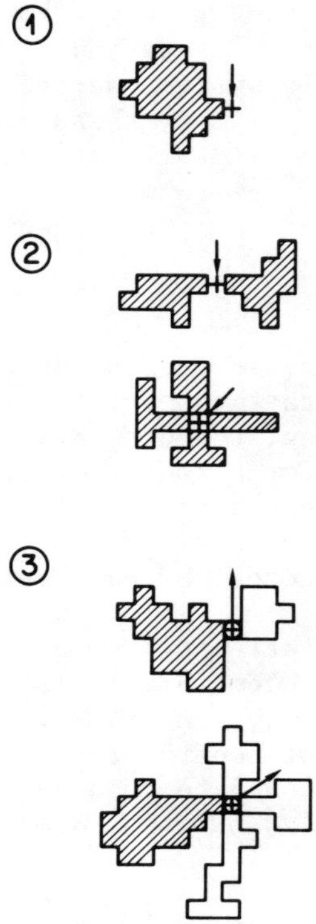

Figure 26. Possible creation processes of a cluster of $n = 20$ spins.

an ideal cluster gas. Invoking detailed balance, the equation of motion then becomes (14)

$$\frac{\partial P(n)}{\partial t} = D(n)\, P_{eq}(n) \left\{ \frac{P(n+1)}{P_{eq}(n+1)} - \frac{P(n)}{P_{eq}(n)} \right\}$$

$$- D(n-1)\, P_{eq}(n-1) \left\{ \frac{P(n)}{P_{eq}(n)} - \frac{P(n-1)}{P_{eq}(n-1)} \right\} . \qquad (45)$$

$D(n)$ is a diffusion constant, depending on the cluster size only. To first order, this difference equation is equivalent to

$$\frac{\partial P(n)}{\partial t} = \frac{\partial}{\partial n} \left\{ D(n)\, P_{eq}(n) \frac{\partial \left[\frac{P(n)}{P_{eq}(n)} \right]}{\partial n} \right\} . \qquad (46)$$

The expression in the bracket corresponds to a current

$$- J(n) = D(n)\, P_{eq}(n) \frac{\partial[P(n)/P_{eq}(n)]}{\partial n} \quad . \tag{47}$$

In terms of this current we may rewrite equation (46) in the form of a continuity equation

$$\frac{\partial P(n)}{\partial t} = - \frac{\partial}{\partial n} J(n) \quad . \tag{48}$$

In analogy to the droplet model, we have

$$P_{eq}(n) = P_0\, e^{-\frac{\Phi(n)}{k_B T}} \quad , \tag{49}$$

where $P_{eq}(n)$ increases for $n > n_c$. As a consequence the clusters $n > n_c$ would grow indefinitely. This restricts the range of validity of the continuity equation to clusters $n \leq n_c$. To overcome this difficulty, Becker and Döring assumed (13) that all clusters reaching the critical size n_c, are removed from the system and replaced by an equivalent amount of $n = 1$ clusters. Here $\partial P(n)/\partial t = 0$ and thus equation (46)

$$D(n)\, P_{eq}(n) \frac{\partial}{\partial n} \frac{P_s(n)}{P_{eq}(n)} = - J_s(n) = \text{constant} \quad , \tag{50}$$

where $P_s(n)$ is the steady-state cluster distribution and $J_s(n)$ the steady-state nucleation rate, i.e., the net number of clusters growing from size n to $n + 1$ per unit time. Figure 27 shows that the steady-state cluster distribution $P_s(n)$ no longer exhibits the large cluster catastrophe, namely, that $P_s(n)$ increases for $n > n_c$. Δ (Figure 27) characterizes the width extending from the point where the steady-state distribution deviates from the equilibrium one to the upper limit, where the fraction $P_s(n)/P_{eq}(n)$ becomes negligibly small. One obtains (14)

$$\frac{1}{\Delta} \approx \sqrt{+ \frac{1}{8} \left. \frac{\partial^2 \ln P_{eq}(n)}{\partial n^2} \right|_{n = n_c}}$$

$$\approx \sqrt{- \frac{1}{8 k_B T} \left. \frac{\partial^2 \Phi(n)}{\partial n^2} \right|_{n = n_c}} \tag{51}$$

Of particular interest is the relaxation of the system to the steady state. From equation (48), Zeldovich (34) derived for the associated relaxation time the estimate

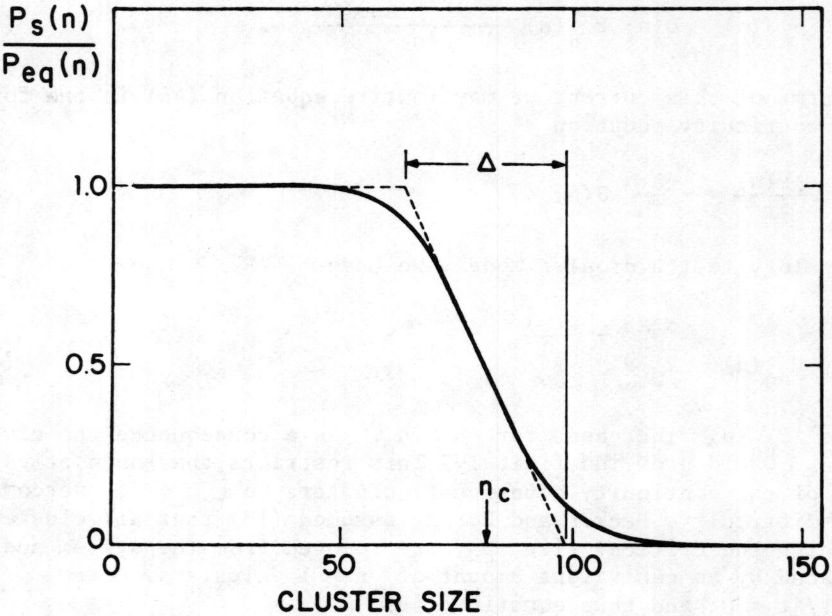

Figure 27. Steady state - divided by equilibrium cluster distribution as a function of cluster size, according to Feder et al. (14).

$$\tau \sim \frac{P_{eq}(n_c)\,\Delta}{J_s(n_c)} \quad , \tag{52}$$

where J_s is the steady-state current. Recalling that $J_s(n_c)$ is the net number of clusters growing from n_c to $n_c + 1$ per unit time, it becomes evident that $J_s(n_c)$ is proportional to the inverse lifetime of a metastable state

$$J_s \sim 1/\tau_R \quad . \tag{53}$$

Combining equations (52) and (53) we find

$$\frac{\tau}{\tau_R} \sim P_{eq}(n_c)\,\Delta \quad . \tag{54}$$

Using for $\Phi(n)$ the expression of droplet model, as given by equation (28) $(H < 0)$ we find with the aid of equations (49), (51) and (54)

$$\frac{\tau}{\tau_R} \sim |H|^{-3/2} \quad \exp\left[\frac{\text{const}}{|H|}\right] \quad , \tag{55}$$

where τ_R denotes the lifetime of a metastable state and τ the relaxation time to the steady state, respectively.

For small applied fields and $T \ll T_c$, where the Becker-Döring approach is expected to hold, equation (52) predicts long-lived metastable states, as revealed by our simulations (Figure 8(b)). Moreover, τ_R is expected to remain finite for non-zero applied fields.

If the amount of supercooling is great, however, the system is far from equilibrium and for temperatures $T \lesssim T_c$ the cluster gas is no longer ideal. In those situations, the basic assumptions of conventional nucleation theory no longer hold. In fact, the motion picture clearly reveals that in those situations, the interaction between the clusters is no longer negligible (processes 2 and 3 sketched in figure 26). The sequence of snapshots (Figures 20-25) also reveals the importance of the cluster border anisotropy, which is neglected in the conventional droplet model and the conventional nucleation theory.

To summarize, we may say that the approximate descriptions of the nucleation process are qualitatively consistent with the motion picture, as long as $\bar{\mu}(t)$ exhibits a flat region (Figure 22). In fact the existence of a flat region in $\bar{\mu}(t)$ or in $\Phi_{\mu}^{0,\Delta H}(t)$ supports a quasi-steady-state regime. The appearance of a discontinuity in the derivative of $\bar{\mu}(t)$ with respect to time, as revealed by our simulations (Figure 22) indicates, however, that processes 2 and 3 shown in figure 26 become dominant. As a consequence, conventional nucleation theory, neglecting these processes, cannot account for the decay process of the steady state.

We acknowledge stimulating discussions with G. Benedek, K. Binder, P. Heller, P. Meier, K.A. Müller, St. Sarbach and H. Thomas.

REFERENCES

1. See for example A.B. Pippard, Classical Thermodynamics, (Cambridge, University Press, 1957) Chap. 8.

2. D. Dahl and M.R. Moldover, Phys. Rev. Letters, 27, 1421 (1971).

3. O. Penrose and J.L. Lebowitz, J. Stat. Phys., 3, 211 (1971).

4. B. Chu, F.J. Schoenes and M.E. Fisher, Phys. Rev., 185, 219

(1969).

5. M.E. Fisher, Physics, 3, 255-83 (1967).

6. A.F. Andreev, Sov. Phys.-JETP, 18, 1415 (1964).

7. C. Domb, J. Phys. Chem., 6, 39 (1973).

8. O.E. Lanford and D. Ruelle, Commun. Math. Phys., 13, 194 (1969).

9. D.S. Gaunt and G.A. Baker, Phys. Rev., B 1, 1184 (1970).

10. P. Schofield, Phys. Rev. Letters, 22, 606 (1969).

11. P. Schofield, J.D. Litster and J.T. Ho, Phys. Rev. Letters, 23, 1098 (1969).

12. E. Stoll and T. Schneider, Phys. Rev., A 6, 429 (1972).

13. J. Frenkel, Kinetic Theory of Liquids, New York, Dover Publications, 1955) Chap. VII.

14. J. Feder, K.C. Russell, J. Lothe and G.M. Pound, Adv. in Phys., 15, 117 (1966).

15. C.S. Kiang, D. Stauffer, G.H. Walker, O.P. Puri, J.D. Wise, Jr. and E.M. Patterson, J. Atmospheric Sci., 28, 1112 (1971).

16. J.L. Katz, J. Stat. Phys., 2, 137 (1970).

17. A. Eggington, C.S. Kiang, D. Stauffer and G.H. Walker, Phys. Rev. Letters, 26, 820 (1971).

18. R.J. Glauber, J. Math. Phys., 4, 294 (1963).

19. K. Kawasaki, in Phase Transitions and Critical Phenomena, edited by C. Domb and M.S. Green, New York, Academic, 1972), 2, p. 443.

20. M. Suzuki and R. Kubo, J. Phys. Soc. Japan, 24, 51 (1968).

21. N. Ogita, A. Ueda, T. Matsubara, H. Matsuda and F. Yonezawa, J. Phys. Soc. Japan, 26 S, 146 (1969).

22. H. Müller-Krumbhaar and K. Binder, J. Stat. Phys., 8 (May 1973).

23. T. Schneider, E. Stoll and K. Binder, Phys. Rev. Letters, 29, 1080 (1972).

24. E. Stoll, K. Binder and T. Schneider, to be published.

25. H. Yahata and M. Suzuki, J. Phys. Soc. Japan, 27, 1421 (1969).

26. H. Yahata, J. Phys. Soc. Japan, 30, 657 (1971).

27. N. Metropolis, A.W. Rosenbluth, M.N. Rosenbluth, A.H. Teller and E. Teller, J. Chem. Phys., 21, 1087 (1953).

28. R. Abe and A. Hatano, Progr. Theoret. Phys. (Kyoto), 41, 941 (1969).

29. B.I. Halperin, P.C. Hohenberg and S. Ma, Phys. Rev. Letters, 29, 1548 (1972).

30. K. Binder and H. Müller-Krumbhaar, to be published.

31. J.S. Langer, Ann. Phys. (N.Y.), 41, 108 (1967).

32. K. Binder, E. Stoll and T. Schneider, Phys. Rev. B 6, 2777 (1972).

33. H. Riesz, J. Stat. Phys., 2, 83 (1970).

34. J.B. Zeldovich, J. Expt. Theor. Phys., 12, 525 (1942).

CONTINUOUS VERSUS DISCONTINUOUS PHASETRANSITIONS

G. Alefeld

Physik-Department der Technischen Universität
München, 8046 Garching

ABSTRACT

It will be argued that with lattice defects as additional degree of freedom melting or structural phasetransitions, which according to Landau must be discontinuous, may become continuous. Under this aspect detailed consideration will be given to martensitic transformations of β-brass type alloys, including a short discussion of the phenomena of shapememory.

INTRODUCTION

In 1937 Landau (1) published two fundamental papers in which he showed that a large group of structural phasetransitions cannot be continuous, it must be of first order. His arguments are based on symmetry considerations. (For a detailed discussion see N. Boccara (2)!). Such e.g. a fcc lattice cannot be transformed into a hcp lattice by an infinitesimal small deformation. Also on the basis of symmetry arguments the phasetransition solid liquid should always be first order.

In this communication we like to show very briefly for melting and more elaborately for martensitic transformations that Landau's arguments do not hold if lattice defects are admitted to exist. We like to show that phasetransitions, which according to Landau must be discontinuous may occur in a continuous way due to the existence of a further degree of freedom of the system, and thus the possibility to use a different order parameter than used by Landau.

MELTING

In Landau's argumentation the perfect solid is assumed to exist in equilibrium with the liquid. Yet with increasing temperature in all solids lattice defects exist in thermal equilibrium, in metals especially lattice vacancies. Not the perfect crystal has lowest free energy but the imperfect crystal. If one admits vacancies one has also to admit in some temperature and pressure regime clusters of vacancies, dislocations and also grainboundaries. Due to entropy the polycrystalline solid may have lower free energy than the monocrystal. A polycrystal and a liquid differ only quantitatively and not qualitatively. Therefore a continuous transition between a solid and a liquid seems possible. There may be substances for which the phase separation line solid-liquid ends in a critical point.

MARTENSITIC PHASETRANSITIONS

For simplicity we will first consider the transition fcc-hcp. Both structures are closepacked structures. The essential difference is the packing sequence in [111] direction. For the fcc lattice the sequence is ABCABC... with A,B,C denoting the three possible positions in a closepacked layer. For the hcp lattice the sequence is ABAB... . Therefore the fcc lattice can be considered as a hcp lattice with a periodic arrangement of stacking faults and vice versa. Removing these stacking faults (or introducing new stacking faults) transforms one lattice into the other. Therefore a continuous transition between a fcc and hcp structure may be possible. We do not want to discuss in details how such a transformation would occur. Yet it is e.g. conceivable that stacking faults existent in a fcc structure on all four {111} planes condense into one set of planes, thus transforming the crystal to a hcp structure.

At the fcc-hcp transition of Co a high stacking fault density is being observed (3). Yet this transition shows large hysteresis effects and also intermediate phases in which the stacking faults assume periodic arrangements (4). We will therefore briefly consider the martensitic transformations of β-brass type alloys, which are close to be hysteresis free under suitable conditions (5). These alloys are essentially bcc-structures which transform into an orthorhombic structure. In Fig. 1 the transformation is shown for the (001) plane of the cubic crystal which coincides with that of the new orthorhombic crystal. In Fig. 1 the transformation of the cubic unit cell is shown separately. As long as no volume change occurs in the transformation the areas shown in the insert in Fig. 1 are equal.

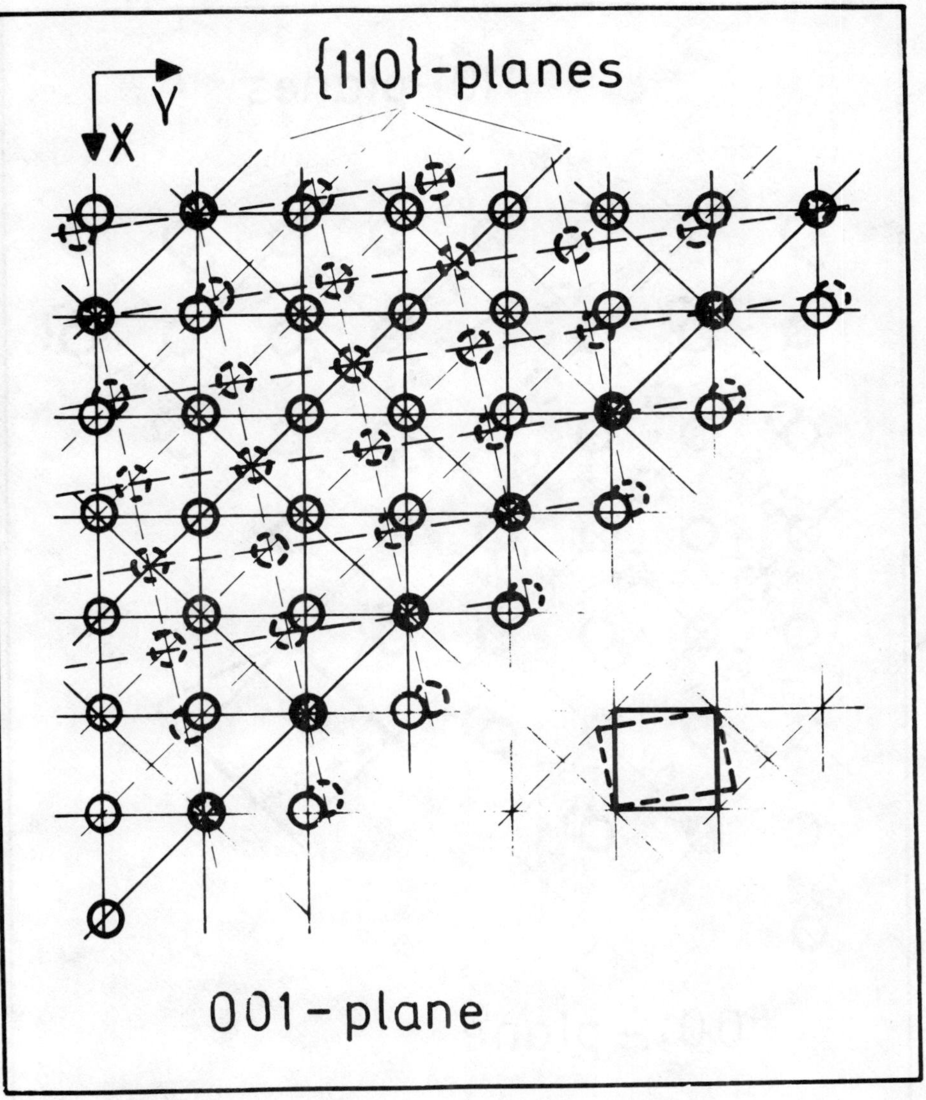

Fig. 1 Transformation of a cubic lattice (O) to an orthorhombic lattice (◌) by shear on {110} planes

Furthermore one diagonal of the cell remains equal to a $\sqrt{2}$, but the 90° angle is changed to about 87°. Thus the length of one side of the cell is increased, whereas the other side shrinks. The transformation is easier understood by considering the de-

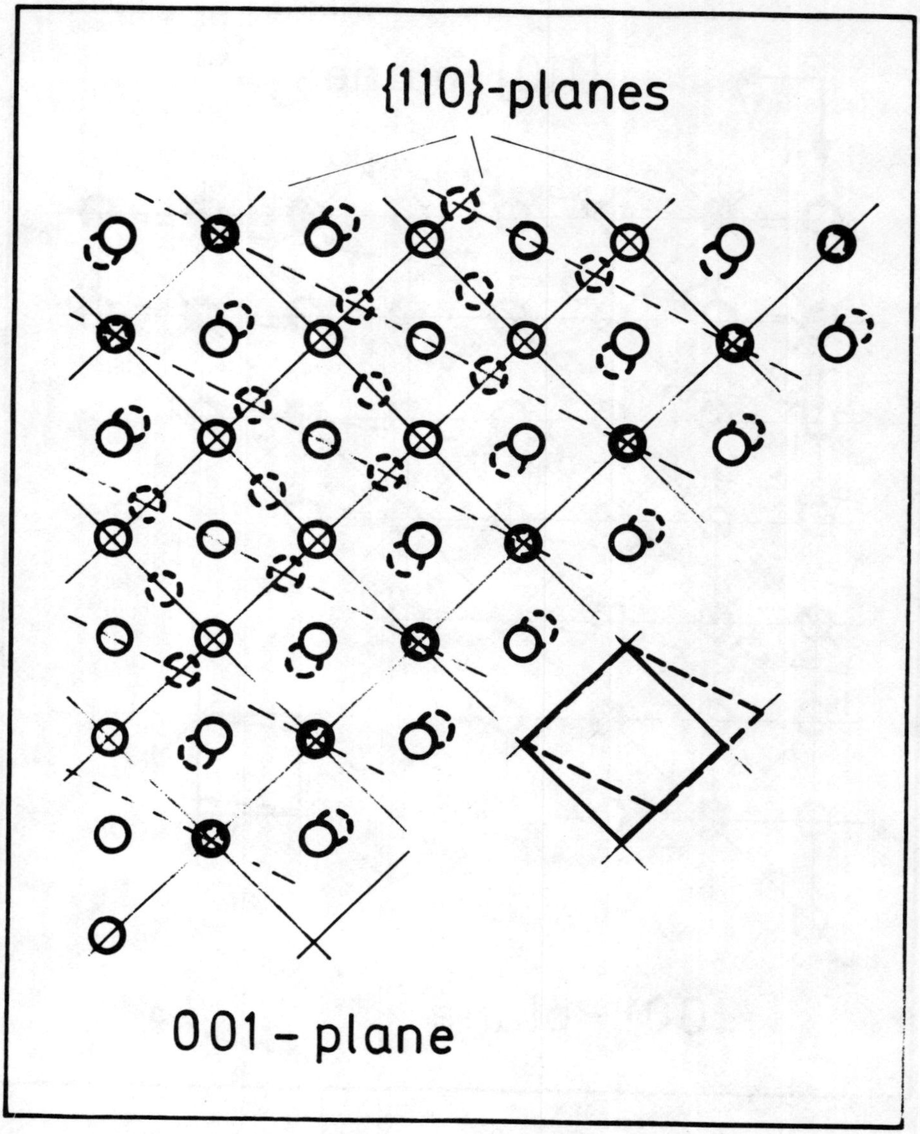

Fig. 2 The transformation of Fig. 1, but using a larger unit cell

formation of a larger unit cell with the axes parallel to the [110] and [$\bar{1}$10] directions (Fig. 2). The deformation of this cell can be described by one shear component, namely a shear parallel to the [110] direction. The 90° angles are deformed to 90 ± α with tan α = 1/3. After performing exactly this shear deformation a new orthorhombic cell can be drawn with the axis parallel to the [001], [$\bar{1}$10] and [110] directions of the cubic lattice and the lengths a, a$\sqrt{2}$ and 3 a $\sqrt{2}/2$. Fig. 1 shows the orthorhombic lattice and its orientation relationship to the cubic lattice. The shear deformation can thus be described as follows: the cubic lattice shears on the (110) plane in [$\bar{1}$10] direction. On each plane one portion of the crystal is shifted relative to the other part by 1/6 [$\bar{1}$10] thus transforming cell by cell from cubic to orthorhombic.

There is no reason to assume that this transition must be first order, in spite of the finite discontinuous transformation of the individual cell. We are using as an order parameter the number of planes shifted, or the number of cells transformed instead of the degree of deformation as suggested by Landau's treatment. Such an order parameter has become possible by admitting nonperfect lattices to exist in both phases. Shear stress in the (110) plane in [$\bar{1}$10] direction can be considered as the conjugate field coupling to the orderparameter.

Some experimental observations mainly due to Nakanishi and coworkers (5,6) may be summarized:

1) The onset of the phasetransition is accompanied by a strong softening of C_{11}-C_{12}, i.e. a transverse soft mode corresponding to the shear on the (110) plane in [$\bar{1}$10] direction.

2) Immediately above T_M double hysteresis curves are observed similar to double hysteresis P-E curves of ferroelectric crystals. Apparently the spontaneous shearing leading to the phasetransition at and below T_M can be achieved above T_M by an external shear stress. On removal of the stress the shearing disappears spontaneously before the stress is zero.

3) Below T_M the stress-strain curves show strong hysteresis loops similar to those of ferroelectric or ferromagnetic crystals.

4) One of the most remarkable properties of these alloys is the socalled "shape-memory" effect. If a crystal is deformed below T_M and then heated above T_M the original macroscopic shape recovers almost completely. For complete recovery the strains should not be too large ($\varepsilon < 0{,}3$), but nevertheless strains of this order of magnitude are usually never considered to be reversible. Having the shear mechanism of the phasetransition in mind, the shape memory effect and the strong hysteresis loops below T_M may be interpreted as follows (see Fig. 3): In cooling through T_M, the crystal may be sheared in 1/6 [$\bar{1}$10] as well as 1/6 [1$\bar{1}$0] direction, as long as no external stress field is applied, similarly to the orientation of the magnetization without external magnetic field. The crystal is made up of domains, the

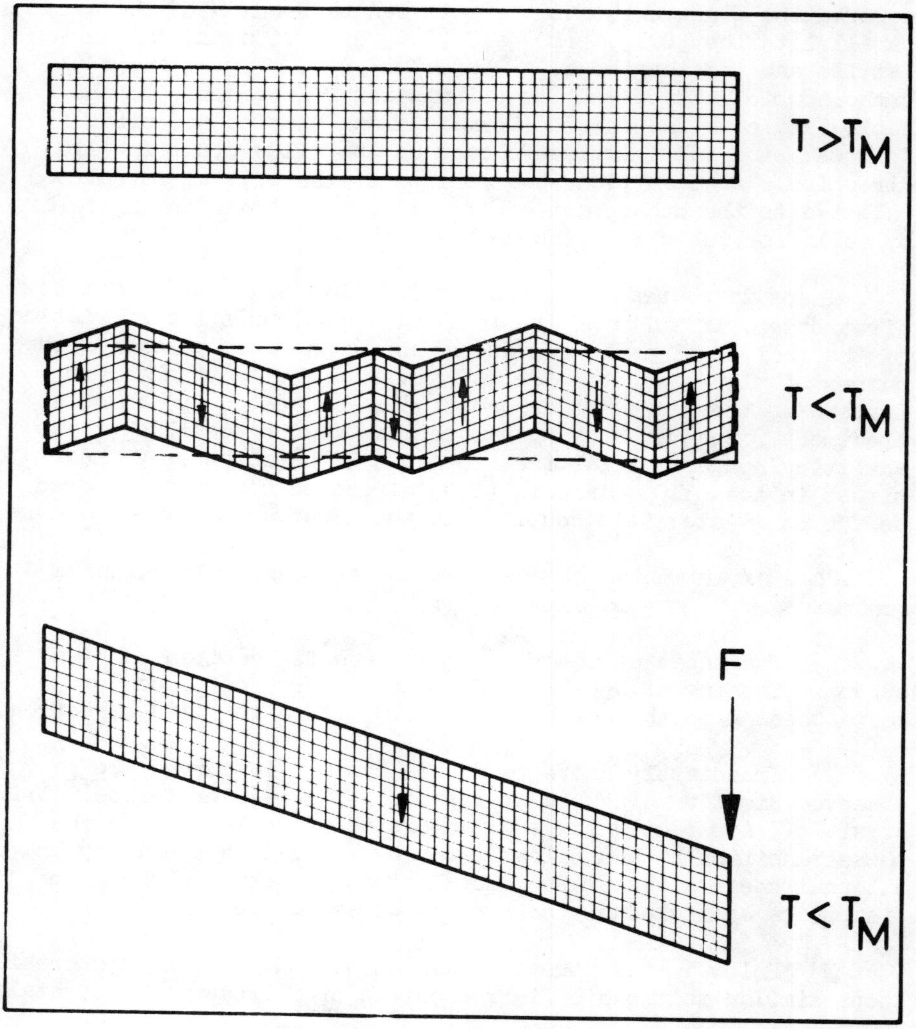

Fig. 3 The basic mechanism for the "shape memory effect"

domain boundaries are twin boundaries of the orthorhombic lattice. The macroscopic shape of the crystal is essentially unchanged, as long as the domains are small. By applying an external field the specimen can be deformed by shifting the domain boundaries and eventually planes, which have not yet sheared. The phenomena are similar to moving Blochwalls and eventually aligning additional spins. In the process of "deformation" the specimen will finally end up as a monodomain crystal. Heating a such deformed crystal to a temperature above T_M, recovers the original shape due to a complete reversible shear on all planes by the definite amount of 1/6 [$\bar{1}$10]. In common plastic deformation the shear on a plane occurs by multiples of a vector of the elementary cell. No mechanism is built in for remembering how large the shear on a plane has been. In the martensitic transformation described each plane shears by an amount smaller than half a vector of the elementary cell. Therefore the amount and the direction of shear is memorized, each plane remembers exactly its original position relative to the neighboring ones.

It should be pointed out that the mechanism described above which certainly is an idealization also works for polycrystalline materials or for such martensitic materials, in which the transformed phase nucleates in platelets in the cubic matrix. By properly arranging the domain boundaries the geometrical shapes of the transformed region can be adjusted to boundaries required (7).

Martensitic transformations between hcp and fcc structures, if performed on perfect single crystals and if occuring without volume change, should show no shape memory effects. For the removal or creation of a stacking fault in one of these structures the ambiguity of two equivalent shear vectors exists. Nevertheless in polycrystals e.g. the coherence of grains sets up boundary conditions which can serve as a memory-mechanism for the macroscopic shape, as observed for Co and Co-Ni alloys (4). Finally it should be pointed out that the pioneering studies on shape memory have been performed on the system Au-Cd (8).

The shape memory effect is in our opinion one feature of the more general group of properties which have been summarized under the name "Ferroelasticity" (9, 10, 11). Fig. 3 clearly demonstrates the bistability of the low temperature phase.

CONCLUSION

The intention of this paper was to point out, that Landau's symmetry arguments which imply discontinuity for certain structural phasetransitions, must be used with proper caution. The existence of lattice defects in thermal equilibrium may spoil the applicability of these arguments. We have only touched the discussion, as to how such transformation may proceed in a continuous way by shearing on successive planes. In this connection we have discussed a mechanism for the shape memory effect.

The author gratefully acknowledges discussions with Dr. Warlimont, Max-Planck-Institut Stuttgart, on the questions of martensitic transformations.

REFERENCES

1. L.D. Landau, Phys. Z. Sowjet 11, 26 (1937) and 11, 545 (1937)

2. N. Boccara, Annals of Physics 47, 40 (1968)

3. S. Kajiwara, Japan. J. Appl. Phys. 9, 385 (1970)

4. A. Nagasawa, Phys. stat. sol. (a) 8, 531 (1971)

5. N. Nakanishi, Y. Murakami and S. Kachi, Scripta Metallurgica 5, 433 (1971)

6. N. Nakanishi, Y. Murakami, S. Kachi, T. Mori and S. Miura, Phys. Letters 37A, 61 (1971)

7. See J.W. Christian, "The Theory of Phase Transformation in Metals and Alloys", Pergamon Press, New York, 1965

8. See D.S. Lieberman in Phase Transformations, American Society for Metals, Metals Park, Ohio (1970)

9. K. Aizu, J. Phys. Soc. Japan 27, 387 (1969)

10. G. Alefeld, G. Schaumann, J. Tretkowski and J. Völkl, Phys. Rev. Lett. 22, 697 (1969)

11. S.C. Abrahams, Mat. Res. Bull. 6, 881 (1971)

DYNAMICS OF FLUIDS NEAR THE CRITICAL POINT: LIGHT SCATTERING INVESTIGATIONS*+

H.Z. Cummins** and H.L. Swinney

Department of Physics, New York University,
New York, New York 10003, U.S.A.

1. INTRODUCTION

The critical point of the liquid-vapor phase transition has been studied extensively for more than a century, and the results of experimental observations of critical phenomena in simple fluids and binary fluid mixtures have frequently provided a model for analyzing critical phenomena in other systems.

Experimentally, many thermodynamic properties of fluids are found to exhibit anomalies near the critical point of the form $\varepsilon^{-\psi}$ where $\varepsilon = |T-T_c|/T_c$ is the reduced temperature and ψ is a "critical exponent".(1) The fact that the exponent describing a particular property (such as the susceptibility) has the same value for widely different systems such as fluids, ferromagnets, binary mixtures, etc., is a consequence of the increasing range of correlations near the critical point which results in the long wavelength fluctuations (which tend to average out local structure) dominating critical behavior.

The consequences of critical phenomena being dominated by a characteristic length ξ which diverges as the critical point is approached were shown by Widom and by Kadanoff to suggest algebraic relations between the critical exponents known as "Static Scaling Laws".(1) Recently, the scaling transformations introduced by Kadanoff have been extended by Wilson, who has been able to deduce a preliminary numerical value for the exponent γ characterizing the susceptibility divergence.(2)

About ten years ago, interest developed in exploring the dynamical properties of the fluid critical point, stimulated in part by the availability of lasers and the new technique of light beating spectroscopy.[3] As the results of these experiments appeared, a theoretical program based on the work of Fixman was developed by Kawasaki and by Kadanoff and Swift which has come to be known as the mode-mode coupling theory. This theory gave specific predictions for the dynamical properties of fluids in the critical region which could be compared directly with experimental results.

By 1970, the mode-mode coupling theory was in qualitative agreement with the results of many experiments,[4] but several serious problems remained:

(1) The exponent describing the relaxation rate for long-wavelength order parameter fluctuations varied from 0.59 to 0.75 for various fluids and binary mixtures, while for SF_6, the exponent had been reported as 0.63 on the coexistence curve and 1.26 on the critical isochore. (Theoretically this exponent should have the same value for all systems, and the same exponent should characterize the behavior along both the coexistence curve and the critical isochore.)

(2) The mode-mode coupling theory had several unfinished aspects: (a) the theoretical results included an ambiguous parameter η_s^* defined as the high frequency shear viscosity. (b) the calculations did not include vertex corrections and the extent to which they would influence the results was unknown.

By now some of these problems have been resolved. In this lecture we will review the development of the theory and present a summary of the results of experiments.

2. DENSITY FLUCTUATIONS AND CRITICAL OPALESCENCE[5]

As a fluid approaches the critical point, the isothermal compressibility κ_T diverges as $\varepsilon^{-\gamma}$; hence the amplitudes of the long wavelength density fluctuations also diverge. When light of wavevector $\vec{k}_o$ passes through the fluid, the intensity of the light scattered at an angle θ from the forward direction is given by:

$$I_s(q) = A\langle(\rho_q)^2\rangle = A'S(q) \tag{1}$$

where A and A' are constants, $\langle(\rho_q)^2\rangle$ is the mean square amplitude of the $\vec{q}^{\,th}$ Fourier component of the density fluctuation, $|\vec{q}| = |\vec{k}_o - \vec{k}_s| \simeq 2k_o \sin(\theta/2)$, and S(q), the structure factor, is related to g(r), the radial distribution function, by:

$$S(q) = 1 + \rho \int \exp\,[i\vec{q}\cdot\vec{r}]\,[g(r)-1]\; d^3r, \tag{2}$$

where ρ is the number density. This formulation does not include the direct coupling to temperature fluctuations discussed in the lecture of Professor Klein. In most fluids, however, the effect of such coupling is negligible. For very long wavelength fluctuations, the compressibility theorem, $1+\rho\int[g(r)-1]\,d^3r = \rho kT\kappa_T$, when combined with Eq. (2), gives $S(q)_{q\to 0} \propto \rho^2 kT\kappa_T$, which is just Einstein's 1910 result showing that the mean-square density fluctuations $\langle(\rho_q)^2\rangle$ and the light scattering intensity should diverge as κ_T which qualitatively explains the observed phenomenon of critical opalescence.(6) If the Ornstein-Zernike form for g(r) is assumed,

$$g(r)-1 = (1/4\pi\rho R^2)\,\frac{\exp(-r/\xi)}{r}\;, \tag{3}$$

then Eq. (2) becomes

$$S(q) \propto (\rho^2 kT\kappa_T)/(1+q^2\xi^2). \tag{4}$$

Plots of the inverse scattering intensity vs. q^2 for fluids and binary mixtures at different temperatures give a series of straight lines,(4) and measurement of the intercepts and slopes of these lines gives κ_T and ξ, hence γ and ν where $\xi = \xi_o\varepsilon^{-\nu}$. (Experimentally, $\gamma \simeq 1.25 \simeq 2\nu$).

3. DYNAMICS OF THE DENSITY FLUCTUATIONS: HYDRODYNAMICS

Measurements of the light scattering intensity $I_s(q)$ determine $\langle(\rho_q)^2\rangle$ which depends on the equilibrium properties of the fluid, κ_T and ξ. The dynamics of the density fluctuations are contained in the correlation function $C_q(t) = \langle\rho_q(0)\;\rho_q(t)\rangle$, which can be deduced from measurements of either the spectrum or the intensity autocorrelation function of the scattered light.(3,4)

The dynamics of long wavelength density fluctuations far enough from the critical point for $q\xi\ll 1$ to apply are governed by the classical equations of hydrodynamics. When those equations are Fourier analyzed following Mountain,(7) one finds the density autocorrelation function:

$$C_q(t)/\langle(\rho_q)\rangle^2 = \frac{c_p-c_v}{c_v}\exp[-(\Lambda/\rho c_p)q^2|t|] + \\ + \frac{c_v}{c_p}\;\exp -\Gamma_s q^2|t|]\;\cos\,(C_o qt) \tag{5}$$

where c_p and c_v are the specific heat at constant pressure and volume, ρ is the mass density, Λ is the thermal conductivity, and C_o and Γ_s are the speed and damping of sound. The Fourier transform of $C_q(t)$ is $S(q,\omega)$; Eq. (5) transforms to give the triplet which was discussed by Professor Sjölander.

The first term in Eq. (5), which dominates the critical dynamics, is of the form $C_q(t) = \langle(\rho_q)^2\rangle\, e^{-\Gamma_q t}$, with:

$$\Gamma_q = (\Lambda/\rho c_p)q^2 = \chi q^2 \tag{6}$$

The "critical slowing down" concept of Van Hove[8] suggests that the relaxation rate for density fluctuations goes to zero with the generalized restoring force. In this approach the transport coefficient Λ is assumed to be well-behaved in the critical region, so $\chi \sim c_p^{-1} \sim \varepsilon^{\gamma}$ (where χ is the thermal diffusivity), since c_p has the same critical anomaly as κ_T.

When the first experiments were initiated about ten years ago, it was generally expected that the experiments would give $\chi = \chi_o \varepsilon^{\phi}$ with $\phi = \gamma \simeq 1.25$. As we shall discuss in section 6, however, the experiments now clearly indicate that $\phi \simeq 0.63$. Thus, contrary to earlier expectations, the thermal conductivity Λ must diverge strongly with an exponent $\psi = \gamma - \phi \simeq 0.6$.

4. FIXMAN'S THEORY

The analysis of hydrodynamics in the critical region is extremely complicated since the growing range of correlation destroys the strictly local nature of the equations of motion. In order to extend the hydrodynamic equations, the non-locality is formally incorporated into the transport and response coefficients by making them q-dependent, and the solutions to the linearized local equations above are retained.

The occurrence of critical anomalies in the transport coefficients had been predicted by Fixman on the basis of the interaction between the transport currents and the spontaneous density fluctuations,[9] and some evidence for the existence of such anomalies was found in thermodynamic measurements.[10]

Fixman had first considered the constant volume specific heat c_v. Since large density fluctuations involve large fluctuations in the energy density, c_v may also become anomalous since in general $c_v = (1/kT^2)\,\langle(\delta E)^2\rangle$. Fixman related δE to the density fluctuations, assumed the Ornstein-Zernike form for $\langle(\rho_q)^2\rangle$, integrated over $\vec{q}$, and found an anomalous contribution to c_v:[9,11]

$$\Delta c_v \sim \xi\left(\frac{\partial}{\partial T}\ \xi^{-2}\right)^2 \sim \varepsilon^{3\nu-2}. \tag{7}$$

He then used the "classical" exponent for ξ, $\nu = 1/2$, which gives $\Delta c_v \sim \varepsilon^{-1/2}$ which is too strong a divergence. However, the divergence in c_v is characterized by an exponent α, ($c_v \sim \varepsilon^{-\alpha}$ with $\alpha \simeq 0.05$), so Fixman's derivation without the last step yields $\alpha = 2-3\nu$ which is just a static scaling law![1,12]

Fixman analyzed the critical behavior of the shear viscosity η_s which would arise from "The frictional dissipation of energy which occurs when the shear flow interacts with the spontaneous composition fluctuations".[9] For any transport problem, there is a current $\vec{J}$ flowing down a potential gradient $\nabla\mu$, which causes energy dissipation and therefore entropy production given by:

$$T\dot{S} = -\int \langle \vec{J}\cdot\nabla\mu \rangle \, dV. \tag{8}$$

Since $\vec{J} = \theta\nabla\mu$, where θ is a transport coefficient, we have

$$T\dot{S} = \frac{1}{\theta} \int \langle J\cdot J \rangle \, dV = \frac{V\langle J^2 \rangle}{\theta} = V\theta \, \langle (\nabla\mu)^2 \rangle.$$

Fixman defined $\theta \equiv T\dot{S}/V(\nabla\mu)^2$ and then computed $T\dot{S}$ from Eq. (8), assuming that the fluctuations in μ result from the spontaneous composition or density fluctuations.

5. MODE-MODE COUPLING THEORY

In 1966, Kawasaki[13] reconsidered Fixman's idea of critical anomalies in the transport coefficients, but proposed a different method of calculation, starting from correlation function expressions for the transport coefficients (Kubo Formulas).[14] The currents $J(t)$ in the correlation function $\int \langle J(0)J(t) \rangle dt$ were then expanded in a power series in the macroscopic variables A_k whose equations of motion, as well as the coefficients in the power series expansion, must be deduced from the macroscopic equations of motion. (This method is discussed in some detail in the lecture of Professor Schwabl.) Kadanoff and Swift[15] extended Kawasaki's work, incorporating static scaling results, and were able to deduce the temperature dependence of the transport coefficients. In particular, they found that in the hydrodynamic region the dominant singularity in Λ is $\sim \xi^{-1}$ so that $\Lambda/\rho c_p \sim \varepsilon^{\gamma-\nu}$. Since $\gamma \simeq 2\nu$, they predicted $\Lambda/\rho c_p \sim \varepsilon^{\nu}$ which agreed, at least qualitatively, with all the experiments with the exception of SF_6 on the critical isochore.[4]

Subsequently, Kawasaki continued the development of this mode-mode coupling theory, extending it to also describe behavior in the non-hydrodynamic regime $q\xi \gtrsim 1$.[16-20]

We digress for a moment to consider a different problem - the lattice dynamics of anharmonic crystals.[21] Here one begins with solutions to the harmonic problem, truncating the potential energy expansion after the terms quadratic in the ion displacements. The anharmonic potential terms can then be expressed in terms of the normal coordinates, and then in terms of creation and annihilation operators for the harmonic phonons.

Green's functions for the exact solution (physical phonons) can then be found - in principle - from the Green's functions for the harmonic solutions by systematic application of perturbation theory, using the anharmonic potential terms as coupling constants.

Diagrammatically, the perturbation calculation is:

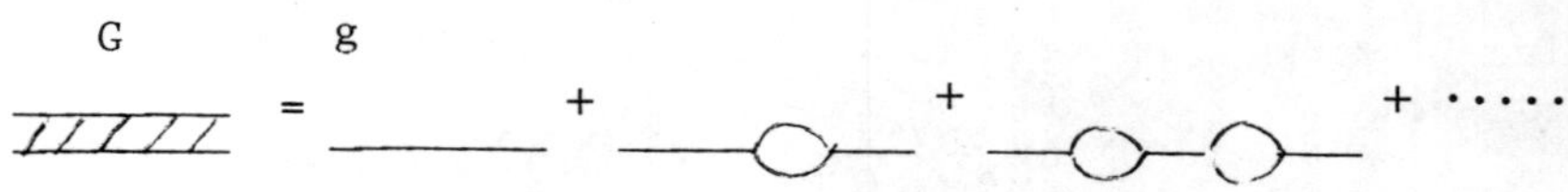

Since there are an infinite number of diagrams to be calculated the equation cannot be solved. But a Partial summation can be achieved by working with the lowest order diagram but making the intermediate states physical rather than bare. This "self-consistent phonon" approximation gives a single integral equation for the self-energy Σ which must then be solved self-consistently in the sum over intermediate states.[22] (This method sums up all the bubble diagrams but misses the vertex corrections.)

In the mode-mode coupling theory, a similar procedure is applied. The linearized hydrodynamic equations are first solved (with transverse velocity retained) to give the macroscopic normal modes. These are of the form $A_q^j(t) = A_q^j(0)\exp(-S_j t)$ where the different j-modes are:

(1) Heat flow $S_T = (\Lambda/\rho c_p)q^2$

(2) Viscous flow $S_\eta = (\eta_s/\rho)q^2$

(3) Sound $S_s = \pm\, iCq + D_s q^2/2$

These modes serve as the phenomenological "bare propagators" of the theory, and a perturbation calculation similar to the anharmonic lattice calculation can be performed. Since no microscopic theory exists, however, the form of the coupling constants is not obvious, and they were found by a projection operator technique. But physically they must represent the leading non-linear terms in the hydrodynamic equations.

Kadanoff and Swift found that the dominant contribution to the anomaly in Λ comes from intermediate states involving one heat mode and one viscous flow mode:[23]

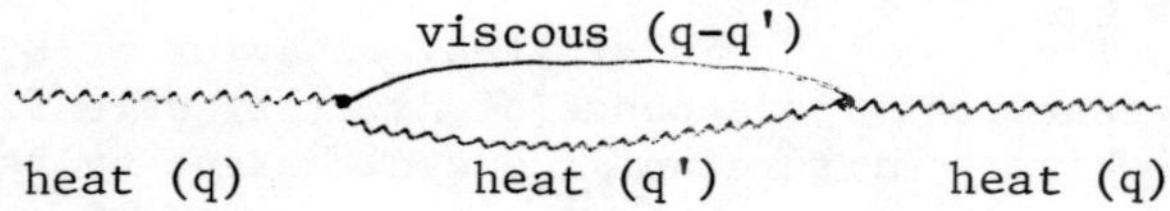

Their result for the contribution of this diagram to the thermal conductivity was:

$$\Lambda(q,\ s) = \frac{kT}{3}\int\frac{d^3q'}{(2\pi)^3}\ \frac{c_p(q')}{S_T(q') + S_\eta(q-q') - s}\ . \tag{9}$$

They then argued that the main contribution to the integral comes from $q \simeq \xi^{-1}$, and that the denominator is dominated by the viscous relaxation rate. For $q\xi<<1$ this leads to the result

$$\Lambda(q,\ s)_{q\xi<<1} \sim (\rho c_p \xi^{-1}/\eta^*)\ kT, \text{ and } \Lambda/\rho c_p \sim (\xi^{-1}/\eta^*)\ kT, \tag{10}$$

where η^* is the shear viscosity for a viscous flow mode with wavevector $\sim\xi^{-1}$, which is expected to be non-anomalous.

Kawasaki, carrying out a similar analysis, found[17,18]

$$\frac{\Lambda(q)}{\rho c_p}\ q^2 = \Gamma_q = \frac{kT}{(2\pi)^3\ \eta_s^*}\int d\vec{k}\ [(\frac{q}{k})^2 - (\frac{\vec{q}\cdot\vec{k}}{k^2})^2]\ \frac{c_p(\vec{q}-\vec{k})}{c_p(q)}, \tag{11}$$

which when integrated with the Ornstein-Zernike form $c_p(q) \sim (q^2+\xi^{-2})^{-1}$, reduces to:

$$\Gamma_q = (kT/6\pi\eta_s^*)\ \xi^{-3}K_o(q\xi), \tag{12a}$$

where

$$K_o(x) = (3/4)\ [1 + x^2 + (x^3 - x^{-1})\ \tan^{-1} x] \tag{12b}$$

A form which is consistent with the predictions of dynamical scaling as discussed in the lecture of Professor Schwabl. Kawasaki's result, Eq. (12), predicts the decay rate of order parameter fluctuations over the entire range from the hydrodynamic region ($q\xi<<1$) to the extreme critical region ($q\xi>>1$).

In the hydrodynamic and extreme critical regions, Eq. (12) becomes:

$$\Gamma_q\ (q\xi<<1) = (kT/6\pi\eta_s^*)\ \xi^{-1}\ q^2, \tag{13a}$$

$$\Gamma_q \; (q\xi >> 1) \;=\; (kT/16\eta_s^*)\, q^3. \qquad (13b)$$

Note that Eq. (13a) can be interpreted as ordinary diffusion of a "droplet" with radius = ξ.

Experimentally, the extreme critical behavior of Eq. (13b) as well as the functional dependence of Eq. (12) were first observed in the binary mixture aniline-cyclohexane by Bergé et al.[16,24]

Equation (12), however, cannot represent the full mode-mode coupling theoretical result as it stands since the viscosity η_s^* is still ambiguous, no vertex corrections are included, and, finally, intermediate states with three or more modes (representing higher-order nonlinearities) are not included. We return to these points later.

6. EXPERIMENTS IN THE HYDRODYNAMIC REGION

The time dependence of density fluctuations in fluids and binary mixtures in the critical region has been studied in many different systems since the first experiments in 1965.[25,26] The experimental technique generally employed utilizes a laser beam which traverses the sample in a carefully controlled environment, and the light scattered into a small solid angle at an angle θ from the forward beam is focussed on a photomultiplier. The output current of the photomultiplier is then analyzed either with an electronic spectrum analyzer, or in the most recent experiments, with a digital photoelectron correlator. (For reviews of the experiments see references 3, 4, 27 and 28).

The photoelectron correlation function (or the photocurrent spectrum) measured at different temperatures and scattering angles is then analyzed to find the density autocorrelation function $C_q(t)$ which is of the form $C_q(t) = \langle(\rho_q)^2\rangle \exp(-\Gamma_q t)$. Plots of $\chi = \Gamma_q/q^2$ for CO_2, from the 1968 experiment of Swinney and Cummins, are shown in Figs. 1 and 2.[29] Note that in Fig. 1, the thermal diffusivity does approach zero, as expected, but that the downward curvature implies that the exponent ϕ is less than unity. (ϕ is determined from a log-log plot as shown in Fig. 2).

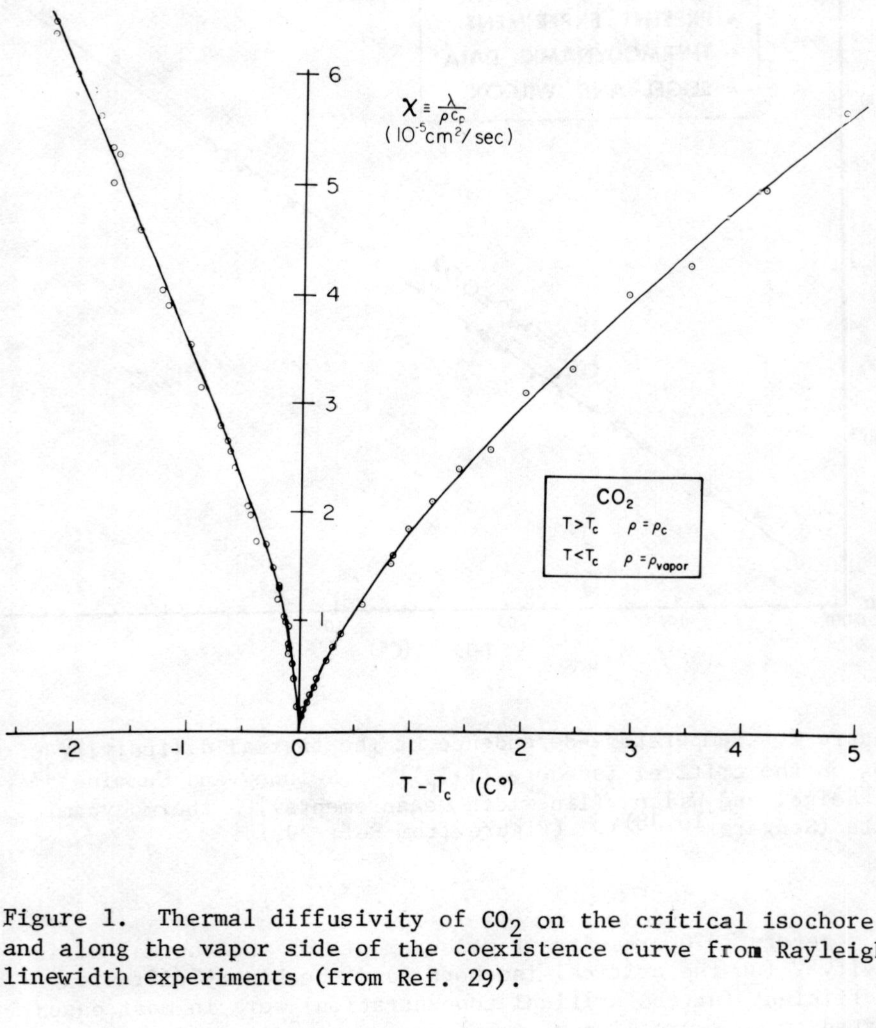

Figure 1. Thermal diffusivity of CO_2 on the critical isochore and along the vapor side of the coexistence curve from Rayleigh linewidth experiments (from Ref. 29).

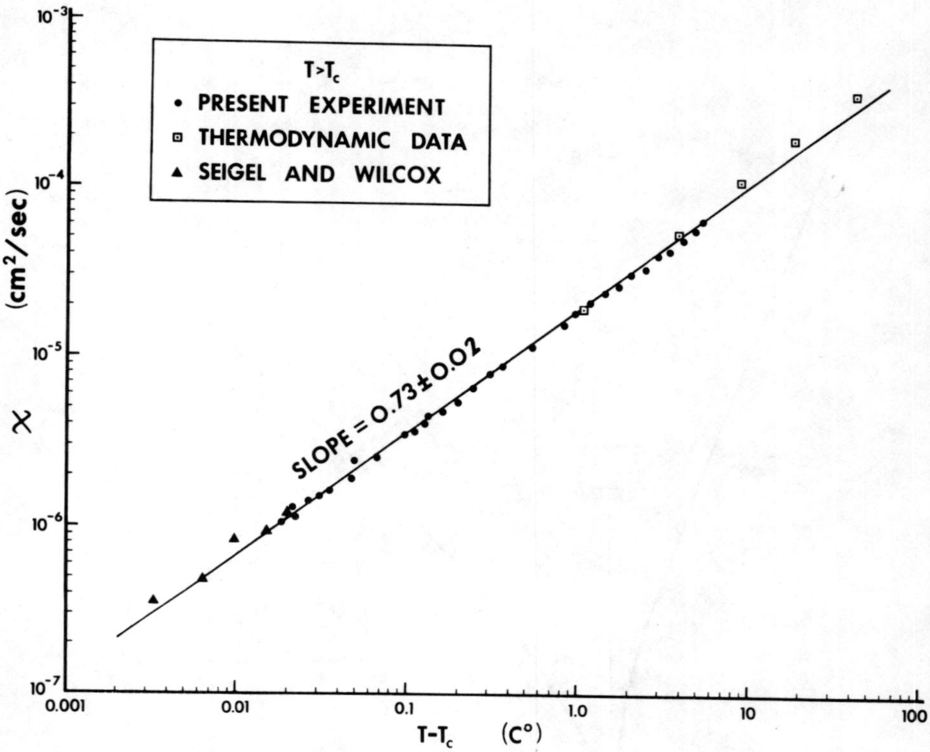

Figure 2. Temperature-dependence of the thermal diffusivity of CO_2 on the critical isochore ($T>T_c$): Swinney and Cummins[29]; Seigel and Wilcox (linewidth measurements); thermodynamic data (Sengers[10,38]). (Figure from Ref. 29.)

Before 1970, results of measurements of the thermal diffusivity χ (on the critical isochore) and the binary diffusion coefficient (at the critical concentration) were in most cases fitted to an expression of the form $\chi = \chi_o \varepsilon^{\phi}$ in analogy with the simple exponential laws known to describe the behavior of static properties of fluids. (See Table I.)

Table I. - Rayleigh linewidth determinations of the quantities $\chi = \chi_o \varepsilon^{\phi}$; $D = D_o \varepsilon^{\phi}$. ($\varepsilon = |T - T_c|/T_c$.)

System	Author	Path	χ_o or D_o (cm^2/s)	ϕ
SF_6	Benedek (3b,42)	$T>T_c$, $\rho=\rho_c$ $T<T_c$, gas $T<T_c$, liquid	$(1.26\pm0.04)\times10^{-2}$ $(1.79\pm0.01)\times10^{-3}$ $(1.75\pm0.02)\times10^{-3}$	1.26 ± 0.02 0.632 ± 0.002 0.635 ± 0.003
SF_6	Braun (30)	$T>T_c$, $\rho=\rho_c$ $T<T_c$, gas $T<T_c$, liquid	$(1.89\pm0.08)\times10^{-3}$ $(5.62\pm0.06)\times10^{-3}$ $(3.92\pm0.18)\times10^{-3}$	0.89 ± 0.07 0.88 ± 0.02 0.83 ± 0.04
CO_2	Swinney (29)	$T>T_c$, $\rho=\rho_c$ $T<T_c$, gas $T<T_c$, liquid	$(1.176\pm0.032)\times10^{-3}$ $(1.57\pm0.13)\times10^{-3}$ $(2.14\pm0.15)\times10^{-3}$	0.73 ± 0.02 0.66 ± 0.05 0.72 ± 0.05
Xe	Henry (31)	$T>T_c$, $\rho=\rho_c$	$(6.94\pm0.21)\times10^{-4}$	0.751 ± 0.004
Isobutyric acid+ water	Chu (32)	$T>T_c$	6.08×10^{-6}	0.68 ± 0.04
n-hexane+ nitrobenzene	Chen (33)	$T>T_c$	$\sim 1.65\times10^{-5}$	0.66 ± 0.02
Aniline+ cyclohexane	Berge (34)	$T>T_c$	$(7.7\pm0.7)\times10^{-6}$	0.588 ± 0.06
Phenol+ water	Pusey (35)	$T>T_c$ $T<T_c$, low ρ $T<T_c$, high ρ	$(1.17\pm0.03)\times10^{-5}$ $(2.67\pm0.08)\times10^{-5}$ $(1.35\pm0.03)\times10^{-5}$	0.63 ± 0.03 0.68 ± 0.03 0.60 ± 0.02

The linewidths measured in binary mixtures were found to be described in the hydrodynamic region by the simple exponential law with $\phi = \gamma - \psi \simeq 0.63$, but the exponents that were obtained for the simple fluids CO_2 and xenon were somewhat higher, $\phi \simeq 0.74$, and for SF_6 the exponent was markedly different, $\phi \simeq 1.26$, contrary to the expected "universality" in critical behavior.

The larger exponents observed for the pure fluids was a matter of serious concern because the concept of universality was well established from numerous measurements of the static properties of many systems near the critical point. It was suggested that perhaps the measurements were affected by impurities in the samples, but extensive systematic studies by Bak and Goldburg,[36] (on phenol-water with hypophosphorous acid as an impurity) and by Bak, Goldburg, and Pusey[37] (on bromobenzene-water with acetone as an impurity) showed that even for fairly high impurity concentrations the critical behavior is unchanged except for a change in T_c.

It was also suggested that the hydrodynamic expression for Γ_q, Eq. (6), might not apply near the critical point, even when $q\xi \ll 1$; however, there is one fluid, CO_2, for which the diffusivity has been determined in the critical region both by linewidth measurements and by conventional thermodynamic techniques, and in the temperature range common to both sets of data the diffusivities obtained by the different techniques are in excellent agreement, thus corroborating Eq. (6). (See Fig. 2).

In 1970 Sengers suggested that the apparently higher exponents observed for the pure fluids could be explained by taking into consideration the nondivergent background contribution to the thermal conductivity.[38] Because of the large contribution of nonanomalous background terms, the behavior of systems in the temperature region readily accessible to experiment may be very different from the true asymptotic behavior which is presumably describable by the simple exponential laws. (See the discussion in the 1967 review by Fisher.)[39] Thus the nonsingular background contributions must first be subtracted if the data are to be analyzed over extended temperature ranges.

Generalizing the hydrodynamic result $\Gamma_q = (\Lambda/\rho c_p)q^2$ to include a q dependence in Λ and c_p, and separating Λ and c_p into background and singular parts, we have

$$\Gamma_q = [\Lambda^b/\rho(c_p)_q]\, q^2 + \Gamma_q^s\, [(c_p)_q^s/(c_p)_q], \tag{14}$$

where $\Gamma_q^s \equiv [\Lambda_q^s/\rho(c_p)_q^s]\, q^2$ is the singular part of the linewidth.

If the Ornstein-Zernike form is assumed for the q dependence of $(c_p)^s_q$ and $(c_p)_q$, then Eq. (14) becomes:

$$\Gamma_q = (\Lambda^b/\rho c_p)\, q^2\, (1+q^2\xi^2) + \Gamma^s_q\, [(c_p)^s/c_p], \tag{15}$$

where the absence of a subscript q on c_p indicates the $q = 0$ (thermodynamic) quantity.

The partition of a transport coefficient into background and critical parts is clearly a crucial part of the data analysis in any experimental investigation of the dynamics of a system near the critical point. In such experiments, which include, for example, measurements of the spin diffusion rate in magnets and the sound velocity and attenuation in fluids as well as measurements of the viscosity, conductivity and diffusivity, meaningful comparison between theory and experiment can be made only if a systematic procedure can be developed for estimating the bare transport, or Onsager kinetic coefficient. For the thermal conductivity and shear viscosity of a pure fluid Sengers and Keyes have developed a method for estimating Λ^b and η^b_s using data obtained far from the critical point.[40] The procedure is based on the empirical result, frequently used in the engineering literature, that the "excess" thermal conductivity,

$$\tilde{\Lambda}(\rho) = \Lambda(\rho, T) - \Lambda(0, T) \tag{16}$$

[where $\Lambda(\rho,T)$ is the thermal conductivity at a density ρ and temperature T, and $\Lambda(0, T)$ is the thermal conductivity in the dilute gas limit], is independent of temperature for temperatures and densities up to approximately twice the critical temperature and density. The Sengers-Keyes ansatz is that the background thermal conductivity in the critical region is given by

$$\Lambda^b\, (\rho, T) = \tilde{\Lambda}(\rho) + \Lambda(0, T), \tag{17}$$

where $\tilde{\Lambda}(\rho)$ is determined using data obtained away from the critical region. A similar expression is assumed to hold for the background viscosity.

Linewidth measurements indicate that the background contribution to the conductivity is far less important for mixtures than for pure fluids, and de Gennes has argued that this is plausible on physical grounds.[41]

Sengers and Keyes[40] found that when the CO_2 data were analyzed with the background thermal conductivity taken into account, the exponent ϕ was reduced from 0.73 to 0.62; in a similar analysis of the xenon data we found that ϕ was reduced from 0.75 to 0.64.[28] But Benedek, et al. found that the thermal conductivity background correction did not bring their

SF_6 data into agreement with the results of other fluids.[42] (See also Refs. 30 and 43). The SF_6 puzzle has recently been solved by three new independent experiments (Langley and coworkers;[44,45] Lim and Swinney;[45] and Feke, Hawkins, Lastovka and Benedek),[46] all in agreement with one another and in strong disagreement with the previous SF_6 data. The new SF_6 linewidth data, after subtraction of the background terms, yield $\phi = 0.61 \pm 0.04$; hence SF_6 does indeed exhibit the same critical behavior as other fluids. (Separation of the total decay rate into critical and background parts is illustrated for SF_6 in Fig. 3, where the singular part is calculated using the mode-mode coupling theory expression of Kawasaki and Lo.[20])

Thus the experimental results for all systems studied to date indicate that in the hydrodynamic region ($q\xi \ll 1$), the critical contribution to the decay rate Γ_q is of the form q^2/ξ, in agreement with the predictions of Kadanoff and Swift.[15]

7. EXTENSION OF EXPERIMENTS INTO THE NONHYDRODYNAMIC REGION

In 1969 Berge, Calmettes, Volochine, and Laj[24] extended their linewidth measurements on aniline-cyclohexane into the extreme nonhydrodynamic regime where $q\xi \gg 1$, and subsequently this region has been investigated for other mixtures and simple fluids. The measured decay rates, after the subtraction of the background, have been found to be described well by the Kawasaki expression, Eq. (12). However, for most systems this comparison was made taking η_s^* and ξ (or at least ξ_o) as adjustable parameters since they were not known from independent measurements (see, e.g., Refs. 24, 31 and 47). In those systems for which independent measurements of ξ and η_s were performed (see, e.g. Refs. 28, 48 and 49) the theory was found to yield linewidth values near T_c larger than those measured if the background viscosity were used for η_s^*, while the predicted linewidths near T_c would be too small if the theory was interpreted using for η_c^* the full measured macroscopic values of the shear viscosity, indicating that the mode-mode coupling theory in the form indicated by Eq. (12) is inadequate.

Since 1970 there have been three significant modifications to the original Kawasaki result (Eqs. 11 and 12), which we discuss briefly before presenting our final comparison of the data with the most complete version of the mode-mode coupling theory currently available.

a) The viscosity η_s^* appearing in Eq. (11) is ambiguous since it represents an effective weighted average over all viscous modes appearing in intermediate states. The correct interpretation of

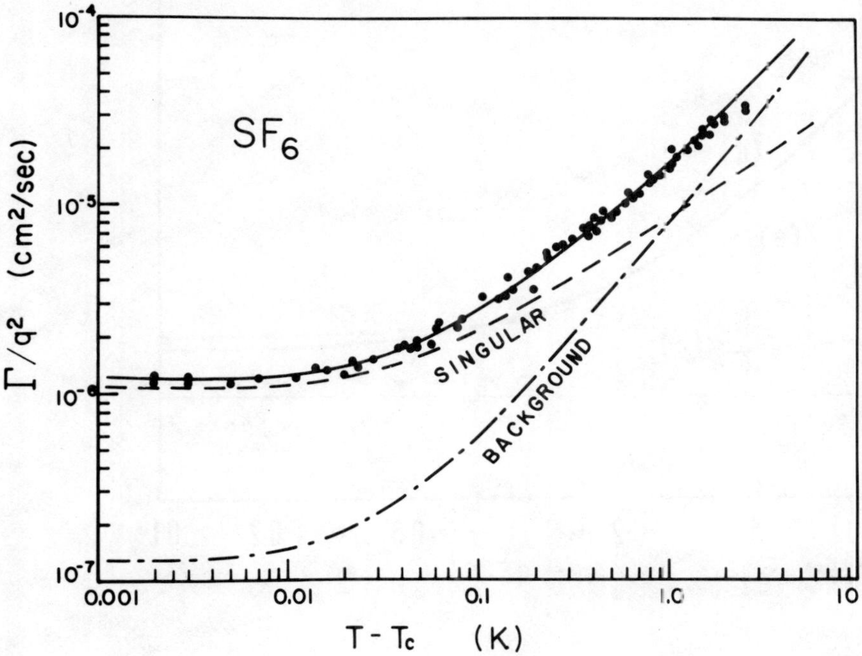

Figure 3. Separation of the SF_6 decay rate into singular and background parts. The solid curve is the sum of the singular and background contributions. (Ref. 45).

η_s^* requires self-consistent evaluation of both the order-parameter and viscous flow mode propagators. Recently, Kawasaki and Lo have succeeded in partially solving the self-consistency equations to the point of deducing an expression which relates η_s^* to the macroscopic shear viscosity η_s, thus removing the ambiguity in the viscosity.[20] They find:

$$\eta_s^* = [K_o(q\xi)/K(q\xi)]\ \eta_s \tag{18}$$

where $K(q\xi)$ is given numerically by an integral. $K(q\xi)/K_o(q\xi)$ is shown in Fig. 4, curve (a) (taken from Fig. 3 of Kawasaki and Lo). Note that η_s^* differs from η_s even far from T_c; in that region $\eta_s^* = \eta_s/1.055$. [$K_o(q\xi)$ is given by Eq. (12).] The problem of the wavenumber-dependent viscosity has also been considered by Perl and Ferrell with somewhat different results.[50]

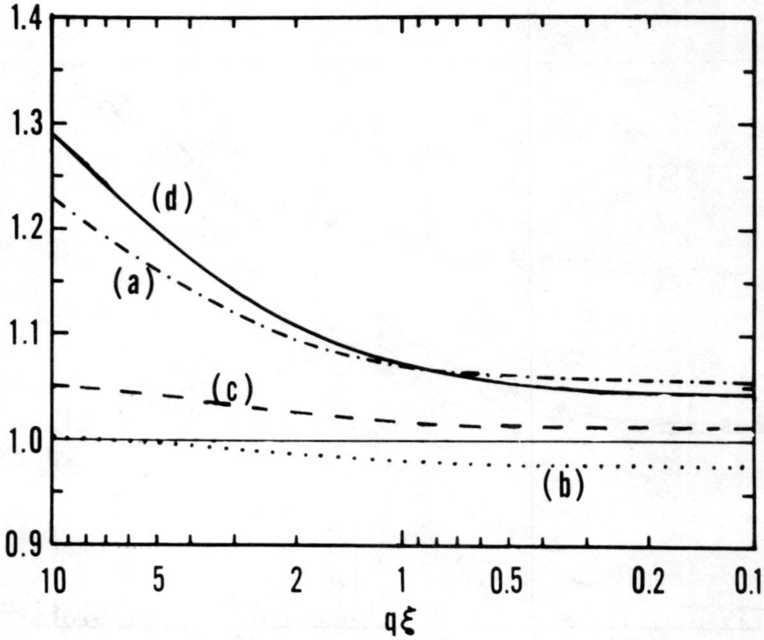

Figure 4. Modifications to the Kawasaki theory which are discussed in the text. (a) - self consistent solution for the viscosity η_s^*;[20] (b) - lowest-order vertex corrections;[19] (c) - correlation function correction using the Fisher-Burford form for the correlation function.[54] Curve (d) includes all three corrections and gives the ratio of the modified decay rate to that of Eq. 12.

b) Equation (12) was derived in "self consistent phonon approximation" and contains no vertex correction terms to the self energy. (This same result was also derived by Ferrell[51] from a direct evaluation of the current correlation function defining Λ. Ferrell's "Decoupled-Mode Approximation" is equivalent to the Kawasaki result without vertex corrections because the absence of internal lines between the two intermediate state propagators allows them to be factored within the Kubo integral.)[52]

Recently, Lo and Kawasaki have computed the self energy corrections resulting from the simplest vertex correction terms.[19] They found that their effect is to reduce Eq. (12) by 2.44% for $q\xi \ll 1$ and increase Eq. (12) by 0.40% for $q\xi \gg 1$. The vertex correction $V(q\xi)$, and the ratio of the corrected to the uncorrected decay rate, is shown by curve (b) in Fig. 4, which was obtained by connecting the limiting values of $V(q\xi)$, $V(\infty)$ and $V(0)$, by a smooth curve. [A calculation of this small modification to the theory for intermediate values of $q\xi$ would require the evaluation of a complicated integral expression - Eq. (2.10) in Ref. 19.] The vertex correction to the linewidth is frequency dependent, with the above values for $q\xi \ll 1$ and $q\xi \gg 1$ applying only in the zero frequency limit. Because of the frequency dependence of the vertex correction, the observed spectral line will in principle deviate from the Lorentzian lineshape, (or equivalently, the correlation function will deviate from simple exponential decay) but the correction is so small that the predicted deviations would be very difficult to observe.

c) The integral expression Eq. (11) for the decay rate was evaluated by Kawasaki using the Ornstein-Zernike form for the correlation function, leading to the result, Eq. (12). But scattering experiments and the theoretical investigations of the Ising model by Fisher and Burford(53) have shown that there are small departures from Ornstein-Zernike behavior near the critical point. The correct asymptotic form for the correlation function at the critical point is expected to be $r^{-(1+\eta)}$ with $\eta \simeq 0.05$ to 0.1, while for the Ornstein-Zernike theory $\eta = 0$.

Fisher and Burford found that correlations in the Ising model are accurately described by:

$$G_{FB} \propto (\xi^{-2} + \phi^2 q^2)^{\eta/2} / [\xi^{-2} + (1+\phi^2\eta/2)\, q^2] \tag{19}$$

where $\phi = 0.15 \pm 0.01$, independent of the type of lattice. Swinney and Saleh(54) have evaluated the decay rate integral using G_{FB} and the result for the decay rate ratio,

$$C(q\xi) \;=\; \Gamma_q^s\,(G_{FB},\, q\xi)/\Gamma_q^s\,(G_{OZ},\, q\xi), \tag{20}$$

is given by curve (c) in Fig. 1 for $\eta = 0.1$. (The decay rate integral was also evaluated by Swinney and Saleh and by Chang _et al._(55) for other forms of the correlation function which have been used in the analysis of data from scattering experiments; however, G_{FB} is more satisfactory theoretically since, as explained in Ref. 53, it leads to the correct asymptotic behavior at large r both at the critical point and away from the critical point.)

With the viscosity, vertex, and correlation function modifications included, the linewidth expression Eq. (12a) in the mode-mode coupling theory becomes

$$\Gamma_q^s = (kT/6\pi\eta_s\xi^3)\ K_o(q\xi)\ H(q\xi), \tag{21}$$

where η_s is the macroscopic shear viscosity, and the correction factor $H(q\xi)$, which most analyses of linewidth data have heretofore assumed to be unity, is given by:

$$H(q\xi) \equiv [K(q\xi)/K_o(q\xi)]\ V(q\xi)\ C(q\xi) \tag{22}$$

and is plotted as curve (d) in Fig. 4. Although the vertex correction has not been evaluated for intermediate values of $q\xi$ and the correlation function correction is somewhat uncertain because the correct form for the correlation function is not well established, these two corrections are nevertheless both small, and we will consider them as well as the nonlocal shear viscosity correction in our data analysis.

These recent refinements of the mode-mode coupling theory warrant a new analysis of the existing data, particularly since recent independent measurements of the shear viscosity and correlation length for several systems now permit a direct comparison between the measured linewidths and the theoretical predictions using no adjustable parameters. In the present paper we analyze all the linewidth data for fluids for which independent η_s and ξ data exist, testing particularly the prediction of the mode-mode coupling theory that a particular dimensionless combination of the measured quantities, $(6\pi\eta_s\Gamma_q^s/kTq^3)$, where Γ_q^s is the singular part of the linewidth, should be described by the same universal function of $q\xi$ for all simple fluids and mixtures.

The equation for the critical part of the linewidth Eq. (21) can be rewritten as:

$$\Gamma^*_{th} = \left(\frac{1}{q\xi}\right)^3 K_o(q\xi)\ H(q\xi), \tag{23a}$$

where the "scaled" linewidth Γ^*_{exp} is defined as [with Γ_q^s defined in Eq. (15)]:

$$\Gamma^*_{exp} \equiv \left(\frac{6\pi\eta_s}{kT}\right)\left(\frac{\Gamma_q^s}{q^3}\right). \tag{23b}$$

Thus the theory predicts that the data for different temperatures and scattering angles, obtained for various simple fluids and binary mixtures, should all fall on a single universal curve when the (dimensionless) quantity Γ^* is plotted as a function of $q\xi$. This single curve is predicted to describe the critical behavior not only along the critical isochore and the coexistence curve, but also along any other thermodynamic path in the critical region.

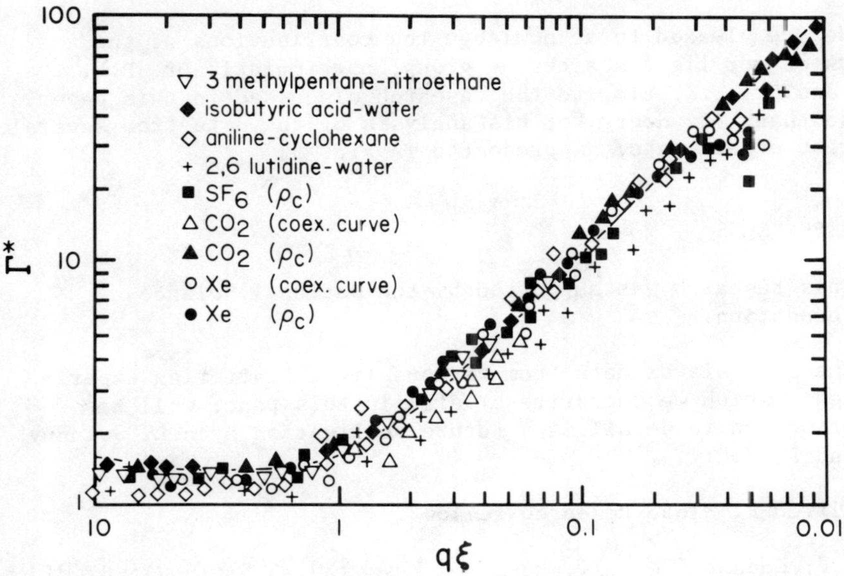

Figure 5. Comparison of experimental data for different systems and thermodynamic paths "scaled" according to Eq. (23b) with the prediction of the modified Kawasaki theory, Eq. (23a). The comparison is shown with no adjustable parameters.

A universal plot of the existing data for fluids and binary mixtures, reduced to the form of Eq. (23) is shown in Fig. 5. We emphasize again that there are no adjustable parameters in this plot. (A full discussion of the data analysis will appear in a forthcoming paper by Swinney and Henry.)

The conclusion is that the theory now accounts for all available data within the experimental accuracy. Subsequent refinements will be required in the experiments, particularly those providing thermodynamic parameters, before a more rigorous test of the theory will be possible, and the effects of additional corrections (such as higher order nonlinear coupling terms) can be evaluated.[56]

8. ACKNOWLEDGMENTS

We are pleased to acknowledge the contributions of the members of our light scattering group, particularly Dr. D.L. Henry and Dr. T.K. Lim, to the research discussed in this paper. We also thank Dr. Henry for his analysis of the data from several different experiments, as presented in Fig. 5.

9. REFERENCES

* This research was supported by the National Science Foundation.

+ The analysis of data from various light scattering experiments which we summarize briefly in this paper will be presented in detail in a future publication by H.L. Swinney and D. Henry.

** Alfred P. Sloan Research fellow.

1. L.P. Kadanoff et al., Rev. Mod. Phys. 39, 395 (1967); B. Widom, J. Chem. Phys. 43, 3898 (1965).

2. K.G. Wilson, Phys. Rev. Letters 28, 548 (1972).

3. H.Z. Cummins and H.L. Swinney, in: Progress in Optics, Vol. 8, edited by E. Wolf (North-Holland, Amsterdam, 1970), p. 133; G.B. Benedek, in Polarization Matiere et Rayonnement, Livre de Jubilé en l'Honneur du Professeur A. Kastler, edited by the French Physical Society (Presses Universitaires de France, Paris, France, 1969), p. 49.

4. See review by H.Z. Cummins, in: Critical Phenomena, International School of Physics "Enrico Fermi", LI Course (1970), edited by M.S. Green (Academic Press, New York, 1971), p. 380.

5. The material in this section is presented in greater detail in Ref. 4.

6. A. Einstein, Ann. Phys. 33, 1275 (1910).

7. R.D. Mountain, Rev. Mod. Phys. 38, 205 (1966).

8. L. Van Hove, Phys. Rev. 95, 1374 (1954); see also L.D. Landau and I.M. Khalatnikov, Dokl. Akad. Nauk. SSSR 90, 469 (1954).

9. M. Fixman, Adv. Chem. Phys. 6, 175 (1964).

10. J.V. Sengers, in Critical Phenomena, edited by M.S. Green and J.V. Sengers (National Bureau of Standards Miscellaneous Publication 273, Washington, D.C., 1966), p. 165.

11. W. Botch and M. Fixman, J. Chem. Phys. 42, 196 (1965).

12. The fact that Fixman's result recovers the static scaling law was noted by L. Mistura and D. Sette, J. Chem. Phys., 49, 1419 (1968). Mistura (loc. cit. Ref. 4) has generalized Fixman's derivation and found a complex frequency-dependent specific heat from which the attenuation and dispersion of sound can be predicted.

13. K. Kawasaki, Phys. Rev. 150, 291 (1966).

14. See, for example, R. Zwanzig, Ann. Rev. Phys. Chem. 16, 67 (1965).

15. L.P. Kadanoff and J. Swift, Phys. Rev. 166, 89 (1968); L.P. Kadanoff, J. Phys. Soc. Jap., 26 (supplement), 122 (1969).

16. K. Kawasaki, Phys. Letters 30A, 325 (1969).

17. K. Kawasaki, Ann. Phys. (N.Y.) 61, 1 (1970).

18. K. Kawasaki, Phys. Rev. A 1, 1750 (1970).

19. S. Lo and K. Kawasaki, Phys. Rev. A 5, 421 (1972).

20. K. Kawasaki and S. Lo, Phys. Rev. Letters 29, 48 (1972).

21. See, for example, R.A. Cowley, in: Phonons in Perfect Lattices and Lattices with Point Imperfections, edited R.W.H. Stevenson (Plenum Press, N.Y., 1966), and the paper of R. Klein in this volume.

22. R.D. Mattuck, A Guide to Feynman Diagrams in the Many-Body Problem (McGraw-Hill, N.Y. 1967?) Sections 10, 11; R. Silberglitt, Solid State Commun. 11, 247 (1972).

23. Our schematic discussion of mode-mode coupling theory is close to the spirit of Ref. 17. The approach of Kadanoff and Swift leads to the decay rate directly. (Ref. 15)

24. P. Bergé et al., Phys. Letters 30A, 7 (1969); Phys. Rev. Letters 23, 693 (1969).

25. S.S. Alpert, Y. Yeh and E. Lipworth, Phys. Rev. Letters 14, 486 (1965).

26. N.C. Ford and G.B. Benedek, Phys. Rev. Letters 15, 649 (1965).

27. B. Chu, Ann. Rev. Phys. Chem. 21, 145 (1970).

28. H.L. Swinney, D.L. Henry and H.Z. Cummins, J. de Phys. (Paris) 33, C1-181 (1972).

29. H.L. Swinney and H.Z. Cummins, Phys. Rev. 171, 152 (1968).

30. P. Braun, D. Hammer, W. Tscharnuter and P. Weinzierl, Phys. Letters 32A, 390 (1970).

31. D.L. Henry, H.L. Swinney and H.Z. Cummins, Phys. Rev. Letters 25, 1170 (1970).

32. B. Chu and F.J. Schoenes, Phys. Rev. Letters 21, 6 (1968).

33. S.H. Chen and N. Polonsky-Ostrowsky, Optics Communications 1, 64 (1969).

34. P. Bergé, P. Calmettes, C. Laj, M. Tournarie and B. Volochine, Phys. Rev. Letters 24, 1223 (1970).

35. P.N. Pusey and W.I. Goldburg, Phys. Rev. Letters 23, 67 (1969).

36. C.S. Bak and W.I. Goldburg, Phys. Rev. Letters 23, 1218 (1969).

37. C.S. Bak, W.I. Goldburg and P.N. Pusey, Phys. Rev. Letters 25, 1420 (1970).

38. J.V. Sengers, loc. cit. (Ref. 4), p. 445.

39. M.E. Fisher, Reports Prog. Phys. 30, 615 (1967).

40. J.V. Sengers and P.H. Keyes, Phys. Rev. Letters 26, 70 (1971).

41. P. de Gennes, unpublished lecture, Second International Conference on Light Scattering in Solids, Paris, 1971.

42. G.B. Benedek, J.B. Lastovka, M. Giglio and D. Cannell, in: Critical Phenomena, edited by R.E. Mills, E. Ascher and R.I. Jaffey (McGraw-Hill, New York, 1971), p. 503.

43. N. Theodorakopoulos, Ph.D. Thesis, Brown University, 1972 (unpublished).

44. R. Mohr and K.H. Langley, J. de Phys. (Paris) 33, C1-97 (1972).

45. T.K. Lim, H.L. Swinney, K.H. Langley and T.A. Kachnowski, Phys. Rev. Letters 27, 1776 (1971).

46. G.T. Feke, G.A. Hawkins, J.B. Lastovka and G.B. Benedek, Phys. Rev. Letters 27, 1780 (1971).

47. W.I. Goldburg and P.N. Pusey, J. de Phys. (Paris) 33, C1-105 (1972).

48. R.F. Chang, P.H. Keyes, J.V. Sengers and C.O. Alley, Phys. Rev. Letters 27, 1706 (1971).

49. B. Chu, D. Thiel, W. Tscharnuter and D.V. Fenby, J. de Phys. (Paris) 33, C1-111 (1972).

50. R. Perl and R.A. Ferrell, Phys. Rev. Letters 29, 51 (1972); Phys. Rev. A 6, 6 (1972).

51. R.A. Ferrell, Phys. Rev. Letters 24, 1169 (1970).

52. R.A. Ferrell, in: Dynamical Aspects of Critical Phenomena, edited by J.I. Budnick and M.P. Kawatra (Gordon and Breach, New York, 1971), p. 1.

53. M.E. Fisher and R.J. Burford, Phys. Rev. 156, 583 (1967); see also D.S. Ritchie and M.E. Fisher, Phys. Rev. B 5, 2668 (1972).

54. H.L. Swinney and B.A. Saleh, Phys. Rev. A 7, 747 (1973).

55. R.F. Chang, P.H. Keyes, J.V. Sengers and C.O. Alley, Bunsen-Ges. Phys. Chem 76, 260 (1972).

56. In the extreme critical region ($q\xi >> 1$) there is some evidence for a systematic difference between theory and experiments. However, Perl and Ferrell (Ref. 50) have included both the wavevector and frequency dependence of the viscosity in their decoupled-mode derivation of the decay rate in the extreme nonhydrodynamic region, and there the agreement between theory and experiment appears to be somewhat better than that obtained with the theory of Kawasaki and Lo (Ref. 20), who considered the effect of nonlocality on the shear viscosity, but neglected the frequency dependence. This comparison, along with a full discussion of the analysis of all the experimental data will be discussed in a future publication by H.L. Swinney and D. Henry.

NEUTRON SCATTERING IN SOLID AND LIQUID H_2

K. Carneiro and M. Nielsen

Danish Atomic Energy Commission
Research Establishment Risø
Roskilde, DK 4000 Denmark

ABSTRACT

As an introduction a brief review is given of the variety of general physical problems that have been investigated by studying the para and ortho modifications of H_2 and D_2 in the solid, as well as in the liquid phase. In section II we describe in more detail the properties of H_2 and show how they influence the neutron scattering of this material. This is applied when we describe the lattice dynamics of solid hcp H_2 in section III and the dynamics of liquid H_2 in section IV.

I. INTRODUCTION

The system that will be described in this lesson, molecular hydrogen in the solid and the liquid phases, is an example of a fairly simple system, whose molecular dynamics can be described in terms of only two parameters which can be systematically varied, so that by putting together the information obtained by different techniques one can get a good understanding of the system. The two parameters are the mass M and the angular momentum quantum number J. The latter will be described in section II. Now we simply notice that for parahydrogen (p-H_2) and orthodeuterium (o-D_2) we have J=0, which means that the molecules interact through isotropic van der Waals forces only, whereas for orthohydrogen (o-H_2) and paradeuterium (p-D_2) the fact that J=1, introduces anisotropic interaction. In Table I we show some of the characteristics of these constituents. Rather than going through this table number by number, we will review some of the experiments that have been performed and try to point out why this particular system is investigated.

Property		Unit	p-H_2	o-H_2	o-D_2	p-D_2
Boiling Temp.	T_B	K	20.27	20.42*	23.57 †	
Melting point	T_M	K	13.81	14.00*	18.6 †	
Transition Temp.	T_c, T_λ	K	0	~ 2	0	~ 4
Mass	M/m_n	-	2	2	4	4
Bond length	d	Å	0.746	0.746	0.742	0.742
Rotational quantum number	J	-	0	1	0	1
Rotational energy	E_J	meV	0	14.6(ΔE)	0	7.1
Molecular Spin	I	-	0	1	0 or 2	1
Neutron scattering cross section (per nucleus).	σ_{coh}	barns	1.77		5.4	
	σ_{inc}	barns	79.9		2.2	

* Obtained by extrapolation. For natural H_2 T_B = 20.38 and T_M = 13.95.

† The data are for natural D_2.

Table I. Some characteristics of the para and ortho modifications of H_2 and D_2, of importance in the solid and liquid phases. Note: 1meV = 11.6 K= 8.06 cm^{-1} = 242 GHz.

In 1967 Egelstaff et.al. (1) reported results from incoherent neutron scattering on liquid o-H_2. Since o-H_2 and vanadium, the latter having a melting point of 1920 ^oC, are the only existing elemental purely incoherent scatterers this was a case to study the molecular "self-motion", in a liquid. Their results show that self -motion in liquids can be well understood when considering only the two extreme cases, namely Brownian motion for long times and free particle motion for short times. The characteristics of the motion then becomes the diffusion coefficient D showing Arrhenius behaviour, and a time constant τ_0. By studying the temperature dependence of τ_0 one can get a hint about the type of diffusion, and the measurements indicate that at low temperatures the diffusion is of the "jump" type like in the solid. As the temperature is increased the diffusive motion approaches free gas-like diffusion.

A complementary approach to that of (1) is to study the liquid phase by coherent neutron scattering from p-H_2, to obtain information on the dynamical behaviour of the liquid (2). Since at present two rather incompatible types of theories have been developed for collective excitations in liquids, and since experimental results on liquid ^4He and liquid Ar are very different, there is an intense discussion at a conceptual level about the existence and the nature of collective excitations in simple liquids, and about the similarities between quantum liquids and classical ones. As we will see in the last section, liquid hydrogen seems to behave "in between" the two extreme cases of liquid ^4He and liquid Ar, indicating that liquid dynamics should be treated as conceptually the same phenomenon in all liquids.

Cooling the liquid one obtains the high temperature hcp phase of the solid. The phonon dispersion relations have been measured in p-H_2 and o-D_2 by coherent neutron scattering (3). In the quantum solids He, H_2, D_2 the zero point motion of the molecules have amplitudes which are not small compared to the intermolecular separations. This introduces a strong anharmonicity, and a modified form of the lattice dynamics applicable to classical van der Waals crystals, has to be established in order to predict the experiments theoretically. We shall return to this in section III in the case of H_2. The Raman active phonon modes (4) were investigated in the four species and in HD. The results are in agreement with the neutron data, and together the two sources provide detailed comparison with the present theories.

Another field of lattice dynamics needing more investigation, is the effect of impurities in crystals. The H_2/D_2 system is a good candidate to bring this field a step further (5). It will be the first hcp-system to be studied by coherent neutron scattering and so far only cubic Bravais lattices have been examined by this method. The simplest theory is based on the impurity being a pure mass defect in the host lattice but a large difference in effective force constants between H_2 and D_2 is expected to be more stringent test of this theory. Further H_2 and D_2 have very different σ_{coh} making it possible not only to measure the perturbation of the phonon dispersion relation but in addition to observe directly the impurity mode.

When the temperature is further decreased the fcc-structure eventually becomes stable if the (J=1)-mole fraction X is larger than 0.5. This was investigated by Stein et.al.(6), who made incoherent neutron scattering in both the hcp- and the fcc-phase, giving e.g. the density of phonon states. The deduced results are in agreement with other measurements. Further this paper investigated in some detail the phase transition where the structure changes to fcc at T_c and an orientational ordering of the (J=1) -molecules at T_λ occurs. To attribute two temperatures to the transition is somewhat artificial. Both T_c and T_λ show hysteresis and cannot be experimentally distinguished, and therefore we set $T_c = T_\lambda$ in Table I. For the theoretical understanding it is important that $T_\lambda < T_c$, i.e. the ordering takes place in the fcc phase. One result of ref.6 was then to measure the total density of states and subtract the known contribution of the phonons. The rest is then attributed to orientational waves (librons) that are analogous to spin-waves in ordered magnetic systems.

The transition is interesting from various viewpoints. The ordering is in close analogy with that of an antiferromagnet and the order parameter (7) has been measured by NMR techniques (8). Since one can vary T_λ by varying X, one can study the order parameter vs T/T_λ for different compositions. Silvera et.al.(9) have studied in more detail the forces involved in the ordering, by Raman scattering. The transition can be theoretically predicted assuming anisotropic interaction between (J=1)-molecules and X> 0.5. The anisotropy was investigated for both small and large concentrations, and assuming only electrical quadrupole - quadrupole (EQQ) interaction in addition to the isotropic van der Waals interaction they were able to explain their results.

Neutron diffraction experiments on powdered o-D_2 have revealed the detailed structure of the fcc-ordered phase (10), but the orientational phase transition is not as easy to investigate as in methane (7) and in the ammonium halides (11) that have been described at this study institute. The transition is characterized by being martensitic, which means that diffusion plays no role in the transition. It is driven by a sudden shearing action within the matrix. The hysteresis makes it difficult to define precisely the transition temperature, and the lower phase is formed in clusters, so that e.g. in fcc D_2 always 10% remains hcp.

II. NEUTRON SCATTERING PROPERTIES OF H_2

Having demonstrated that a variety of problems of general physical interest have been or could be elucidated by mixing the four constituents in table I, we will turn to a more detailed de-

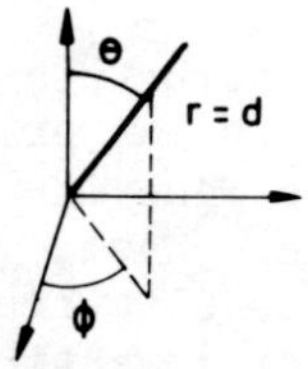

J	Ψ	f	Y	v	
Even	AS	S	S	AS	I=0 Para
Odd	AS	S	AS	S	Ortho I=1

$\Psi = f(r)Y_{Jm_J}(\theta,\varphi)v_I$

Table II. Symmetries of the molecular wavefunction Ψ, in the case of H_2, when the two protons are interchanged. S= symmetric, AS= antisymmetric. At the right is shown in resulting spins of the protons corresponding to the symmetry of the spinfunction v_I.

scription of the neutron scattering properties of solid and liquid H_2. Because of the small moment of inertia of the H_2 molecule, and because the rotations are very little hindered, the rotational quantum number J is well defined below T_B and the rotational states are well separated. In Table II we show how the symmetry of the molecular wavefunction divide the rotational states into two groups, the para-states where J is even and I=0 and the ortho-states where J is odd and I=1.

We will not go into experimental details but notice that one can prepare pure p-H_2 and o-H_2, which below T_B will be in the J=0 and J=1 states respectively. The scattering cross-sections for neutrons that do not undergo energy changes large enough to change the rotational state then becomes (12):

1: (J=0) → (J=0) scattering:

$$\frac{d^2\sigma}{d\Omega d\omega} = N_o \frac{k}{k_o} \frac{\hbar}{\pi} \sigma_{coh}\, j_0^2\left(\frac{\varkappa d}{2}\right) S(\varkappa, \omega), \tag{1}$$

where

$$S(\varkappa, \omega) = \frac{1}{2\pi\hbar}\int_{-\infty}^{\infty} dt \int d\underline{r}\left[e^{i(\varkappa \cdot \underline{r} - \omega t)} G(\underline{r},t)\right] ,$$

2: (J=1) → (J=1) scattering:

$$\frac{d^2\sigma}{d\Omega d\omega} = N_1 \frac{k}{k_o} \frac{\hbar}{\pi}\left(\sigma_{coh} + \frac{2}{3}\sigma_{inc}\right)\left\{ j_0^2\left(\frac{\varkappa d}{2}\right) + 2j_2^2\left(\frac{\varkappa d}{2}\right)\right\} S_i(\varkappa,\omega) \tag{2}$$

where

$$S_i(\varkappa,\omega) = \frac{1}{2\pi\hbar}\int_{-\infty}^{\infty} dt \int d\underline{r}\, e^{i(\varkappa \cdot \underline{r} - \omega t)} G_s(\underline{r},t) ,$$

If one, however, makes the neutron energy large enough to make the para-ortho transition the cross-section is:

3: (J=0) → (J=1) scattering:

$$\frac{d^2\sigma}{d\Omega d\omega} = N_o \frac{k}{k_o} \frac{\hbar}{\pi} 3\sigma_{inc}\, j_1^2\left(\frac{\varkappa d}{2}\right) S_i\left(\varkappa, \omega - \frac{\Delta E}{\hbar}\right) , \tag{3}$$

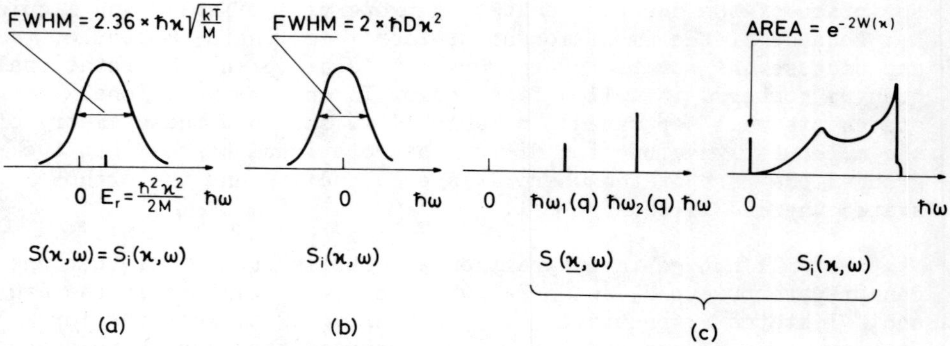

<u>Figure 1</u>. Scattering laws for some systems, shown for constant κ's
(a) Scattering from noninteracting nuclei.
(b) Scattering from a system with simple self-diffusion.
(c) $S(\underline{\kappa}, \omega)$ and $S_i(\kappa, \omega)$ for one-phonon scattering in a harmonic cubic solid.

In (1), (2), and (3) we have not yet defined the momentum transfer $\hbar\ \underline{\kappa} = \hbar\ (\underline{k}_0-\underline{k})$ and the energy transfer $\hbar\ \omega = \hbar^2(k_0^2-k^2)/(2M)$, where $\underline{k}_0$ and $\underline{k}$ are the incoming and outgoing wavevectors respectively. $G(\underline{r},t)$ and $G_s(\underline{r},t)$ are the total and the self correlation function respectively, N_J is the number of molecules in the J'th state, and j_i is the i'th spherical Bessel function.

Having in mind that we are now going to study the two scattering laws S and S_i obtained through neutron scattering we show on Fig.1 these functions for some ideal systems (13). The lineshape of the ideal hydrodynamic system is shown elsewhere (14). It should be noticed that neutron scattering at present cannot with-

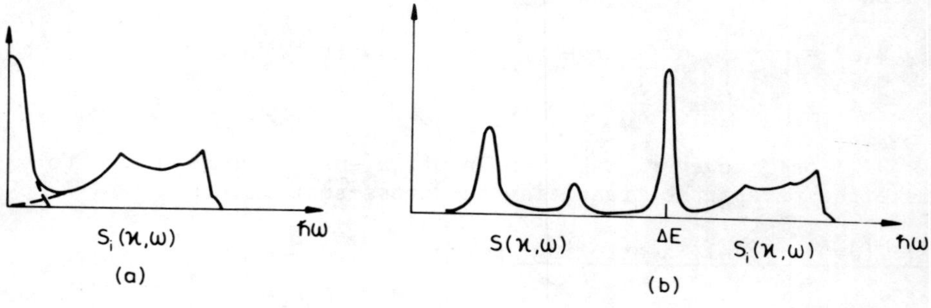

<u>Figure 2.</u> Total scattering from molecular hydrogen.
(a) Scattering from o-H_2 according to eq. (2).
(b) Scattering from p-H_2 according to eqs. (1) and (3).

out great difficulty reveal the detailed shapes of narrow peaks in S and S_i. Experiments can verify the position of a peak, its area, and if the natural linewidth is big enough, also this width can be extracted.

Anticipating the results of the next section we conclude by showing on Fig.2 the scattering from molecular hydrogen. Fig.2a shows the scattering from o-H_2 according to eq.(2) that qualitatively describe the way in which Egelstaff et.al. and Stein et.al. obtained their information. At Risø we have used the properties of p-H_2, and Fig.2b shows how the rotational energy ΔE separates the coherent and the incoherent spectrum in this case, as described by eqs. (1) and (3).

III. LATTICE DYNAMICS OF SOLID H_2 (ref.3)

As mentioned above solid H_2 belongs to the class of quantum solids, for which the energy of zero-point motion of the particles is comparable with the potential energy, and the amplitudes of vibrations are significant fractions of the distance between neighbouring particles.

As shown in Fig.3 the classical lattice dynamics based on the situation shown on Fig.3a breaks down in two ways. Firstly because the zero-point motion expands the crystal so that the nearest neighbour distance corresponds to a distance with imaginary force constants. Taking into account that the vibrations sample the potential over a large region one gets real frequencies, but because of the finite distribution of the ground state wave function there is a finite probability that one particle enters the hard core region of the pair potential, i.e. two particles make an unphysi-

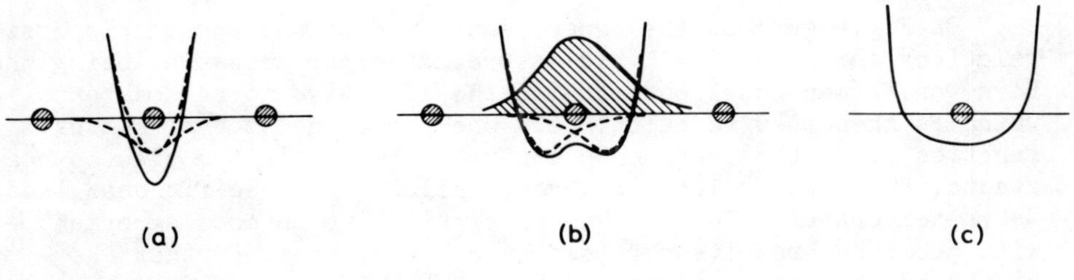

Figure 3. Potential wells (full lines) for a one-dimensional van der Waals crystal with nearest neighbor Lennard-Jones 6-12 interaction (dashed lines)

(a) No zero-point motion.
(b) Large zero-point motion.
(c) Effective potential after quantum corrections.

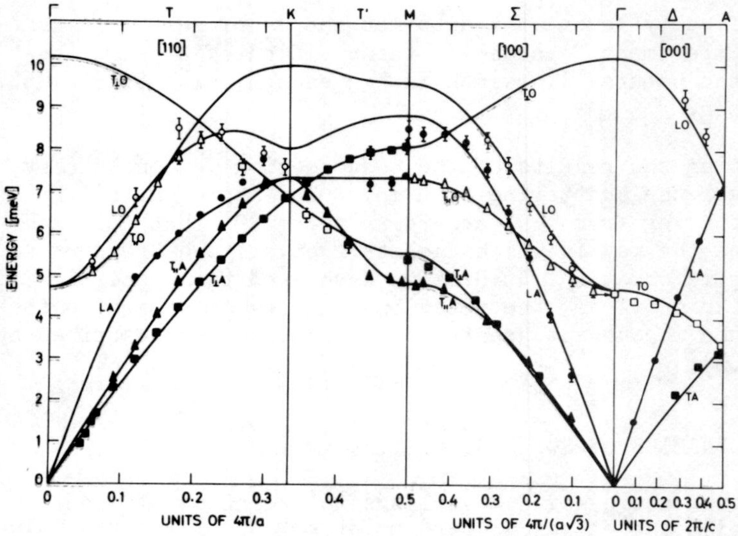

Figure 4. Phonon dispersion relations for p-H_2 at 5.4 K and at saturated vapour pressure. The full lines are results of the Born -von Karman fit where a third nearest neighbor general force model is used. The arrow at Γ shows the frequency of the TO mode as de-derived from Raman spectroscopy.

cally tight collision. To overcome this second problem, a short range correlation function can be introduced to give an effective potential as shown on Fig.3c. Based on this potential a classical "harmonic" theory of quantum lattice dynamics can be established (16). Although this approach may seem very crude, it reproduces the experimental evidence fairly well, although the phonon energies calculated by different methods differ by up to 40%.

On Fig.4 we show the experimentally obtained phonon dispersion relations for H_2 at 5.4 K and saturated vapour pressure. Using the Born-von Karman model one obtains the effective force constants, which is then used to calculate a one-phonon density of states function $Z(\omega)$, the one-phonon eigenvectors $\sigma(q)$'s, the elastic constants, the sound velocities, compressibility, specific heat, and Debye temperature. The derived properties are in good agreement with published results obtained by other methods. Further the widths of the measured groups show that the phonon lifetimes are large for all frequencies. The incoherent scattering described by (3) and Fig.2b has been used to measure directly the Debye-Waller -factor $< \exp(i\, \underline{\kappa} \cdot \underline{u}) >^2 = \exp(-2W(\kappa))$ which in the cubic approximation is equal to $\exp(-1/3\, \kappa^2 u^2)$. Using this expression we get $< u^2 > = 0.48$ Å^2 in agreement with the result obtained from the density of states function, determined by the effective force constants.

Although the consistency in the above mentioned results is surprising, remembering the potential of Fig.3c, this picture of H_2 as a harmonic crystal does not hold in all respects. Two exceptions will be mentioned, the anomalous variation of the intensity of the measured phonon peaks, and the difference between the calculated and the incoherently measured density of states.

The scattering law for one phonon creation scattering is (13):

$$S_{+1}(\underline{q},\omega) = \frac{(2\pi)^3}{v_o}\,\frac{1}{2M}\sum_{\underline{\tau}} e^{-2W(\kappa)}$$

$$\times \sum_{j,\underline{q}} \frac{|\underline{\kappa}.\ \underline{\sigma}j(q)|^2}{\omega_j(q)}\,\{n(\omega_j)+1\}\,\delta(\omega-\omega_j(\underline{q}))\cdot\delta(\underline{\kappa}-\underline{q}-\underline{\tau}) \qquad (4)$$

From (4) one gets the intensity of a particular phonon peak

$$I_{+1}^{j\underline{q}}(\underline{\kappa}) = \frac{(2\pi)^3}{v_o}\,\frac{1}{2M}\,e^{-2W(\underline{\kappa})}\,\frac{|\underline{\kappa}\cdot\underline{\sigma}^j(q)|^2}{\omega_j(q)}\,\{n(\omega_j)+1\} \qquad (5)$$

We have compared the experimentally obtained intensities I_{exp} corrected for instrumental sensitivity and the trivial factors of eq.(1) with this expression, and on Fig.5 we show the ratio I_{exp}/I_{+1} as a function of κ, for a number of phonon branches. The ob-

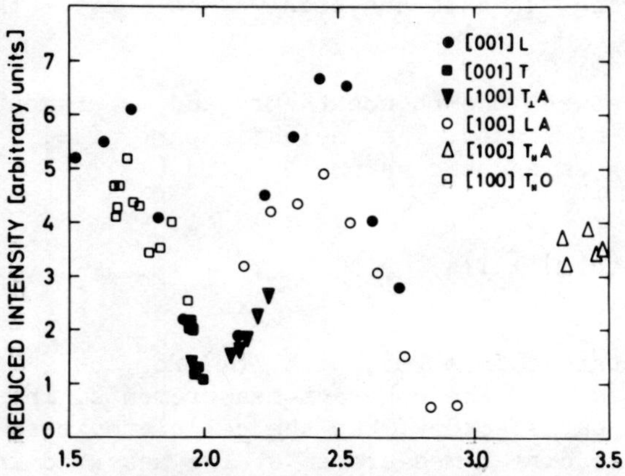

Figure 5. Reduced intensities of the measured neutron groups in p-H_2, shown for the dispersion curves for which the inelastic structure factor is slowly varying with κ. If the harmonic approximation applied the reduced intensities would be independent of κ.

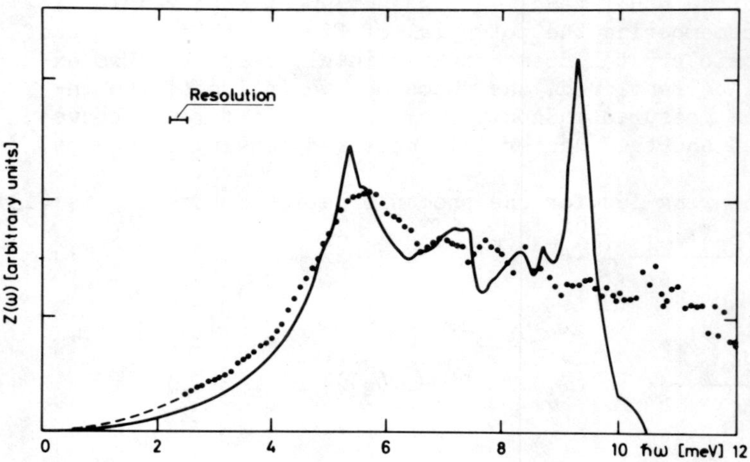

Figure 6. Density of states for solid p-H_2. Full line is Z(ω) calculated from the effective force constants obtained by coherent neutron scattering (3). Dots are experimental points from incoherent scattering (6).

served strong variation in the reduced intensities is ascribed to anharmonic effects, and similar results have been obtained in solid He. The anharmonicity, although of special origin in the quantum solids, seems to be well explained in terms of interference between one- and two-phonon scattering, as this has been worked out for the alkali halides (15).

A similar comparison between one-phonon theory and experiment can be made for the density of states. The inelastic part of S_i is for one-phonon scattering in a cubic harmonic solid (13):

$$S_i(\underline{\kappa},\omega) = \frac{\kappa^2}{2M} e^{-2W(\underline{\kappa})} \frac{Z(\omega)}{\omega} \{n(\omega) + 1\} \quad , \tag{6}$$

On Fig.6 we compare Z(ω) obtained from (6), with Z(ω) for one -phonon scattering calculated from the coherent measurements. The agreement for $\hbar\omega < 8$ meV is satisfactory, but the calculated peak at 9.5 meV is not seen in the direct measurements. The measured intensity $\hbar\omega > 10$ meV might be due to two-phonon scattering. Whether the second observation is consistent with the first mentioned has not been fully verified.

IV. COLLECTIVE DYNAMICS OF LIQUID H_2(ref.2)

Using the same technique as in the solid phase we have performed neutron scattering on liquid H_2. Since no dynamical model like the Born-von Karman model for solids, exists for liquids one cannot in the same way extract significant parameters from the data. The qualitative behaviour of $S(\kappa,\omega)$ and $S_i(\kappa,\omega)$ becomes therefore important.

Two different approaches have been made to model the dynamics of a liquid. The first to be mentioned is based on the intuitive picture of the liquid as a non-stationary, disordered solid (17), and the resulting $S(\kappa,\omega)$ then shows well-defined but rather broad peaks as a function of ω. Since these models fail to describe the measurements on liquid Ar(18), the concept of memory-functions has been adopted. Compared with the solid-like models, memory-functions techniques may seem rather unphysical, but the formalism is known to satisfy the exact moment relations (sum rules) that are violated by some of the models mentioned above. The $S(\kappa, \omega)$ that results from memory-functions techniques shows a steadily decreasing function for increasing κ's. An intrinsic deficiency of the memory-functions is that they only include longitudinal excitations (14). Although transverse waves in liquids have been neglected for various reasons until quite recently, Alder et.al.(19) have shown that such waves with long lifetimes may exist. $S(\kappa,\omega)$ for liquid Ar does not provide a good test case because the rather structureless picture may easily be misinterpreted.

In the solid phase the quantum character of H_2 caused some fundamental problems. This is not the case in liquid H_2, because a quantum liquid may be described as being simply a very cold liquid. Since $S(\kappa,\omega)$ is seen to show rather detailed structure this liquid might be a better test case for dynamical theories.

On Fig.7 we show $S(\kappa,\omega)$ for liquid H_2 . It demonstrates that a well defined peak exists in $S(\kappa,\omega)$ for $\omega \neq 0$, signifying the existence of a collective excitation in the density autocorrelation function. In other liquids such as argon and neon such peaks are not seen in $S(\kappa, \omega)$ but only in the velocity autocorrelation function $\omega^2 S(\kappa, \omega)$, which of course has a peak at finite energy, even if the density autocorrelation function is overdamped. The total spectra have been tested for the ACB one-phonon sum rule. For $\kappa \leq 2.3$ Å^{-1} our results are consistent with one-phonon scattering, with a mean square displacement $\langle u^2 \rangle = 1.0$ Å^2. At larger κ's, however, $S(\kappa,\omega)$ shows a significant multiphonon contribution. The magnitude of the Debye-Waller factor indicates that multiphonon scattering should predominate for $\kappa > 1.5$ Å^{-1}, but this scattering seems to be concentrated mainly at energies above 8 meV, a region not accessible in this experiment.

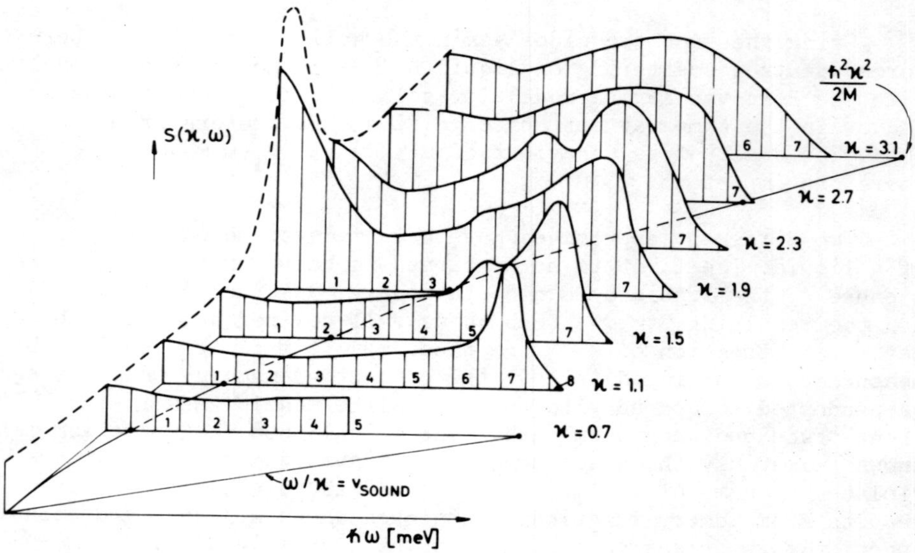

Figure 7. Scattering law $S(\kappa,\omega)$ for liquid parahydrogen at T=14.7 K. In the base plane the recoil energy curve is shown, and the extension of the position of the Brillouin peak is indicated.

From a preliminary analysis, several qualitative features are evident. One of these is the similarity with the dynamics of a solid. Adherence to the ACB sum rules was mentioned above. Furthermore the characteristic excitation energies are in some respects solid-like. For instance, the longitudinal phonon energy at the first zone boundary in the solid at a density equal to that of liquid H_2 would be 7.5 meV, as estimated by the Grüneisen relation and the data of ref.3. This is to be compared with the peak in our $S(\kappa, \omega)$ at $\kappa = 1.0$ Å^{-1} which is at 7.2 meV. At higher κ's, corresponding to the second and third Brillouin zones in the solid, intensity appears in the liquid at lower energy reminiscent of the neutron scattering law calculated for a polycrystalline powder.

In the latter case this feature is related to modes which possess appreciable transverse character, suggesting that transverse modes may contribute to the scattering in the liquid, although it is not clear how the neutron can couple to transverse modes in the absence of a reciprocal lattice. In addition to the maximum at finite frequency there is also a portion of the spectrum centered at zero frequency, most pronounced at $\kappa = 2$ Å^{-1}, i.e. near the major peak in the structure factor. In this respect liquid H_2 is similar to classical liquids.

It seems instructive to compare $S(\kappa,\omega)$ in H_2 with that of ^4He and Ar. It is well known that superfluid ^4He shows a sharp phonon -roton curve at low temperatures. No scattering occurs, however, at zero frequency. Measurements show that thermal broadening becomes significant below the λ-point and that, although heavily damped, the excitations persist in the normal phase. The characteristic energies are, however, much greater in H_2, and here thermal broadening is not as severe even at the boiling point. In the case of Ar, $S(\kappa, \omega)$ shows no structure as a function of κ, indicating that no long living collective excitations can exist in this liquid. This is perhaps a reflection of the fact that Ar is a dynamically warm liquid, for which $T_B/\Theta_D= 1.0$, where Θ_D is the Debye temperature of the solid at the melting point. The corresponding numbers for ^4He and H_2 are 0.16 and 0.20 respectively.

ACKNOWLEDGEMENT

We acknowledge the collaboration of J.P. McTague during the liquid work. Further, we are grateful to H. Bjerrum Møller and B.M. Powell for valuable discussions and comments on the manuscript.

REFERENCES

1. P.A. Egelstaft, B.C. Haywood, and F.J. Webb, Proc.Phys.Soc 90, 681 (1967).
2. K. Carneiro, M. Nielsen, J.P. McTague, Phys.Rev.Let. 30, 481 (1973).
3. M. Nielsen and H. Bjerrum Møller, Phys.Rev. B 3 4383 (1971), and M. Nielsen, Phys.Rev. B 7 1626 (1973).
4. I.F. Silvera, W. Hardy, and J.P. McTague, Phys.Rev. B 5 1578 (1972).
5. B.M. Powell, To be published.
6. H. Stein, H. Stiller, R. Stockmeyer, Jour. Chem.Phys. 57 1786 (1972).
7. W. Press and A. Hüller, These Proceedings.
8. H. Meyer, F. Weinhaus, B. Maraviglia, and R.L. Mills. Phys. Rev. B 6 1112 (1972).
9. I.F. Silvera, W. Hardy, and J.P. McTague, Phys.Rev. B 4 2724 (1971).
10. R.L. Mills, J.L. Yarnell, and A.F. Schuch, Los Alamos Report LA-DC-72-894.
11. W.B. Yelon, These Proceedings.
12. G. Sarma, in "Inelastic Scattering of Neutrons in Solids and Liquids" (IAEA, Vienna, p.397,1961).
13. W. Marshall and S. Lovesey, "Theory of Thermal Neutron Scattering" (OXFORD, 1971).
14. G. Niklasson and A. Sjöander, These Proceedings.

15. R.A. Cowley and W.J.L. Buyers, J.Phys. C. 2 2262 (1969).
16. N.R. Werthamer, Amer. J.Phys. 37 763 (69), and references therein.
17. J. Hubbard and J.L. Beeby, J.Phys. C 2 556 (1969).
18. K. Sköld, J.M. Rowe, G. Ostrowski, and P.D. Randolph, Phys. Rev. A 6 1107 (1972).
19. B.J. Alder and W.E. Wainwright Phys.Rev. A 1 18 (1970).
20. R.A. Cowley and A.D.B. Woods, Can.J.Phys. 49 177 (1971).

NEUTRON SCATTERING BY A LIQUID CRYSTAL AT TEMPERATURES CLOSE TO ITS MELTING POINT

R. Pynn and T. Riste

Institutt for Atomenergi, Kjeller, Norway

ABSTRACT

Neutron diffraction data have been obtained for a deuterated sample of the nematic liquid crystal para-azoxyanisole (PAA) at temperatures close to the melting point of this material. It is found that, as the melting point is approached from below, Bragg intensity decreases continuously in the manner usually associated with the approach of a second-order phase transition. Further, at temperatures close to the melting point, the Bragg intensity is superimposed on a diffuse background which has an angular distribution reminiscent of scattering from the fluid phases of PAA. We interpret these observations as evidence for pretransitional effects associated with the melting of PAA and present a possible explanation of the data. This explanation suggests that the melting of PAA may be associated with the softening of a librational mode.

INTRODUCTION

Somewhat unoriginally we shall begin by pointing out that melting (or its converse, solidification) must have been one of the first phase transitions to impress itself upon mankind. Indeed, one can easily imagine the expression of incredulity on the face of the caveman who tried to dive into Ustevann (1) in winter. Not withstanding its historical impact, the phenomenon of solidification is still incompletely understood although results like those of Professor Cotterill are beginning to force back the barriers of our ignorance. The major difficulty in making definite statements about melting arises, of course, from the totally different symmetries of phases related by such a transition. Since this symmetry difference is obliged by the definition of melting as the disappearance of some form of long-range order, one may argue that the development of a theory of melting is difficult a priori.

In view of the last remark one may ask why we chose to study the melting of a liquid crystal when so many simpler systems are available. Our primary reason for this choice was that, in liquid crystalline systems, long-range translational and orientational ordering of the constituent molecules disappear at different temperatures. Thus independent study of these two types of melting is possible. In the solid phase, long, rod-like molecules are arranged on a regular lattice and have specified orientations in space. For the material we have studied, the solid phase melts (i.e. becomes fluid) to yield a nematic phase in which the molecular long axes tend to be aligned in a particular preferred direction (2). However, there is, in the nematic, no long-range correlation between positions of molecular centres of mass. As the nematic material is heated the degree of orientational ordering decreases somewhat and then disappears abruptly at a particular temperature known as the clearing point (3). The disappearance of orientational order yields a material which is an isotropic fluid. The situation which we have just described is summarized by the following diagram (Figure 1).

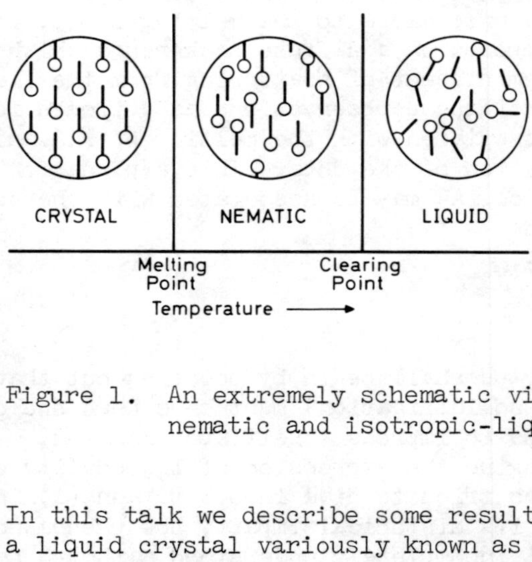

Figure 1. An extremely schematic view of the solid, nematic and isotropic-liquid phases.

In this talk we describe some results which have been obtained with a liquid crystal variously known as para-azoxyanisole (PAA) or di-methoxy-azoxybenzene. A molecule of this substance is drawn schematically in figure 2. In practice, even in the fluid phases of PAA, the molecules remain elongated as in the figure but they

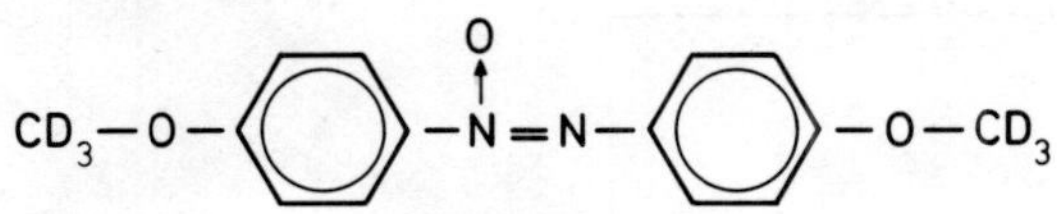

Figure 2. A molecule of deuterated PAA.

probably lose much of their planar quality.

Our measurements have been made with a fully deuterated (4) sample of PAA and we shall restrict ourselves in this talk to a discussion of neutron diffraction patterns. We have made some measurements of the frequency dependence of the neutron-scattering law, but without startling results. The apparatus used in our experiment was a conventional powder diffractometer operating with incident neutrons of wavelength 1.863 Å. The sample, of area $(40 \times 40) mm^2$ and thickness 5 mm, was contained in a double-walled aluminium container equipped with an electrical heater and a temperature control device. In order to define the preferred direction of the nematic phase, one of two magnets, each giving a field of about 850 G was used. These magnets were arranged so that the field $\vec{H}$ (and hence also the preferred direction in the nematic) could be chosen either parallel or perpendicular to the neutron scattering vector, $\vec{Q}$. One may denote these configurations by the symbols $\vec{Q} \parallel \vec{H}$ and $\vec{Q} \perp \vec{H}$ respectively.

DIFFRACTION PATTERNS OBTAINED WITH PAA

Diffraction patters of PAA, typical of the phases for which they were recorded, are displayed in figure 3. In the upper part of the figure we indicate, by vertical bars, the positions of intense Bragg reflections obtained with a solid sample at room temperature. At this temperature the background between Bragg peaks is roughly independent of scattering angle, whereas just below the melting point noticeable structure begins to appear in the background. This situation is depicted in part A of figure 3. In nematic and isotropic-liquid phases (parts B through E of figure 3) the diffraction patterns are composed of one or two broad peaks. The distinctive feature of the nematic phase is that either of these two peaks can be suppressed by a suitable choice of the magnetic-field direction. In the isotropic-liquid phase (part E of figure 3) the magnetic field has no effect on the diffraction pattern.

The point we want to make at this stage is that diffraction patterns from the three phases of PAA are different and that these

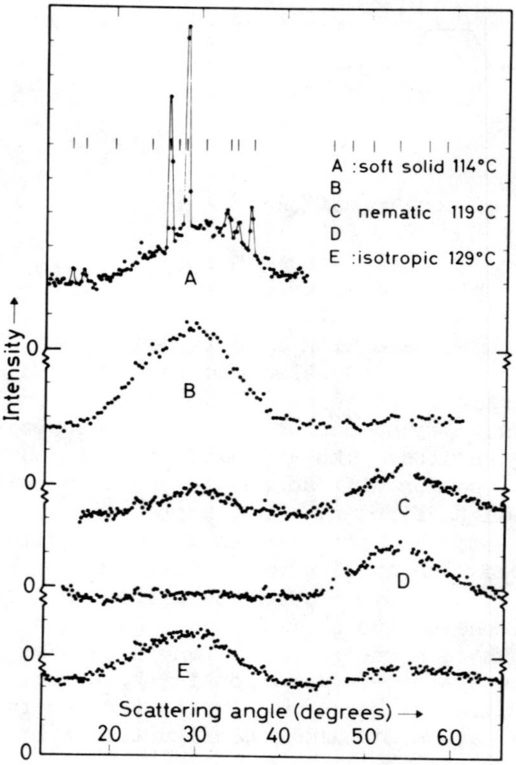

Figure 3. Neutron diffraction patterns for PAA.
Parts A and B: $\vec{Q} \perp \vec{H}$
Parts C and E: $\vec{H} = 0$
Part D: $\vec{Q} \parallel \vec{H}$

patterns serve to identify the phase of the sample. Alternatively, one may say that coherent neutron scattering provides a tool for the study of the transitions between solid, nematic and isotropic-liquid phases. Of course, it is somewhat inconvenient (not to mention thoroughly confusing) to examine a complete diffraction pattern at each temperature of interest. Fortunately, however, the patterns can, for our purposes, be characterized by the intensity of Bragg reflections (in the solid phase) and by the intensi-

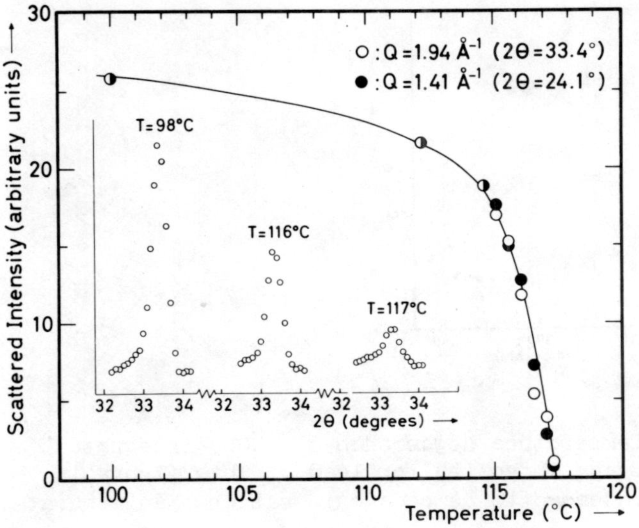

Figure 4. Variation of integrated Bragg intensity with temperature for two PAA Bragg peaks: the curve through the points is intended only as a guide to the eye. The insert shows profiles of one of the peaks at three temperatures

ties of the two broad peaks centred at $2\theta \sim 30^\circ$ and $2\theta \sim 55^\circ$ in the fluid phases. In the remainder of this talk we shall discuss the temperature variation of these intensities.

THE SOLID-NEMATIC TRANSITION

In order to study the solid-nematic transition we have made measurements of the (integrated) intensity of a number of Bragg reflections as a function of temperature. Results for two such peaks are shown in figure 4: the insert of the figure is meant to demonstrate that Bragg peaks do not change significantly in width or position even close to the melting temperature: that is, there is no anomalous expansion or softening of the sample.

Lest you should find figure 4 unsurprising, the following diagram shows a result equivalent to that of figure 4 obtained

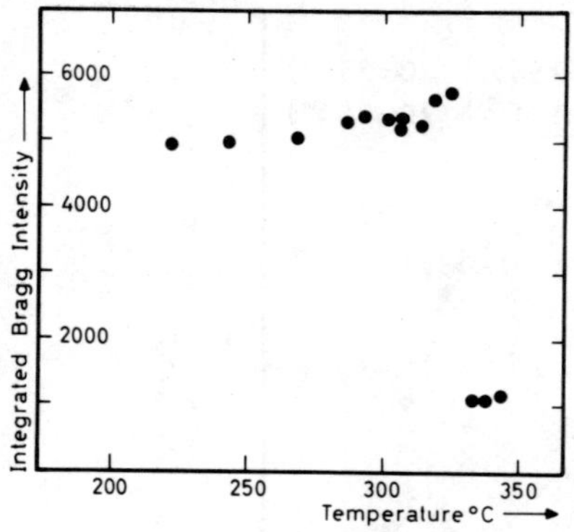

Figure 5. Temperature dependence of (200) integrated Bragg intensity for lead. (This figure is intended to be pedagogic rather than accurate.)

with a lead sample. Clearly in the latter case melting is signified by the abrupt disappearance of Bragg intensity whereas for PAA the system seems to have some foreknowledge of the transition.

For the benefit of the "naturally suspicious" amongst you we should say that figure 4 is not the result of poor temperature control or of thermal gradients in the sample (5). Nor can the figure be explained in terms of a partial melting of the sample: one may go from point to point below the melting temperature T_m quite reproducibly and reversibly.

Of course, scattered intensity cannot just disappear at one point in $\vec{Q}$ space without reappearing somewhere else. In our case the intensity lost from the Bragg reflections shows up as structure in the background between these reflections. We have already shown you an example of this in part A of figure 3 where one sees that, as the melting temperature is approached, the Bragg reflections ride on a background which is reminiscent of diffraction patterns obtained in the fluid phases of PAA.

In order to make this last remark somewhat more specific we show in figure 6 the variation of scattered intensity at the two Q values which correspond to the maxima of the broad peaks in parts B through E of figure 3. One sees, once again, a definite pretransitional increase in intensity; an increase which is apparently independent of $\vec{H}$ except for temperatures very close to T_m.

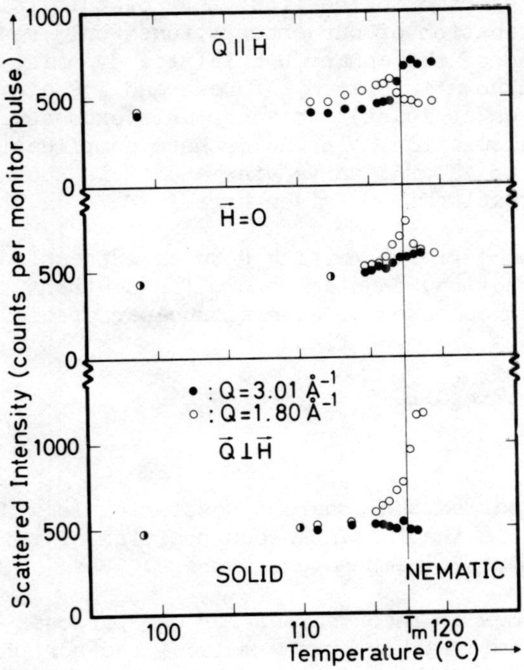

Figure 6. Variation of scattered intensity at $Q = 1.80$ Å^{-1} ($2\theta \sim 30^{\circ}$) and $Q = 3.01$ Å^{-1} ($2\theta \sim 55^{\circ}$) with temperature for three configurations of $\vec{Q}$ and $\vec{H}$.

Incidentally, perhaps it ought to be mentioned at this stage that we have searched around in $\vec{Q}$ space (down to $Q = 0.05$ Å^{-1}) for critical scattering associated with the melting of PAA but without success.

AN EXPLANATION

This section is deliberately entitled "AN explanation (of the experimental data reported in the first part of this talk)". Clearly a thorough explanation of our observations would require an exhaustive discussion of the anharmonic lattice dynamics of solid PAA. However, unknown interatomic forces and 132 atoms per unit cell furnish (for us at least) two acceptable excuses for not contemplating such a calculation. In fact, these complications essentially oblige one to adopt a naive view and it is such a view which constitutes the "explanation" given here.

It is convenient to begin by writing down an expression for the diffraction cross-section, denoted $d\sigma/d\Omega$, for neutrons scattered by a target system composed of several nuclear species. Under suitable conditions (7), one finds

$$\frac{d\sigma}{d\Omega} = \sum_{m,n} a_m\, a_n \langle e^{-i\vec{Q}\cdot\vec{r}_m}\, e^{i\vec{Q}\cdot\vec{r}_n} \rangle_T \tag{1}$$

In this equation a_m is called the coherent scattering length of the m'th nucleus and $\vec{r}_m$ is the instantaneous position of this nucleus. $\langle \cdots\cdots \rangle_T$ denotes a temperature average.

Equation (1) is somewhat inconvenient for our purposes because it provides an atomic rather than a molecular description of the scattering. That is, the equation ignores the fact that (in PAA) groups of 33 atoms move around together as tightly bound molecules. To remedy this deficiency and, at the same time, to reduce the complexity of our problem by an order of magnitude, let us assume the molecules to be so tightly bound as to be rigid. In this case one has

$$\vec{r}_m = \vec{R}_i + \underline{\underline{D}}(\alpha_i\, \beta_i\, \gamma_i)\cdot \vec{u}_j \tag{2}$$

where the nuclear index m has been separated into an index i which refers to a molecule and an index j which labels a nucleus within a molecule. In equation (2) $\vec{R}_i$ is the position of a molecular centre-of-mass and $(\alpha_i\, \beta_i\, \gamma_i)$ are the Euler angles which specify the orientation of this molecule. The matrix $\underline{\underline{D}}(\alpha_i\, \beta_i\, \gamma_i)$ causes the intermolecular coordinate $\vec{u}_j$ (defined in a coordinate frame tied to a molecule) to be rotated through the appropriate Euler angles.

Plugging (2) into (1) gives immediately:

$$\frac{d\sigma}{d\Omega} = \sum_{i,i'} \langle \alpha_i(\vec{Q})\, \alpha^*_{i'}(\vec{Q})\, e^{-i\vec{Q}\cdot\vec{R}_i}\, e^{i\vec{Q}\cdot\vec{R}_{i'}} \rangle_T \tag{3}$$

where $\alpha_i(\vec{Q})$ is a molecular form-factor which depends on the orientation of the i'th molecule: thus

$$\alpha_i(\vec{Q}) = \sum_{j=1}^{n} a_j \, e^{-i\vec{Q}\cdot \underline{\underline{D}}_i \cdot \vec{u}_j} \qquad (4)$$

In this equation $D(\alpha_i\ \beta_i\ \gamma_i)$ has been abbreviated to D_i and the molecules have been assumed to contain n atoms.

In order to make further progress let us make two drastic assumptions:

1) We (i.e. the authors) know of no evidence which supports the assertion that melting is accompanied by the softening of a translational mode of vibration. This observation implies that our experimental results are not to be explained by a mystic temperature variation of a (translational) Debye-Waller factor. Let us therefore exclude this possibility entirely by fixing the coordinates $\vec{R}_i$ in (3) to a regular, immobile array of sites.

2) Let us assume the librational motions of different molecules to be uncoupled.

With these gross simplifications one may easily find

$$\frac{d\sigma}{d\Omega} = \sum_{i,i'} e^{-i\vec{Q}\cdot\vec{R}_i} \, e^{i\vec{Q}\cdot\vec{R}_{i'}} \, |\langle\alpha(\vec{Q})\rangle_\Omega|^2$$

$$+ N[\langle|\alpha(\vec{Q})|^2\rangle_\Omega - |\langle\alpha(\vec{Q})\rangle_\Omega|^2] \qquad (5)$$

where $\langle\cdots\cdots\rangle_\Omega$ denotes an average over molecular orientations. Equation (5) tells us that the total scattering is composed of Bragg intensity (the $\exp[-i\vec{Q}\cdot(\vec{R}_i - \vec{R}_{i'})]$ term) and a part which in some way represents the mean-square angular displacement of molecules from their official, equilibrium orientations. One may call this last contribution to the intensity "the diffuse scattering": it is the background in, for example, figure 3A.

It is possible to evaluate equation (5) for PAA for various hypothetical distributions of molecular orientations: details are given in reference (8). The best result (i.e. that closest to experiment) seems to be obtained for a model in which:

a) the solid sample is assumed to be polycrystalline (which is undoubtedly correct (6))

b) all molecules within a particular crystallite have the same equilibrium orientation

and c) molecules are allowed to librate independently about their long axes (i.e. the long axes are fixed) with a frequency which decreases markedly with temperature in the few degrees below T_m.

With this model the diffuse scattering is found to increase with temperature (cf. (c) above) and to be distributed in Q-space as in figure 7. As one can see by examining figures 2, 6 and 7 the calculated and observed intensity distributions are in fair agreement, a fact which suggests, to us at any rate, that we have at least used the correct molecular coordinate in the calculation. In addition, since $|\langle\alpha\rangle|^2$ decreases sharply with decreasing frequency of libration, the model proposed here also explains the variation of Bragg intensity displayed in figure 4.

The model we have described above suffices to explain all the data presented in this talk. However, the model also predicts that the intensity of a Bragg peak should decay with temperature at a rate which depends on the peak observed. In particular, one should be able to find peaks which have $\vec{Q}$ parallel to the long molecular axes and which behave in a manner similar to that portrayed in figure 5. We have been unable to observe this effect although some of our data might support the contention that Bragg peaks decay at different rates. The main obstacle to obtaining conclusive data is, of course, Murphy's law (10): Bragg peaks in PAA are weak and, for reasonable Q values, are also extremely close together.

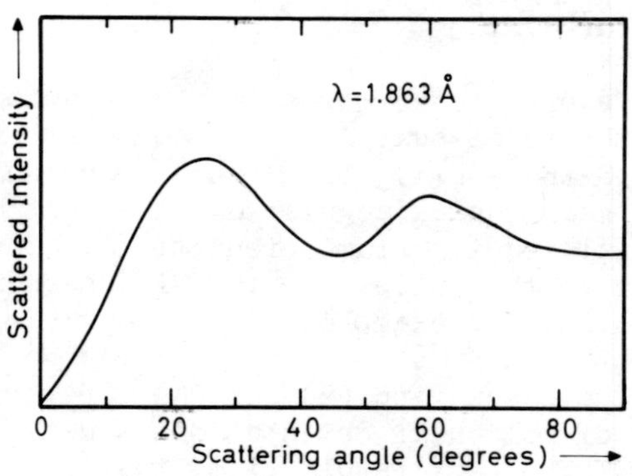

Figure 7. Diffuse scattering from deuterated PAA calculated from the model discussed in the text.

CONCLUSION

In conclusion, we would like to suggest a possible model for the melting of PAA which is, in some sense, an extension of Lindemann's well known law to a system containing large molecules. We propose that as PAA approaches it's melting temperature, librational motions of molecules about their long axes increase substantially in amplitude (i.e. this mode becomes soft). Since PAA adopts an imbricated structure (11), such wild librational motions must cause translation of the molecules. In the limiting case of free rotation for example, the system can probably no longer achieve the packing density required for solid binding. Thus, our suggestion is that PAA melts when the librational motions (rather than the translational motions discussed by Lindemann) reach a critical amplitude. It is not difficult to rephrase this statement in the language used by Professor Cotterill in his lectures: the critical librational amplitude is a measure of the number of molecules which has sufficient energy to reorientate and to cause a packing defect.

Finally one may ask (and we shall not attempt to answer) the question: is the behaviour reported here typical of materials possessing a liquid crystalline phase?

REFERENCES AND FOOTNOTES

1. Ustevann is the nearest lake to the conference hotel.

2. This direction can be defined in practice by applying an external perturbation such as a magnetic field.

3. So called because the nematic phase appears turbid while the isotropic phase is clear.

4. The sample was deuterated so that coherent scattering could be observed: see for example, W. Marshall and S.W. Lovesey, Theory of Thermal Neutron Scattering, (Oxford, Clarendon Press, 1971).

5. This point is discussed in more detail in reference 6.

6. T. Riste and R. Pynn, Solid State Comm., 12, 409 (1973).

7. In particular, the nuclei of the target system must scatter purely coherently and behave as Boltzmann particles if equation (1) is to be strictly valid.

8. R. Pynn, J. Phys. Chem. Solids, 34, 735 (1973).

9. If the preferred axes of the crystallites are aligned in this model, the intensity of the $2\theta \sim 30^{\circ}$ peak increases by a factor of about 2.5 while the $2\theta \sim 55^{\circ}$ peak becomes smeared out. These results help to explain the intensity changes at T_m displayed in figure 6.

10. For the benefit of the uninitiated we may state Murphy's law as: "All interesting natural phenomena are a priori difficult to observe".

11. W.R. Krigbaum, Y. Chitani and P.G. Barber, Acta Crystallogr., B26, 97 (1970).

X-RAY STUDIES OF THE SMECTIC STRUCTURE AND THE SOLID-SMECTIC TRANSITION

M. Lambert, A.M. Levelut

Laboratoire de Physique des Solides (associé au CNRS)
Université Paris-Sud, 91405 ORSAY, France

I - CLASSIFICATION OF SMECTIC MESOPHASES

Smectic liquid crystals have been identified since a long time by Herrmann (1) and Friedel (2) on the basis of X-ray experiments and are characterized by a periodic packing of molecular layers : the X ray diffraction pattern of the non oriented samples (powder pattern) consists essentially of two rings : an inner ring at small Bragg angles (1-2° for CuKα radiation), an outer ring at larger angles :

- the first ring corresponds to the layer spacing and is very well defined. For some compounds several orders of diffraction are even visible.
- the outer ring is characteristic of the lateral distribution of parallel molecules within each layer and it generally looks like a liquid diffraction ring.

In fact, several polymorphic varieties of thermotropic smectic mesophases have been identified and systematic researches in that field were made mostly by German investigators (3) (4) (5). Sackmann and coworkers distinguish 7 types of smectic modifications which are designated as S_A, S_B, S_C, S_D, S_E, S_F, S_G. The classification of these different phases has been made mostly on the basis of their miscibility relations. One remark can immediately be made concerning the frequency of appearance of these different states : the S_A and S_C phases have been observed in the case of about 100 different compounds, the S_B phase has been identified for 30 different compounds and always appears as an intermediate phase between the crystal and another smectic (S_A, S_C) or nematic phase. Only two compounds present the S_D modification and the S_F and S_G phases were identified in the particular case of only one tetramorphous substance. Very recently, it has been shown by Gray (6) that the S_E modification which was identified by Sackmann in the case of 4 different compounds is in fact relatively frequent and appears in many cases.

The most common smectic phases, S_A and S_C, only differ by the tilting of the long axes of the molecules with respect to the direction of the normal to the layers : in the S_A case, the molecules are perpendicular to the layers and, in the S_C case, there is a tilt angle α which is generally dependent of the sample temperature. In both cases, no order exists either in the packing of the molecules of the same layer or between molecules of adjacent layers and the X-ray patterns are very similar (fig 1). The S_B phase presents a much higher degree of order, and on the corresponding X ray powder pattern of fig 1, the outer ring appears sharp : this observation was the first indication of the existence of some kind of order inside the smectic layers. This was confirmed by experiments performed by Chistyakov (7) which seem to indicate an hexagonal order for this S_B modification.

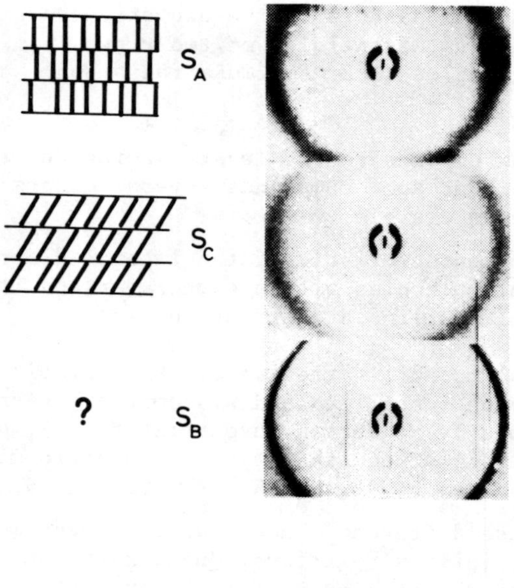

Fig 1 A schematic description of the smectic structures and the corresponding X ray powder patterns (3).

Some further investigations of the S_B modification has been done by De Vries and Fischel (8), Doucet, Levelut and Lambert (9) (10) and a more detailed description of the structure can be given now. Three different compounds have been studied :

1/ E.B.A.C. Ethyl 4-Ethoxybenzal-4'-aminocinnamate

C_2H_5-O-ϕ-CH=N-ϕ-CH=CH-COO-C_2H_5

$$C \underset{\leftarrow}{\overset{82°C}{\rightarrow}} S_B \underset{\leftarrow}{\overset{117°C}{\rightarrow}} S_A \underset{\leftarrow}{\overset{157°C}{\rightarrow}} N \underset{\leftarrow}{\overset{160°C}{\rightarrow}} L$$

2/ T.B.B.A. Terephtal-bis-(4-n-butylaniline)

C_4H_9-ϕ-N=CH-ϕ-CH-N-ϕ-C_4H_9

$$C \underset{\leftarrow}{\overset{113°C}{\rightarrow}} S_B \underset{\leftarrow}{\overset{144°C}{\rightarrow}} S_C \underset{\leftarrow}{\overset{172°C}{\rightarrow}} S_A \underset{\leftarrow}{\overset{200°C}{\rightarrow}} N \underset{\leftarrow}{\overset{237°C}{\rightarrow}} L$$

3/ B.E.A. 4-Butyloxybenzal-4' Ethylaniline

C_4H_9-O-ϕ-CH=N-ϕ-C_2H_5

$$C \underset{\leftarrow}{\overset{41°C}{\rightarrow}} S_B \underset{\leftarrow}{\overset{51°C}{\rightarrow}} N \underset{\leftarrow}{\overset{65°C}{\rightarrow}} L$$

In the first case, E.B.A.C., the molecules are perpendicular to the layers of the S_B modification and there is no change of the molecular orientation at 117°C : the S_B modification transforms **in** a S_A phase. The two other compounds are quite different since, in both cases, the molecules are inclined on the smectic planes in the S_B phase. Although the set of phase transformations differs completely for the two compounds, the structure of their S_B phase is very similar and implies the existence of a large degree of order. The structure of this S_B phase will be described with some details in the case of T.B.B.A. for which some preliminary results of the solid-S_B phase transition are available.

II - STRUCTURE OF THE S_B MODIFICATION

A model (11) of the T.B.B.A. molecule is represented fig 2. It can be seen that the fully extended molecule has a length of 29 Å.

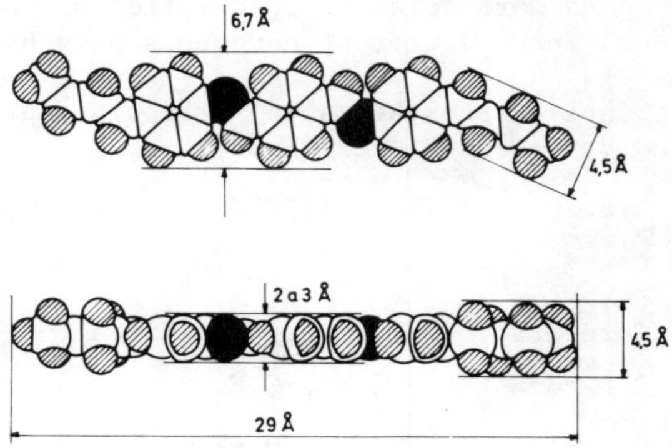

Fig 2 The T.B.B.A. molecule

This compound has 3 different smectic phases and the thermal effects (12) are much more important for the two first transitions ($C \rightarrow S_B$, $S_B \rightarrow S_C$) which correspond to the disappearance of long range order.

The S_B phase presents a mosaic texture as a normal polycrystalline material and a powder X ray pattern (fig 3) shows the presence, at large angle, of several sharp and well defined rings. Nevertheless there exists no diffraction ring at very large angle, as can be seen for the solid phase : the order does not extend to very large distances.

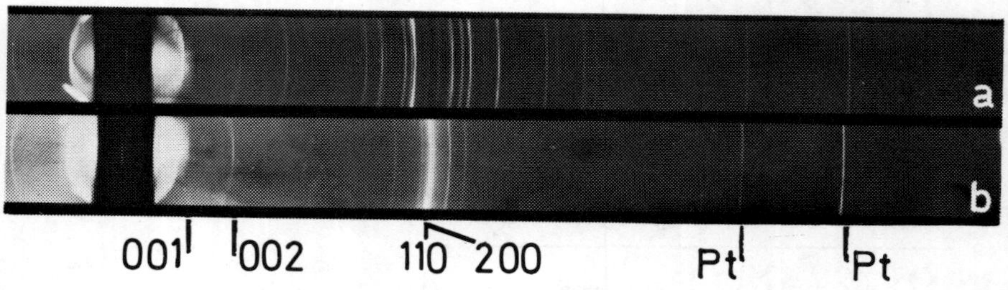

Fig 3 X ray powder patterns of T.B.B.A - CuKα_1 radiation

a) Solid phase

b) S_B phase

To determine the structure of this S_B phase, it was necessary to perform single domain X ray photographs and two of them are shown fig 4.

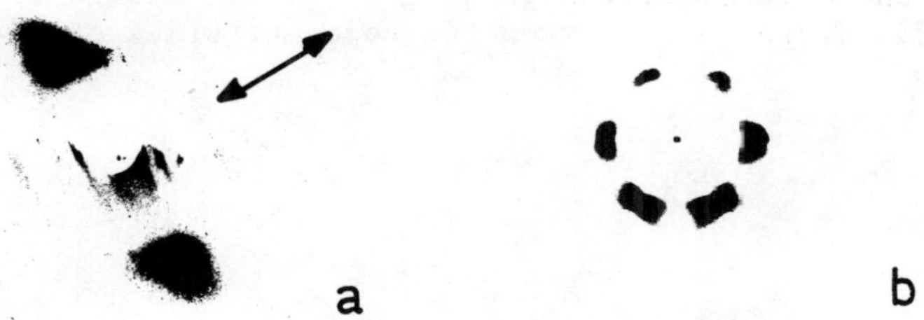

Fig 4 Single domain X ray patterns - CuKα radiation

a) The molecular direction is indicated by the arrow

b) The molecules are perpendicular to the photograph

These two patterns give the basic informations for the determination of the S_B structure :

- in the first case (fig 4 a) the X ray monochromatic beam is perpendicular to the molecular direction and a set of parallel diffuse sheets appears on the photograph : these sheets correspond to reciprocal planes perpendicular to the molecules and their spacing is the inverse of the molecular length 1/29 $\mathring{A}^{-1}$. Such a result indicates the existence in the sample of a longitudinal displacement of parallel strings of molecules. The correlation length along the molecular direction can be measured from the thickness of the diffuse scattering sheets and is of the order of 200 Å (7 molecules) close to the solid S_B transition temperature. From the extent of the diffuse scattering pattern, the fluctuation angle (2°) and the amplitude of the longitudinal motion (2.5 Å) can be deduced (10). The intense spots appearing on the diffuse sheets are the Bragg diffraction spots given by the smectic layers : the tilt angle of the molecules with respect to the normal to the smectic planes can be measured directly on the photograph and is found to be 32° at 115°C.

- the second photograph (fig 4 b) represents the equatorial plane of the reciprocal lattice (perpendicular to the molecular direction) and shows the existence of a pseudo-hexagonal lattice.

Using both the monodomain patterns and the powder pattern of fig 3, it is relatively easy to determine a "local" unit cell characteristic of this S_B phase (fig 5) : this unit cell corresponds to a monoclinic symmetry but, in a plane perpendicular to the molecules, the packing is nearly hexagonal (there is only 1 % difference between two adjacent sides of the hexagone).

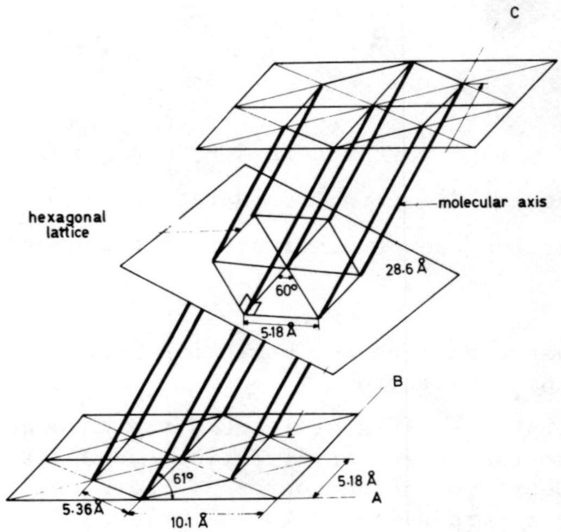

Fig 5 "Unit cell" of the S_B modification of T.B.B.A. at 125°C.

As it can be seen on fig 5, the apparent molecular length is very close to the length of the fully extended molecule (29 Å) and the tilt angle, 29° at 125°C, decreases slowly with temperature and, in the sensibility range of the experiments (13), no discontinuous change could be found at the transition between the S_B and S_C modifications.

From these experimental results the smectic B phase has the following features :

1/ - parallel layers of molecules have parallel hexagonal axes : one layer cannot rotate upon the adjacent one, only a glide motion is possible. Taking into account the difference between strong attractive forces between molecules inside a layer and weak forces between molecules of different layers, a model of the S_B phase has been developped by De Gennes and Sarma (14).

2/ - the nearly hexagonal packing of the molecules suggests the existence of some motion of the molecules about their axes : in that case, the S_B modification would appear as a particular case of a plastic crystal.

3/ - the correlation length along the molecular direction is a function of the sample temperature : for T.B.B.A., it varies from 7 molecules to 4 molecules in the vicinity of the S_B - S_C transition temperature.
This correlation length seems in fact very dependent on the studied compound : it is larger for B.E.A. (10 molecules) and smaller for E.B.A.C. (2 molecules). Much more compounds have to be investigated so that an estimation of the order extent can be made together with a determination of the layer configuration.

III - STRUCTURE OF THE SOLID PHASE

The structure of the solid phase of T.B.B.A. was studied (13) by means of classical X ray diffraction patterns : Weissenberg and precession photographs. The monoclinic unit cell has been determined and is represented on fig 6 together with the "local" S_B unit cell. It may be pointed out that this unit cell is not the most classical one for crystallographers since the β angle (115°) is not the smallest one ; the c axis was taken parallel to the average direction of the molecules in the solid phase. This direction is relatively easy to determine because the longitudinal motion of molecular strings also exists in the solid phase with the same correlation length (7 molecules) as in the S_B phase. The only difference lies in the amplitude of the displacement which is an order of magnitude smaller (0.2 Å).

As it can be seen on fig 6, there is a small change in the average orientation of the molecules between the solid and S_B phases ; the tilt angle is smaller : 25°. The apparent molecular length (half the periodicity along the c direction) seems also smaller : such a small value could be explained by assuming either some interpenetration of molecules belonging to adjacent layers or distorsion of the aliphatic chains associated with some tilting of the molecules relative to the c directions. The complete structure determination has to be solved in order to confirm the configuration change at the transition.

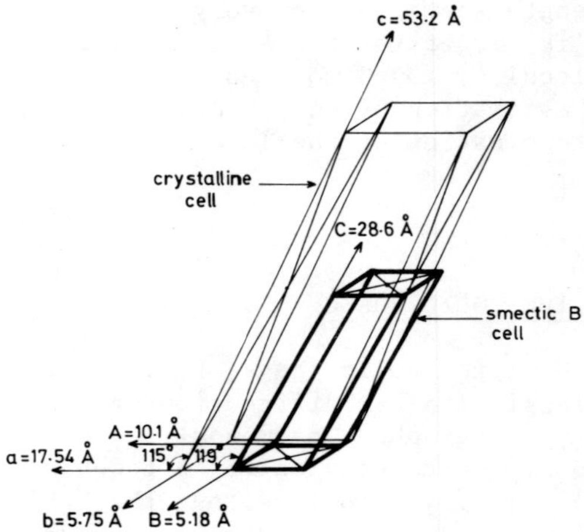

Fig 6 Unit cells of the :
- solid phase 20°C
- S_B phase 125°C

IV - SOLID-SMECTIC TRANSITION

The comparison between the two unit cells of fig 6 concerns the solid phase at 20°C and the S_B phase at 125°C : the essential difference between the two phases lies in the range of the order extent and the S_B phase can be described as a paracrystal (15). In fact, it is interesting to study what happens at the solid-smectic transition point which enables strings of molecules to move longitudinally with such a large amplitude while preserving the lateral order inside the smectic layers.

Some experiments were performed to find out if there is some continuous change in the order parameter in the vicinity of the transition temperature : no important change was found in the X ray diffuse scattering associated with the longitudinal motion of molecules but, contrary to all expectation, a large increase in the intensity of the diffuse scattering connected with the lateral order of molecules was observed in a temperature range of 5° below the transition temperature. The transformation to the S_B phase thus preserves the existence of the molecular strings but involves a modification of the lateral order of molecules which can perhaps be related to a motion about their axes.

Following a suggestion by De Gennes (16), we investigated whether the diffuse scattering in the S_B phase could be attributed to phonons. Measurements were made of the q dependence of the scattered intensity in the vicinity of a Bragg spot. The results are plotted in fig 7 which gives the variation with temperature of the α parameter of the intensity distribution law $I \propto q^{-\alpha}$, where q is the distance to a Bragg spot, measured in the reciprocal lattice (in the case of phonons q is the wave vector and $I \propto q^{-2}$).

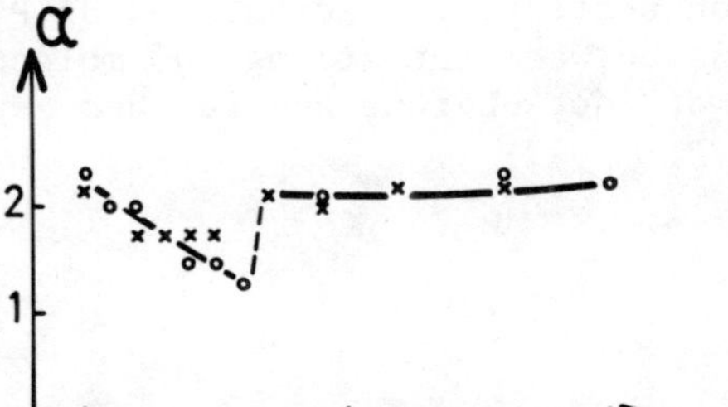

Fig 7 Variation of α with temperature

These experimental results indicate that the dynamics of the S_B modification does not differ very much from the dynamics of a solid phase. The most important anomalies appear as pretransitional effects in the solid phase.

V - DISCUSSION

The S_B phase is characterized both by a periodic order of the molecules inside one layer and some correlations between parallel layers resulting from the correlation in position of parallel adjacent molecules belonging to successive layers. The correlation length differs from one compound to another ; it is smaller for E.B.A.C. where the molecules are normal to the smectic layers and the orientation fluctuations are much more important. In each case, only a local three-dimensional order exists.

An interesting hypothesis was advanced by De Vries (8) proposing that the strings of molecules are essential to the structure of this S_B phase. Consequently it was fundamental to observe if such molecular strings also appear in other liquid crystal modifications and in the solid associated phase. The X ray diffuse scattering experiments performed on different compounds (17) show that such strings of molecules exist in the solid phases, even when they transform, by heating, in a nematic phase (solid P.A.A.). Such correlation effects are also observed for S_C monodomains. Moreover, in all the studied examples, there is no important change of the longitudinal order while the lateral order between adjacent parallel molecules is always affected at the melting point. This result, for P.A.A., is in good agreement with the neutron scattering observations of Pynn and Riste (18). The connections between the strings of molecules and the lateral order range are not obvious and further experimental work is needed.

REFERENCES

1. K. Herrmann, Z. Kristallogr., 92, 49 (1935).

2. E. Friedel, C.R.Acad. Sc. Paris, 180, 269 (1925).

3. S. Diele, P. Brand, H. Sackmann, Mol. Cryst. and Liq. Cryst., 16, 105 (1972).

4. S. Diele, P. Brand, H. Sackmann, Mol. Cryst. and Liq. Cryst., 17, 163 (1972).

5. D. Demus, S. Diele, M. Klapperstück, V. Link, H. Zaschke, Mol. Cryst. and Liq. Cryst., 15, 161 (1971).

6. D. Coates, G.W. Gray, K.J. Harrison, Mol. Cryst. and Liq. Cryst., (1973), in press.

7. I.G. Chistyakov, L.S. Schabischev, R.I. Jarenov, L.A. Gusakova, Mol. Cryst. and Liq. Cryst., 7, 279 (1969).

8. A. De Vries, D.L. Fishel, Mol. Cryst. and Liq. Cryst., 16, 311 (1972).

9. A.M. Levelut, M. Lambert, C.R. Acad. Sc. Paris, 272, 1018 (1971).

10. J. Doucet, A.M. Levelut, M. Lambert, Mol. Cryst. and Liq. Cryst., to be published.

11. Büchi, Dreiding Stereomodeles (normal).

12. T.R. Taylor, S.L. Arora, J.L. Ferason, Phys. Rev. Lett., 25, 722 (1970).

13. J. Doucet, Thèse 3ème Cycle, Orsay, (1972).

14. De Gennes, G. Sarma, Phys. Lett., 38 A, 219 (1972).

15. R. Hosemann, B. Müller, Mol. Cryst. and Liq. Cryst., 10, 273 (1970).

16. P.G. De Gennes, private communication - see also B. Jancovici, Phys. Rev. Lett., 19, 20 (1967).

17. A.M. Levelut, unpublished work.

18. T. Riste, R. Pynn, Solid St. Comm., 12, 409 (1973).

HIGH-FREQUENCY DYNAMICS OF SIMPLE SOLIDS AND LIQUIDS NEAR THEIR MELTING POINTS

P. A. Fleury

Bell Laboratories, Murray Hill, New Jersey, U.S.A.

ABSTRACT

Aspects of the molecular dynamics on either side of the melting transition in Ar, Kr, and Xe have been explored by inelastic light scattering. The simple structure of these materials permits rather clean observation of those higher order light scattering processes sensitive to the material's very short time (10^{-14}-10^{-12} sec) and small scale (1-10Å) dynamics. Measurements of the temperature and density dependences of the spectral lineshapes and strengths have shown that the dense-fluid high-frequency dynamics evolves smoothly and simply from that of the dilute gas, on the one side, and connects continuously to the multiphonon processes in the solid, on the other side. In the solid phase, the temperature dependence of the spectrum has been explored in detail.

INTRODUCTION

In ordinary, or first-order, light scattering one is constrained by momentum conservation to examine individual excitations of wave vector $q \lesssim 2\, k_L$ (where for visible light $k_L \sim 10^5 cm^{-1}$). The frequencies associated with such excitations in simple liquids lie within the hydrodynamic regime and comprise the familiar Rayleigh-Brillouin spectrum. In the solid phases of the classical rare gases, the f.c.c. unit cell contains only one atom and hence sustains no optic phonon modes - such as are usually studied in conventional Raman spectroscopy. The absence of such first-order contributions to the light scattering spectra in the high frequency range (10^{11}-10^{14}Hz) permits observation of those higher order scattering processes which probe molecular motions on the short time (10^{-14}-10^{-12} sec) and length (1-10A) scales lying well outside the hydrodynamic regime. In this paper we summarize experimental results on such higher order scattering processes in classical monatomic materials Ar, Kr and Xe over a range of temperatures and densities which includes the dilute gas, the normal liquid and the solid phases. We find that the spectra, especially at very high frequencies, evolve quite smoothly with changing density and temperature. Thus, the dilute gas $\rightarrow$ super dense liquid range can be described quantitatively by a simple empirical expression, which may be regarded as an extension of the principle of corresponding states to very high frequencies. In addition the development of well-defined high frequency phonon structure in the solid phase at low temperatures, emerges only very gradually upon cooling. These results imply that melting does not involve discontinuous changes in all aspects of a material's dynamics, but only in those at relatively low frequencies. They also support the view that a more fruitful approach to the dynamics of the liquid state is that of a very anharmonic solid, rather than an increasingly dense gas.

To appreciate how light of wavelength $\sim 10^{-5}$cm can probe processes dominated by 10^{-8} cm events, we briefly review the kinematics of second-order Raman scattering. Consider Fig. 1 where an incident photon of frequency and wave vector $\omega_1, \vec{k}_1$ is inelastically scattered into the final photon of $\omega_2, \vec{k}_2$.

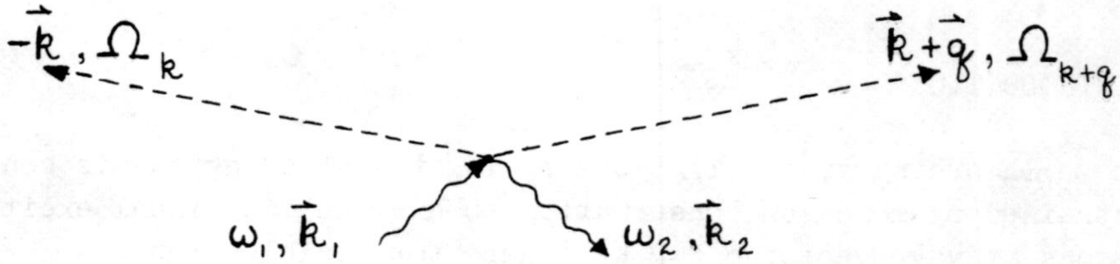

Figure 1. Kinematics of second-order Raman scattering.

In general, regardless of the detailed nature of the excitation responsible for the scattering, momentum and energy conservation require

$$\hbar|\vec{k}_1 - \vec{k}_2| = 2\hbar k_1 \sin\frac{\theta}{2} = \Delta P \equiv \hbar q$$

$$\hbar|\omega_1 - \omega_2| = \Delta E$$

For second-order scattering, the excitation responsible is comprised of a "matched pair" of elementary excitations (e.g. phonons) having individual wave vectors $\vec{k} + \vec{q}$ and $-\vec{k}$, whose vector sum is $\vec{q}$. However, the energy transferred is $\Delta E = \hbar(\Omega_{k+q} + \Omega_{-k}) \approx 2\hbar\Omega_k$. The observed second-order spectrum then is the sum (over k) of contributions from such pairs. In the simplest case (of a well defined dispersion relation Ω_k for all k) this closely resembles the density of states for the excitation. Such second-order scattering (from phonon[1] or magnon[2] pairs) is a familiar process in solids. Recently, these ideas have been applied to liquid He,[3] where the phonon-roton dispersion curve remains well defined out to quite large k values. It has also been suggested that the observed depolarized spectrum in classical fluids (like Ar) may be viewed as a simple extention of this concept to include pairs of density fluctuations which are not

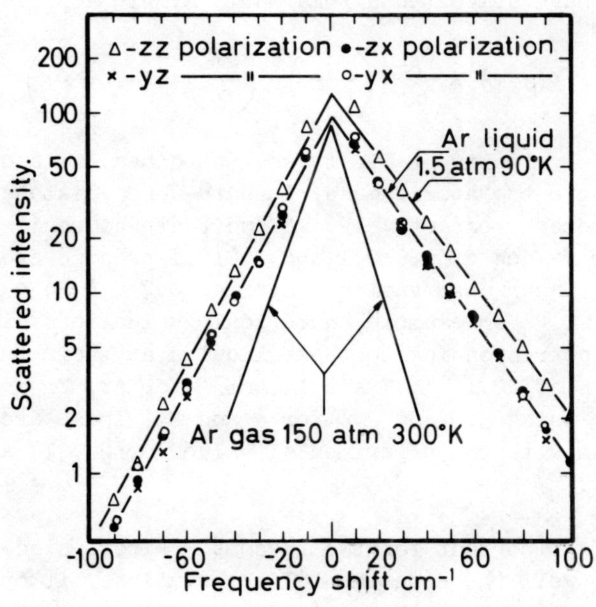

Figure 2. Intermolecular spectra of liquid and dense fluid Ar.

necessarily underdamped.[4] This approach is at least semi-quantitatively successful in accounting for the exponential intermolecular lineshape observed in Ar and other simple fluids.

DYNAMIC CORRESPONDING STATES AND INTERMOLECULAR SCATTERING IN FLUIDS

Fig. 2 shows the observed depolarized spectra from Ar at two different thermodynamic locations.[4] (a) gas at 300°K and 150 amagats; (b) liquid 90°K and 790 amagats. In each case the spectrum exhibits an exponential shape over several decades and extends out to quite high frequencies (corresponding to times $\sim 10^{-14}$ sec). In fact, a series of experiments on Ar, Kr, Xe and Ne covering temperatures between 40°K and 300°K and densities of 10-1800 amagats has been carried out using high pressure techniques.[5] It was found that over a substantial fraction of this range, the spectra can be quantitatively described by $I(\omega) = I_o e^{-\omega/\Delta}$, where Δ^{-1} is some characteristic time descriptive of intermolecular interactions. At present the theory of dynamics in dense fluids is incapable of yielding either Δ or its dependence on density, temperature or material parameters. However, experiments have suggested a remarkably simple empirical relation:[6]

$$\Delta(\rho,T) = \frac{A}{2\pi c}\left(\frac{\varepsilon}{M\sigma^2}\right)^{\frac{1}{2}}\left(\frac{kT}{\varepsilon}\right)^{\frac{1}{2}}\left\{1+\left(\frac{B\sigma^3}{M}\rho\right)^2\right\} \qquad (1)$$

Here c is the velocity of light (so that Δ has units of cm^{-1}), M, σ, and ε are the atomic mass, Lennard-Jones distance and energy parameters respectively. A and B are dimensionless parameters whose values of 3.0±0.1 and 2.0±0.1 produce excellent agreement with our observations covering $0.7 < (\frac{kT}{\varepsilon}) < 4$ and $0.05 < \frac{\rho\sigma^3}{M} < 1$. For example, numerical evaluations of Eq. (1) for ρ and T corresponding to these liquids at their triple points predict Δ_{TP}'s of 22.1, 16.7 and 14.4cm^{-1} for Ar, Kr and Xe, respectively using 3.0 and 2.0 for A and B. These are in excellent agreement with the experimental values of 24, 17 and 14cm^{-1} respectively.[6]

Eq. 1 is important for two reasons. First, since it describes so well the experimental observations, it must provide a bench mark for future theoretical predictions regarding dynamics of simple dense fluids. Second, because the spectral shapes observed are qualitatively all the same (exponential) the expression for $\Delta(\rho,T)$ amounts to an extension of the principle of corresponding states to very high frequency processes.

THE LIQUID-SOLID TRANSITION

Having seen that the spectral changes occur smoothly between the dilute gas and super-dense liquid limits in the fluid state, let us now consider the changes which take place at the liquid-solid transition. The second-order scattering process depicted in Fig. 1 has been calculated in some detail for the classical rare gas solids at zero temperature, and exhibits all of the expected structures of the two phonon density of states for the fcc lattice.[7] However, the spectra shown in Fig. 3, taken on either side of the melting points for Ar, Kr and Xe, reveal that the solids at temperatures close to their melting points exhibit no identifiable phonon structure. Indeed there is no discernible

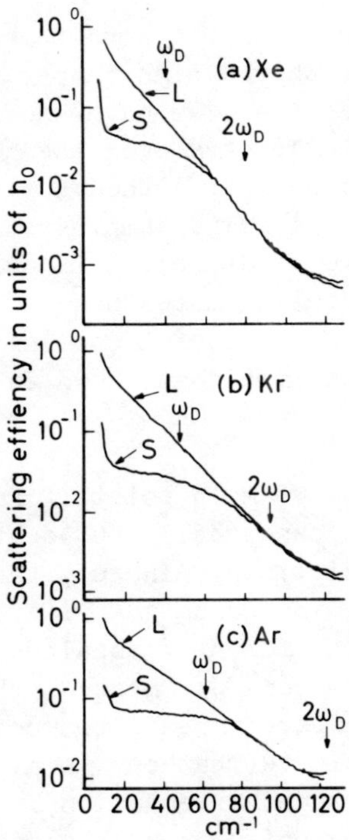

Figure 3. Depolarized spectra of solid (S) and liquid (L) rare gases (a) Xe, (b) Kr and (c) Ar close to their melting points.

difference between the liquid and solid spectra at sufficiently high frequencies ($>\omega_D$, the zero temperature Debye frequency). This means that on the short time and length scales probed by these experiments, melting causes no discontinuous change in the material's dynamics. Notice that the lower the frequency, the larger the discontinuity between liquid and solid behavior; so that in the thermodynamic limit ($\omega \rightarrow 0$) the transition is first order. It should be pointed out here that a similar comparison between liquid and solid spectra for He can be obtained from the work of Slusher and Surko.[8] There the differences between liquid and solid do not become significant until even lower frequencies than those indicated in Fig. 3. Preliminary results of ours on hydrogen indicate that this material exhibits intermediate behavior between the classical Ar and the fully quantum He.

TEMPERATURE DEPENDENCE OF SOLID SPECTRA

In order to make contact with the zero temperature calculations on the one hand, and with the high temperature solid experiments of Fig. 3 on the other, we have studied the temperature dependence of the solid spectra over the entire range, at vapor pressure, of the solid phase. Although with some care it is rather easy to produce rare gas solid samples of excellent optical quality near their melting temperature, it is quite difficult to maintain this quality upon cooling more than ~20% below T_M. The technique finally devised for successful cooling is described in Ref. (6) and produced spectra like that shown in Fig. 4 for Xe at 33°K $\left(\approx \frac{T_M}{5}\right)$.

This figure illustrates two main points. First, there are several sharp features which can readily be identified quantitatively with theoretical predictions. Indeed, the positions indicated by vertical lines, including the 2L(1,0,0) cutoff, were obtained from the calculations for Ar[7] scaled by the Debye frequency (ω_D^{Ar}=62cm^{-1}; ω_D^{Xe}=40cm^{-1}). Second, the differences in relative strengths of the different peaks between polarization combinations (a) and (b) indicate that our sample was predominately a single crystal (of undetermined orientation). Theory[7] predicts that the spectral shape for a polycrystalline sample should exhibit no polarization dependence.

Spectra were obtained at approximately 3° temperature intervals from the melting point down to about 15°K, and evidenced very gradual changes with T. Fig. 5 shows three spectra

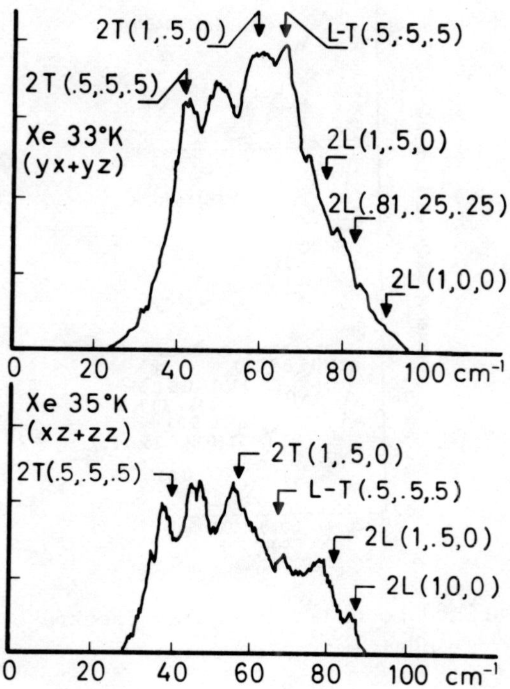

Figure 4. Second-order Raman spectra of solid Xenon. Upper: depolarized scattering. Lower: unpolarized scattering.

selected to illustrate the major stages in this evolution. As T is increased there is a gradual broadening and washing out of the identifiable phonon structure, with the longitudinal contributions submerging by $T/3T_M$ and the transverse contributions (e.g. 2T (0.5,0.5,0.5) at $45cm^{-1}$) persisting to almost $2T/3T_M$. At the same time there is a gradual loss of definition in the high frequency cutoff and the development of a high frequency tail, which for $T \sim T_M$ blends smoothly with the exponential shape observed in the liquid. (See Fig. 4.)

Quite similar behavior was observed in Kr and, over a smaller temperature range, in Ar, the only difference being that all observed frequencies were scaled by the Debye frequency of the appropriate material.

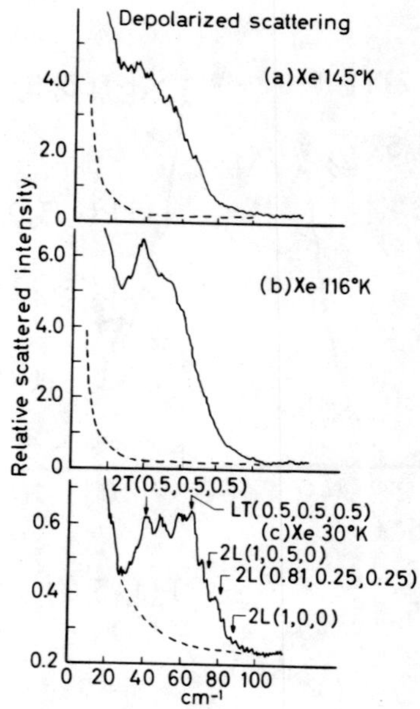

Figure 5. Depolarized second-order Raman spectra of solid Xe at various temperatures.

SUMMARY AND CONCLUSIONS

The experiments reported here have provided a rather broad view of the evolution of high frequency dynamics for simple monatomic materials as they pass from the gas, to the liquid, to the solid phases. Perhaps the most significant result is the remarkable continuity with which this evolution takes place over such a wide range of density and temperature. In the solid phase the low temperature spectra are dominated by well-defined transverse and longitudinal phonons of large wave vector (nearly at the Brillouin zone boundaries). As the temperature is raised, anharmonic effects broaden severely first the longitudinal, then the transverse peaks; until near T_M the zone boundary phonons in the solid are probably all overdamped. This is in agreement with the rather limited direct observations of large wave vector TA and LA phonons at elevated temperatures available from inelastic neutron scattering experiments.[9,10] The view of intermolecular scattering in the liquids as due to pairs of overdamped phonons, then emerges quite naturally. The disappearance of long range static order upon melting has negligible effect on these phonon pair spectra; reminescent incidentally of the persistence of magnon pair spectra well into the paramagnetic phase in antiferromagnets.[11] The suggestion that dynamics in a dense fluid may be more fruitfully approached from the viewpoint of a very anharmonic solid than from that of an increasingly dense gas is then also quite natural.

Finally, some mention must be made regarding the quantitative status of the above discussion. There are two equally important elements in the quantitative relation between the material's dynamics and the observed spectral shapes which we have not emphasized thus far. It is generally true that the quantity measured in a light scattering experiment is $I(\omega) = \int_0^\infty \langle \delta\alpha(t)\delta\alpha(o) \rangle e^{-i\omega t} dt$, the Fourier transform of the auto-correlation function of the polarizability fluctuations; $\delta\alpha$. It is also generally true that $\delta\alpha$ is some function of atomic positions or configurations. In the simplest case, first-order scattering, $\delta\alpha$ is directly proportional to, say, atomic displacement, δu. Then $I(\omega)$ depends simply on $\langle \delta u(t)\delta u(o) \rangle$ and the relation of the spectral shape to the particle dynamics is straightforward and quantitative. For the higher order processes considered in this paper, $\delta_\alpha(t) = \delta u(\vec{r},t)\delta u(\vec{r}',t)\beta(\vec{r},\vec{r}')$ and is simply related to particle motions only if $\beta(\vec{r},\vec{r}')$ - a sort of intermolecular polarizability - has a simple form. Even in this case, $I(\omega)$ measures $\langle \delta u \delta u \ \delta u \delta u \rangle$ a <u>four point</u> correlation function, which especially for anharmonic systems is not simply related to $\langle \delta u \delta u \rangle$.

These two points must be theoretically understood before higher order scattering spectra can be truely quantitatively related to dynamics. Even without this understanding however quite useful semiquantitative information has already been gained. It is worth emphasizing that the development of such understanding - even for simplest systems like argon - is a worthwhile goal since light scattering permits rapid acquisition of data on a complicated correlation function over a range of ρ, T that would require prohibitively long experiments with neutrons.

ACKNOWLEDGEMENTS

I am grateful for the collaboration of J. M. Worlock and W. B. Daniels in several aspects of the work described here. The excellent technical assistance of H. L. Carter is also greatly appreciated.

REFERENCES

1. See for example, J. R. Hardy and A. M. Karo, "Light Scattering Spectra of Solids", edited by G. B. Wright, Springer-Verlag, N. Y., 1969, p.99; and M. Krauzman loc. cit. p.109.

2. P. A. Fleury and R. Loudon, Phys. Rev. 166, 514, 1968.

3. M. J. Stephen, Phys. Rev. 187, 279, 1969.

4. J. P. McTague, P. A. Fleury, and D. B. Dupre, Phys. Rev. 188, 303, 1969.

5. P. A. Fleury, W. B. Daniels, and J. M. Worlock, Phys. Rev. Lett. 27, 1493, 1971.

6. P. A. Fleury, J. M. Worlock and H. L. Carter, Phys. Rev. Lett. 30, 591, 1973.

7. N. R. Werthamer, R. L. Gray, and T. R. Koehler, Phys. Rev. B2, 4199, 1970.

8. R. E. Slusher and C. M. Surko, Phys. Rev. Lett. 27, 1699, 1971.

9. D. N. Batchelder, B. C. Haywood, and D. H. Saunderson, J. Phys. C4, 910, 1971.

10. J. Skalyo, Y. Endoh, and G. Shirane, Bull. Am. Phys. Soc. 18, 313, 1973.

11. P. A. Fleury, Phys. Rev. 180, 591, 1969 and Int. J. Magnetism, 1, 75, 1970.

ANALYSIS OF THE APPROACH TO THE CONVECTIVE INSTABILITY POINT

J.P. Boon (1) and H.N.W. Lekkerkerker

Faculté des Sciences, Université Libre Bruxelles, 1050 Brussels and Fakulteit van de Wetenschappen, Vrije Universiteit Brussel, 1050 Brussels, Belgium

ABSTRACT

A spectral analysis is presented of the fluctuations in a horizontal fluid layer subject to a downward directed temperature gradient, which, for a critical value, drives the system in a convective instability state. It is found that the external force resulting from the combination of the temperature gradient and the gravitation force gives rise to a coupling between the heat diffusion mode and a shear mode. As a result of this mode coupling the damping constant of the heat diffusion mode goes to zero when the temperature gradient increases towards its critical value, i.e. the heat diffusion mode behaves like a "soft mode". The implications of the mode coupling and of the ensuing softening of the heat diffusion mode on the light scattering spectrum are discussed.

INTRODUCTION

In a horizontal fluid layer heated from below a stationary convection mode appears spontaneously for a critical value of the downward directed temperature gradient. This is one of the simplest examples of a hydrodynamic instability, (well known in classical physics as the Bénard problem (2)).

One of the prime objects of investigations in hydrodynamic stability has been the determination of the values of the state parameters at which the transition from the stability regime to the instability domain occurs. So far little attention has been given to the process of initiation of non-equilibrium instabilities (see however references 3 and 4). The reason for this may be the difficulty to observe pretransitional phenomena in classical hydrodynamic experiments. However, it appears that modern experimental methods allow the investigation of the pretransitional states. Indeed, since instabilities are triggered by thermal fluctuations, light scattering spectroscopy which probes these fluctuations, appears as anappropriate tool to investigate the pretransitional fluctuations (5,6).

The Bénard problem, because of its simplicity, was chosen here to investigate pretransitional phenomena. Using hydrodynamic fluctuation theory, we show how the spectral features of the light scattering spectrum are modified when the fluid departs from equilibrium towards the instability critical point. The spectral changes are found to be most important for the central line. An additional Lorentzian component arises because the entropy fluctuations, which in an equilibrium one-component fluid decay in a single heat diffusion mode, decay into two non-propagating modes in the presence of the external force. When the system approaches the instability critical point the damping of one of these modes goes to zero indicating a "soft mode" type of behavior which suggests an analogy between hydrodynamic instability phenomena and structural phase transitions (7).

SPECTRAL ANALYSIS OF THE FLUCTUATIONS

As is well known, the spectral intensity distribution of the polarized component of scattered light is proportional to $S_\epsilon(\underline{k},\omega)$ the spectral density of the $\underline{k}$th spatial Fourier component of the fluctuation $\delta\epsilon$ in the optical dielectric constant. Here $\underline{k}$ is the change in wave vector and ω the change in frequency upon scattering. The dielectric constant can be considered as a function of the thermodynamic state of the system, so that $\delta\epsilon$ can be expressed in terms of the fluctuations α_i in the thermodynamic state variables A_i. Consequently

$$S_\epsilon\ (\underline{k},\omega) = \sum_{i,j} \left(\frac{\partial\epsilon}{\partial A_i}\right)\left(\frac{\partial\epsilon}{\partial A_j}\right) S_{\alpha_i\alpha_j}\ (\underline{k},\omega). \tag{1}$$

The spectral densities of the fluctuations in the thermodynamic state variables occuring on the right hand side of equation (1) will be calculated here by assuming that also in non-equilibrium systems do the fluctuations on the average decay according to the appropriate hydrodynamic equations, i.e. the linearized equations for the fluctuations in the steady state variables. In the Bénard problem the steady state can be described as follows: an adverse linear temperature gradient is maintained steadily; there is no convective motion; and the gravitational force is balanced by the pressure gradient. Labeling the steady state variables with the superscript s one has

$$T^s(z) = T_o - \beta z \; ; \; \underline{v}^s = 0 \; ; \; \partial_z p^s(z) = -g\rho^s(z) \tag{2}$$

where the subscript o denotes the value of a variable at the reference position, which for the sake of convenience is taken at the lower boundary (see figure 1). For the purpose of obtaining the linearized equations for the fluctuations in the steady state variables the equation for $p^s(z)$ and $\rho^s(z)$ need not be solved explicitly as long as $|p^s(z) - p_o| \ll p_o$ and $|\rho^s(z) - \rho_o| \ll \rho_o$ which conditions are usually satisfied for normal fluids and for actual experimental conditions (layer tickness $\sim 0.1 - 1$ cm).

In the present case, we choose to write the thermodynamic equations for the following set of variables:

$$\delta s = s - s^s \; ; \quad \delta p = p - p^s \; ; \; \varphi = \partial_x v_x + \partial_y v_y \; ; \; v_z \; ; \; (\nabla x \underline{v})_z .$$

The above description of the velocity field is a consequence of the symmetry of the system under consideration. It can be shown that φ and v_z have the same transformation properties as δs and δp under the symmetry operations of the system, whereas $(\nabla x \underline{v})_z$ does not and will therefore not couple to δs and δp.

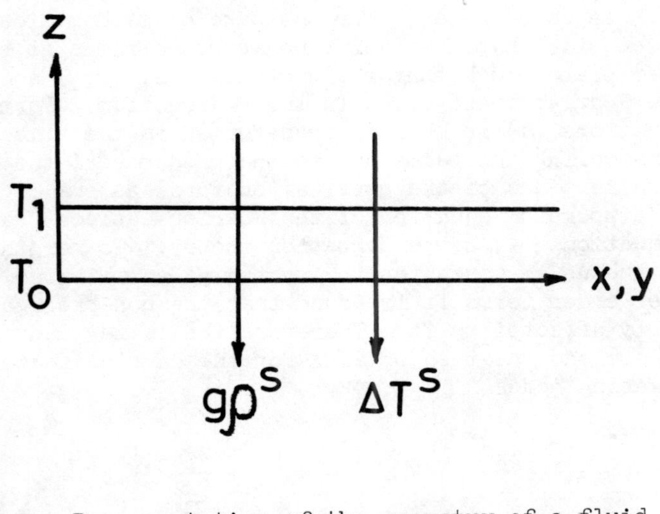

Figure 1. Representation of the geometry of a fluid layer subject to an adverse temperature gradient.

One then obtains the following set of coupled linearized equations

$$(\partial_t - D_T\nabla^2)\delta s - (\alpha/\rho)D_T\nabla^2\delta p - (Cp/T)\beta v_z = 0,$$

$$(\partial_t - (\gamma-1)D_T\nabla^2)\delta p - (\rho/\alpha)(\gamma-1)D_T\nabla^2\delta^s$$

$$+ \varphi/\chi_s - (\alpha\beta-\partial_z)v_z/\chi_s = 0,$$

$$(\partial_t - (\nu+\nu')\nabla^2 + \nu'\partial_z^2)\varphi - \nu'(\nabla^2-\partial_z^2)\partial_z v_z$$

$$+ \rho^{-1}(\nabla^2-\partial_z^2)\delta p = 0,$$

$$(\partial_t-\nu\nabla^2-\nu'\partial_z^2)v_z - \nu'\partial_z\varphi + (g\chi_s + \rho^{-1}\partial_z)\delta p$$

$$- (g\alpha T/C_p)\delta s = 0. \qquad (4)$$

Here is α the thermal expansitivity, χ_S the adiabatic compressibility, γ the ratio of the specific heats at constant pressure (C_p) and at constant volume (C_V), D_T the thermal diffusivity, ν the kinematic viscosity and $\nu' = (\xi + \eta/3)/\rho$ where η is the shear viscosity and ξ the bulk viscosity.

As the system is supposed to be infinite in the x and y directions (i.e. the actual geometry of the system is such that the thickness of the fluid layer is much smaller than its dimensions in the x,y plane) a spatial Fourier transformation can be performed in these coordinates. For the sake of mathematical simplicity we consider hypothetical boundary conditions at the lower and upper plates such that the variables δs, δp, φ and v_z can also be Fourier transformed in the z direction. Further the set of equations (4) is Laplace transformed in the time variable. Considering the terms due to the presence of the temperature gradient and of the gravitation force as small, one can treat the dispersion equation of the Fourier-Laplace transformed coupled equations (4) by perturbation theory. To zeroth order one retrieves the usual modes for the system at equilibrium. From the higher order terms it is found that the non-propagating modes are weakly affected by it. Therefore the latter can safely be neglected here and one then obtains for the damping factors of the non-propagating modes

$$s_\pm = -\frac{k^2}{2}(\nu+D_T) \pm \frac{k^2}{2}[(\nu-D_T)^2 + 4\nu D_T \frac{R}{R_c(\underline{k})}]^{\frac{1}{2}}, \qquad (5)$$

with

$$R/R_c(\underline{k}) = \frac{\alpha\beta g}{\nu D_T} \; \frac{k^2 - k_z^2}{k^6} \; . \qquad (6)$$

Here R is the Rayleigh number and $R_c(\underline{k})$ its critical value for the mode with wave vector $\underline{k}$. Indeed, note that $R/R_c(\underline{k})$ depends strongly on the wavelength of the fluctuation under consideration. This fact has important consequences for probing pretransitional phenomena with light scattering spectroscopy and will be discussed in the next section. For most liquids $\nu > D_T$ and it is easily seen from equation (5) that in that case the external force induces a coupling such that the heat diffusion mode (S_+) behaves like a "soft mode" simultaneously with the broadening of the shear mode (S_-). When considering only the non-propagating modes, the set of equations (4) reduces to two coupled equations. These can easily be solved for $\delta s(\underline{k},s)$ (the Fourier-Laplace transform of δs), from which the appropriate correlation function is constructed to yield the spectral distribution of scattered light. One obtains

$$I^{central}_{\underline{k},\omega} \propto \left(\frac{\partial\epsilon}{\partial s}\right)_p^2 \langle|\delta s(\underline{k})|^2\rangle \left(A_+ \frac{|s_+|}{s_+^2+\omega^2} + A_- \frac{|s_-|}{s_-^2+\omega^2}\right) , \qquad (7)$$

with

$$A_+ = 1 - A_- = \frac{(\nu k^2 - s_+)^2}{(\nu k^2 - s_+)^2 + \nu D_T k^4 R/R_c(\underline{k})} \; .$$

One thus observes that the central peak for a system subject to an adverse temperature gradient consists of two Lorentzians (Note that in the limit $R \rightarrow 0$, i.e. $\beta \rightarrow 0$ one retrieves the single heat diffusion line as expected for a fluid at equilibrium). The spectrum as given by equation (7) is illustrated in figure 2.

DISCUSSION

The most interesting result is undoubtedly the fact that the damping factor of one of the non-propagating modes goes to zero when the instability critical point is approached. From the expression of $R/R_c(\underline{k})$ given by equation (6), it is clear that for a given temperature gradient the effect of the external force will be most important for small values of k. This means that the most dramatic spectral changes are to be observed at small scattering angles. Indeed, the wavelength of the mode which is the first to become unstable is of the order of the vertical dimension of the system.

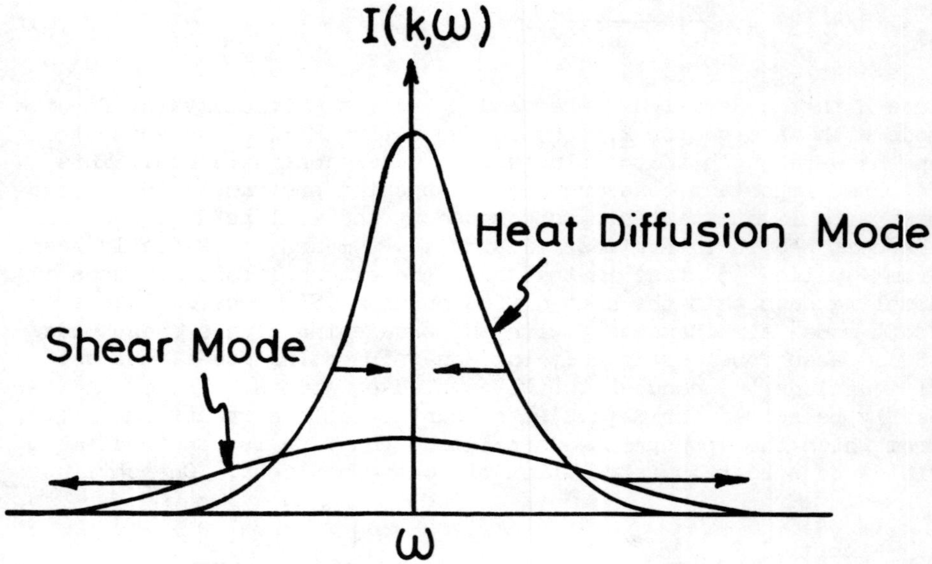

Figure 2. Representation of the central spectral components of a fluid subject to an adverse temperature gradient. The arrows indicate the modification of the width of the spectral components when the temperature gradient increases towards its critical value.

Presently available techniques used in light scattering spectroscopy should allow to probe modes which are expected to be affected in a fluid subject to an external force.

ACKNOWLEDGEMENTS

The award of a NATO Science Fellowship by the Netherlands Organisation for the Advancement of Pure Research for the academic year 1971-1972, when this work was initiated, is acknowledged by HNWL. This research has been sponsored in part by the Internationale Instituten voor Fysica en Chemie, gesticht door E. Solvay.

REFERENCES AND FOOTNOTES

(1) Chercheur qualifié au Fonds National de la Recherche Scientifique (FNRS), Belgium.

(2) S. Chandrasekhar, Hydrodynamic and Hydromagnetic Stability, (Oxford, Clarendon Press, 1961), Chapter 2.

(3) V.M. Zaitsev and M.I. Shliomis, Soviet Physics JETP 32, 866 (1971).

(4) J.P. Boon, Phys. Chem. Liquids 3, 157 (1972).

(5) J.P. Boon and Ph. Deguent, Phys. Letters 39A, 315 (1972).

(6) J.M. Deutch, S. Hudson, P.J. Ortoleva and J. Ross, J. Chem. Phys. 57, 4327 (1972).

(7) H. Thomas, I.E.E.E. Trans. Magn. 5, 874 (1969).

MELTING: THEORIES AND RECENT COMPUTER SIMULATIONS

R.M. Cotterill, E.J. Jensen and W.D. Kristensen

Department of Structural Properties of Materials
The Technical University of Denmark
DK-28 Lyngby, Denmark

ABSTRACT

This brief review of melting is divided into four main parts. It starts with a review of significant phenomenology. This is followed by a description of the essentials of the main theories of the melting transition. An account is then given of recent direct observations of the melting process by computer simulation. In the concluding sections the computer results are compared with the different theories and an attempt is made to present a unified model of the transition.

INTRODUCTION

Although the gross features of the melting transition are understood inasmuch as it is clear that there is a limit to how much thermal agitation a regular crystal structure can tolerate, we still do not have a microscopic picture of exactly what happens when a crystal melts. Melting has it in common with other transitions that it is not particularly accessible experimentally. The obvious approach is to make observations on the solid and liquid states as close as possible to the melting point, and this strategy has now yielded a considerable body of data to which any theory of melting would have to be reconciled. The data falls into two broad classes. On the one hand there are such macroscopic parameters as volume change, latent heat, thermal and electrical resistivity change, the solid-liquid interfacial energy, diffusion constant change, and the marked drop in viscosity, and on the other the more recent X-ray and neutron scattering data on the atomic arrangements in the solid and liquid states. In the case of a liquid a scattering experiment does not of course give a picture of a unique arrangement of atoms but rather a measure of the probability distribution of distances between pairs of atoms. This represents part of the difficulty, but it is not the whole problem. Even working from a knowledge of the exact location of each atom in the liquid, it would still be a formidable task to develop a systematic description of the atomic displacements which are associated with the transition.

There have been many theories of melting, and they have ranged from modest attempts to explain some of the more obvious features of the phenomenon, to more ambitious efforts in which the changes of several parameters across the transition have been quantitatively accounted for. Two of the better known theories are due to Lindemann (1) and Born (2). The first of these says that a solid will melt when the mean vibrational amplitude of the constituent atoms reaches a certain value (a certain fraction of the nearest-neighbour distance say), while the second has it that melting occurs when the rigidity modulus vanishes. More modern theories, which invariably predict not only the melting temperature but also account for the changes in several variables, have often invoked crystal defects. Melting has variously been attributed to the presence of vacancies, interstitials or dislocations.

The main object of this brief review is in fact to discuss the latest evidence for the dislocation theory of melting. However to place it in perspective it is necessary first to consider the phenomenology of melting, both at the macroscopic and microscopic levels. We will find it useful also to consider the Bernal random packing model, which is thought to be a good first approximation to the instantaneous structure of a liquid, and whose difference from the regular packing of a solid must be explained by any melting model. Before considering the different theories, it will also be instructive to examine certain correlations between irradiation damage and melting. After a brief review of past theories we turn to the recent computer simulations, and we close by attempting a synthesis of a unified picture which incorporates certain aspects of several theories into a model in which the dislocation plays a decisive role.

THERMODYNAMICS OF MELTING

Melting is a first-order phenomenon, in the Ehrenfest sense[+], in that the first derivatives of the Gibbs free energy G(P,T) with respect to temperature and pressure suffer a discontinuity across the transition. For a system having a fixed number of atoms we

---

[+] according to which a transition is said to be of nth order if the nth derivative of G(P,T) is the lowest one to show a discontinuity.

have

$$dG(P,T) = VdP - SdT \tag{1}$$

and there are changes of volume and entropy on melting, because the right-hand sides of the following two equations are non-zero

$$\Delta V_m = ((\partial G/\partial P)_T)_{liq} - ((\partial G/\partial P)_T)_{sol} \tag{2}$$

$$\Delta S_m = ((\partial G/\partial T)_P)_{sol} - ((\partial G/\partial T)_P)_{liq} \tag{3}$$

The latent heat of melting, L_m, is related to the entropy change, and the melting temperature, T_m, by the relation

$$L_m = T_m \Delta S_m \tag{4}$$

and it is related to the volume change via the Clausius-Clapeyron equation

$$(dP(T)/dT)_{T_m} = L_m / T_m \Delta V_m \tag{5}$$

Finally, the specific heat at constant pressure of the system, at any temperature, is given by

$$C_p = T(\partial S/\partial T)_P = - T(\partial^2 G/\partial T^2)_P \tag{6}$$

It is instructive to compare the changes which occur during melting and boiling, by refering to the PVT surface, and its projections and sections. Figure 1 shows such a surface for a substance which expands upon melting, (which is by far the more common case). The dashed line traces the changes in the system at a constant pressure below the critical pressure. There are volume changes both upon melting and upon evapouration, although the former is relatively small; (it is never in fact more than a few percent of the solid volume at T_m). Both transitions, being first order, also involve latent heat and changes in entropy. Because the system absorbs latent heat while the temperature is not changing, the specific heat at constant pressure must be infinite for each of these processes, and this in fact is another general characteristic of first order transitions. This behaviour is indicated in figure 15 and will be discussed later in connection with the melting of finite systems. Returning to figure 1, we note that at the critical point the transition from liquid to vapour is such that the system suffers no volume or entropy discontinuity, (and is no longer first order). Thus above this point, (P_c,V_c, T_c),the distinction between these two states vanishes. Their similarly "random" and isotropic natures are consistent with such an occurence. Since crystalline solids, on the other hand, are niether random nor isotropic it is difficult to see how the solid--liquid transition could ever exhibit critical behaviour. No crit-

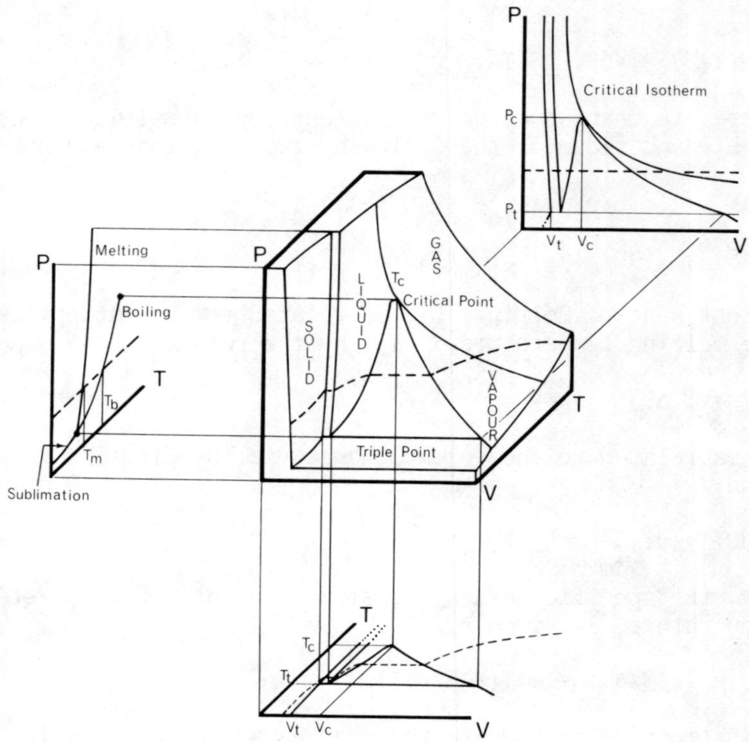

Figure 1. Typical PVT surface for a substance that expands upon melting.

ical melting has indeed ever been observed, and high temperature extrapolations of recent melting data for sodium (3), for instance indicate that ΔS would not go to zero. At the critical point the transition from liquid to vapour is of second order, with an absence of latent heat (and entropy and volume change) and with the specific heat showing a discontinuity (because $\partial^2 G/\partial T^2$ is discontinuous in equation 6) but no tendency to become infinite.

It is found that the majority of elemental substances have PVT surfaces with similar proportions to that shown in figure 1 and this has led to the proposal of several empirical rules. Richard's rule, for instance, has it that L_m/T_m is constant (and roughly equal to 2 cal mole^{-1} $^oK^{-1}$). Lindemann's rule relates the Debye temperature to T_m, and we discuss this in a later section. The Mukherjee (4) and Van Liempt (5) rules relate the vacancy formation energy and the self diffusion energy respectively to T_m. Finally, there is the suggestion (6) that $\Delta V_{sd}/E_{sd} = \Delta V_m/L_m$ where ΔV_{sd} and E_{sd} are the self diffusion activation volume and energy respectively. Although the empirical rules linking diffusion and melting are interesting they do not necessarily show that diffusion actually causes melting. The simple functional relationships might simply occur as a natural consequence of the fact that both processes are linked to the interatomic potential.

CHANGES IN PHYSICAL PROPERTIES UPON MELTING

In this section we review the physical changes that acompany the melting transition, treating first the macroscopically measurable variables and thereafter the differences in instantaneous atomic structure and the dynamics of atomic motion between the solid and liquid states.

By far the most dramatic change is exhibited by the fluidity. It increases approximately by a factor of 10^{19} (from typically 10^{-17} to 10^{2} poise^{-1}). Most other variables show only a relatively subtle change. ΔV_m is usually positive and of the order of a few percent, although there are a few exceptional substances which show small contractions on melting (e.g. Ge,Si,Ga,Sb,Bi, and, of course, H_2O). The solid-liquid interfacial energy is an order of magnitude smaller than either the solid-gas or liquid--gas interfacial energy. The same is true of the latent heat of melting compared with the latent heat of evapouration. The diffusion constant is typically about $10^{-9}cm^2sec^{-1}$ in a crystal just below the melting point, and 10^{-5} cm^2sec^{-1} in a liquid just above the melting point. In the case of crystals the diffusion most commonly proceeds through the agency of vacancies, while in the liquid state it is the vacancy-like holes (whose size and position are the result of statistical fluctuations in the distribution of free volume) that are responsible for the process. The electrical resistivity change on melting has been measured for several metals and it is frequently less than a factor of two. The changes in several macroscopic parameters on melting are listed in Table 1.

Scattering experiments (with X-rays or thermal neutrons) show that the instantaneous structure of a liquid is distinctly different from that of a solid. In the former case the atoms are distibuted on a well-defined lattice. In the limit of zero temperature the lattice is a perfect pattern of delta functions, and these are smeared out at finite temperature into gaussian distributions by amounts proportional to the Debye-Waller factor. The additional "smearing out" observed in a liquid is more than can be explained simply on the basis of the temperature factor. Larger displacements (comparable to, although less than the nearest--neighbour distance) must be involved. The degree of order in either phase is quantified through the radial distribution function (RDF); the probability that given an atom at an arbitrary origin there will also be an atom at a distance r. In a solid at zero temperature the RDF consists of delta functions at the suc-

TABLE 1

MELTING CHARACTERISTICS OF SELECTED ELEMENTS

Element	Melting Temperature T_M [°K]	Latent Heat L_M [cal/mole]	Entropy of Fusion S_M [k_B/Atom]	Volume Change ΔV_M [%]	Electrical Resistiv. $\frac{\rho_\ell}{\rho_s}$	Thermal Conduct. $\frac{\lambda_s}{\lambda_\ell}$	Self-Diffusion $\frac{D_\ell}{D_s}$
Cu (FCC)	1356	3100	1.16	4.5	2.04	1.99	$7.9 \cdot 10^3$
Ag (FCC)	1234	2690	1.10	3.3	2.09	2.03	$4.1 \cdot 10^3$
Au (FCC)	1336	3050	1.15	5.1	2.08	2.35	
Al (FCC)	933	2310	1.25	6.0	2.20	2.22	
Ni (FCC)	1726	4240	1.24		1.30	-	
Li (BCC)	454	690	0.77	1.7	1.64	1.66	
Na (BCC)	371	630	0.86	2.5	1.45	1.36	$2.3 \cdot 10^2$
K (BCC)	335	570	0.85	2.6	1.56	1.80	$2.8 \cdot 10^2$
Ar (FCC)	83	280	1.68	15.2	-	2.27	$1.6 \cdot 10^4$
Kr (FCC)	116	392	1.70	15.0	-	2.65	
Si (DIA)	1683	12150	3.63	-12.0	0.034	-	

cessive neighbour shell positions, while for a liquid, if this were truly random in its atomic arrangement (as for a perfect gas), the RDF would be a parabola. The actual situation in a liquid (argon) is shown in figure 12 (where the pair distribution, PDF, has been plotted; the PDF being simply the RDF divided by $4\pi r^2$). One sees that the distribution is not at all uniform. Inside a certain distance the PDF is zero due to the short-range repulsive forces which become very large for radii below d_o , the nearest-neighbour distance. At a radius slightly less than d_o the liquid PDF shows a distinct first peak, indicating a certain degree of short-range order, while beyond about $5d_o$ it is constant because of the total absence of long-range order.

The PDF is easily obtained from a suitable scattering experiment (7), but it unfortunately does not define the atomic arrangement uniquely. A notable attempt to overcome this difficulty has been made by Bernal et al (8) who showed that a random close-packed assembly of hard spheres has a PDF that is in good agreement with that observed for liquid argon, (see figure 12). From measurements on the model it was also possible to obtain the average coordination number of the "instantaneous hard-sphere liquid", (see figure 2) and this proved to be approximately 9 (instead of 12 as in the close-packed crystal). The volume increase corresponding to the change from regular to random close packing was 15.6% (9), which compares well with the entry for Kr in Table 1. Finally, it is interesting to note that by using the derived PDF together with a Lennard-Jones potential appropriate for Ar, it was possible to compute a latent heat of melting that agreed quite well with the experimental value (10). These simple experiments suggest that the gross features of the instantaneous structure of a liquid are determined by the short-range repulsive forces, and that this structure is essentially random close packing. Bernal has remarked that a solid can be regarded as a pile of atoms whereas a liquid is more like a heap of atoms.

Solids and liquids are also distinctly different in the nature of their atomic motions. Neutron scattering, in particular, has been able to establish this fact because the energy of thermal neutrons is comparable to the atomic translational energy in the two phases. Each atom in a solid vibrates about a fixed point and the assembly of such points forms a time-invariant lattice. The totality of these vibrations is described by the familiar phonon spectrum. Translational migrations that are not necessarily reversed within a vibrational period can occur by diffusion (usually vacancy-assisted ; see later) but the relative occurence of this type of motion only reaches 1 in 10^7 even in the vicinity of the melting point. In a liquid the situation is quite different. The instantaneous positions form, as was noted earlier, not a lattice but a random close-packed assembly, and this assembly changes with time. The atoms still vibrate within their individual (and

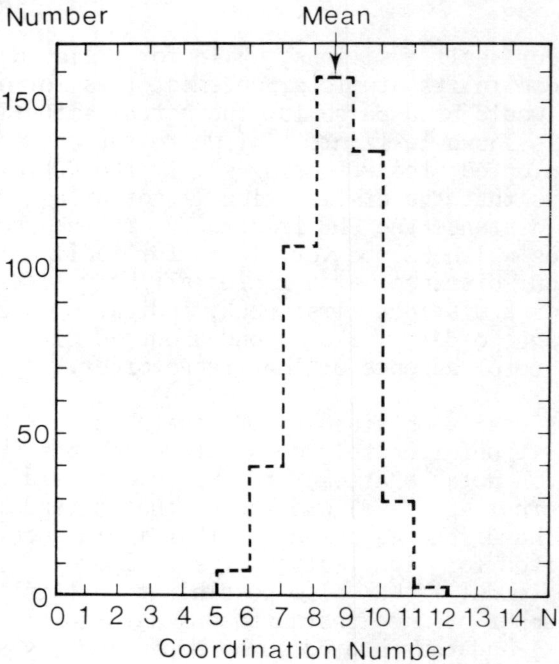

Figure 2. Coordination histogram for a random close--packed assembly of hard spheres (8).

non-identical) coordination shells, but they migrate at the same time. The relative occurences of these two modes of motion are comparable in a typical liquid, the actual distribution being temperature dependent. Just above the melting point vibrations still predominate+ , but as the boiling point is approached the migrations are far more significant. It is a moot point as to whether or not liquids can still be said to exhibit phonon-like behaviour (12).

Before leaving this section it should be noted that we have so far tacitly assumed that the transition from solid to liquid occurs quite suddenly right at the melting point. This might not be strictly true and there has been considerable conjecture regarding the possibility of premonitory effects that are usually referred to as "premelting" (13,14). There is not space here to review the theory of such effects, but there does appear to be experimental support for their occurrence. In particular, the work of Eckert and Drickamer (15) on diffusion in In near the melt-

+ An early theory of liquid viscosity (11) actually assumed that the vibrational frequencies in a solid and a liquid, near to the melting point, are equal.

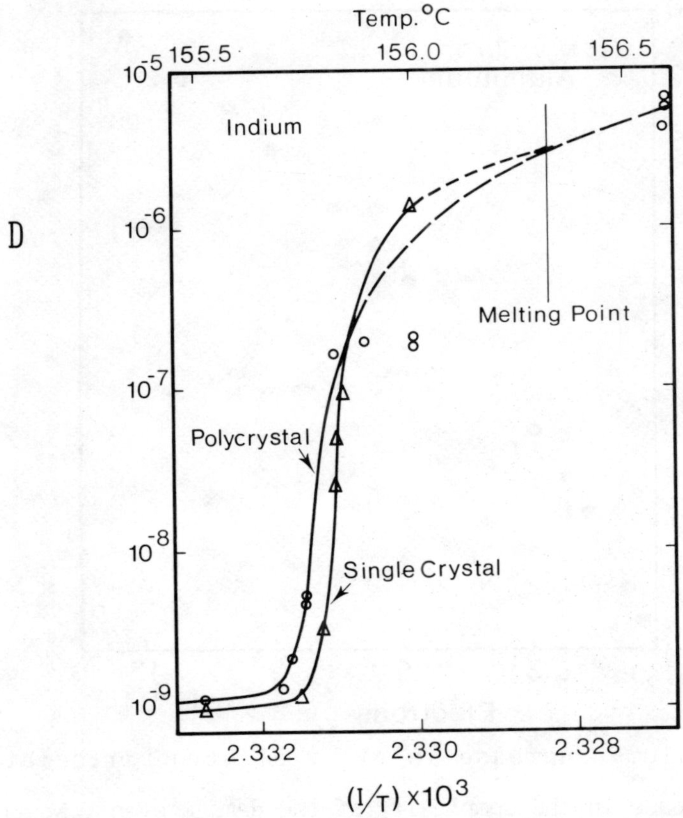

Figure 3. Premelting detected by measurement of the diffusion constant in indium (15).

ing point (see figure 3) indicates that as far as this process is concerned the transition actually occurs gradually over a range of about 0.6 °K. We return to premelting later, and we will offer an explanation for it.

SOME RELEVANT IRRADIATION DAMAGE EXPERIMENTS

During the last twenty years a large amount of information has been collected in experiments in which disorder is introduced in crystals by bombardment with subatomic particles. The simplest form of this damage occurs when the amount of energy imparted to the primary displaced atom is just above the threshold for preventing that atom from spontaneously falling back into its lattice position. The defect thus generated is the single "Frenkel pair" (see also the next section) and it consists of a bound pair of one vacancy and one interstitial. If the energy imparted to the primary struck atom is much larger than the threshold value, further displacements can occur and a "cascade" is produced, with a

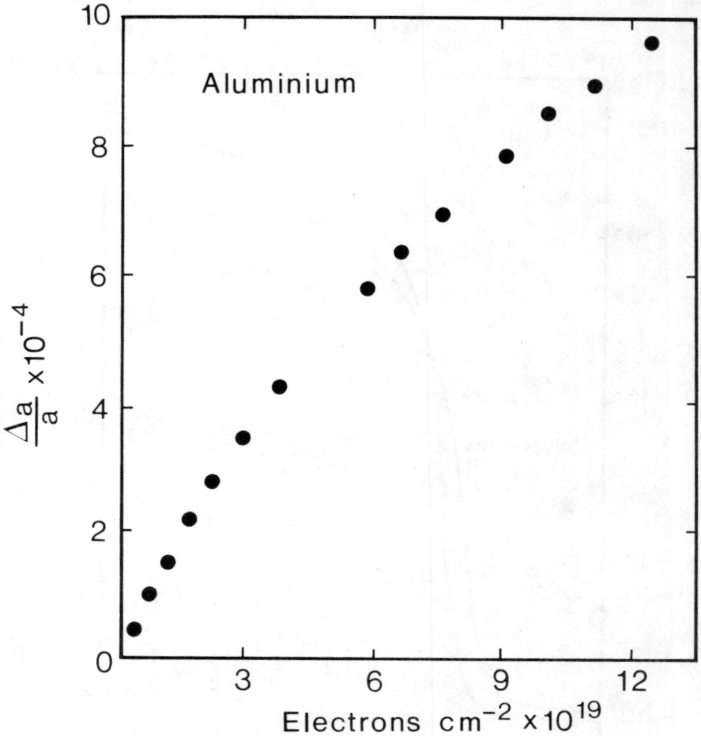

Figure 4. Volume increase in Al by electron irradiation.

resultant increase in the complexity of the damage event. Neutron damage is usually of the cascade type, but electron damage can be carried out with greater control of the incident energy and the single-displacement regime is easily realised.

When a vacancy is formed in a crystal the net effect can be regarded as the sum of two processes. There is the removal of an atom from the interior of the crystal and its replacement on the surface, then there is the relaxation around the vacancy which is essentially a spherically-symmetric shrinking into the resulting hole with a strain which falls off as r^3. The first contribution gives a volume change of Ω_o, where Ω_o is the atomic volume in the perfect lattice, while the second gives $\varepsilon_v\Omega_o$, where ε_v is negative and approximately 0.2 . The volume change per vacancy is therefore positive and roughly $0.8\Omega_o$ in size. When an interstitial is formed an atom is effectively removed from the surface and replaced into one of the interstices of the lattice. The first part now contributes $-\Omega_o$ to the volume change, but the extra atom that is forced into the lattice is in such close contact with the surrounding atoms that the outward expansion $\varepsilon_i\Omega_o$ is quite large. ε_i can in fact easily be + 3.0 , so that the volume change upon generating an interstitial would again be positive and roughly $2\Omega_o$. This would imply a Frenkel pair volume change of $2.8\Omega_o$, which is

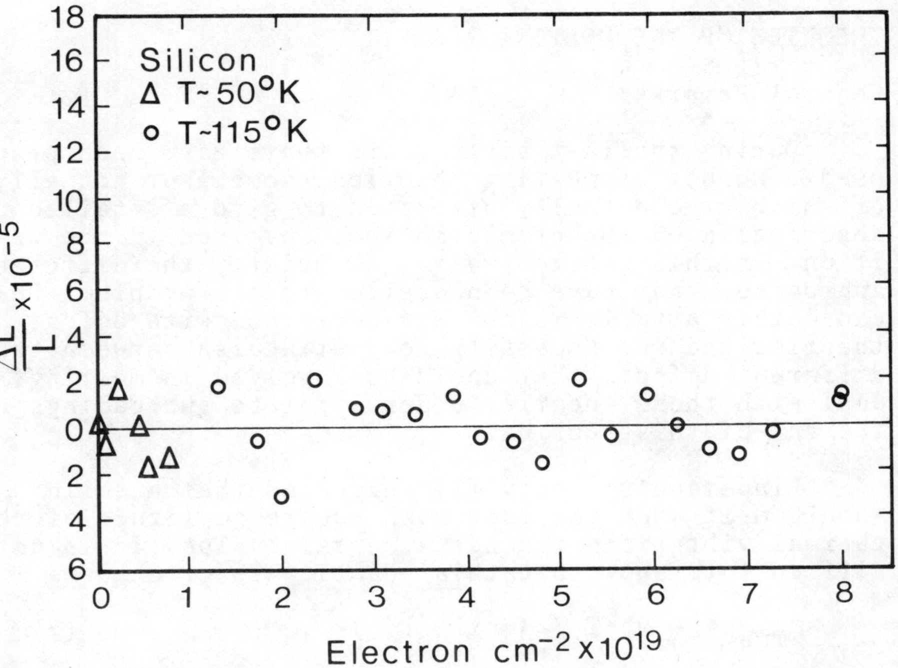

Figure 5. Lack of volume change in Si upon electron irradiation (17).

indeed typical.

It is instructive to compare the volume change (actually it is length change that is measured) in different types of element upon irradiation by electrons. Figure 4 shows that aluminium expands upon irradiation (16). Similar experiments with silicon (17) produce on the other hand no measurable length change, as is shown in figure 5. The aluminium result is what would be expected for such a close-packed (fcc) metal. The interstitial sites have a very small volume and ε_i is consequently large. Silicon has a rather open lattice structure and the relatively large volume of the interstitial sites leads to a small value of ε_i . The experimental result in fact indicates that the absolute magnitude of ε_i is no greater than the absolute magnitude of ε_v since $\varepsilon_v+\varepsilon_i$ is essentially zero. Comparing these results with the corresponding entries in Table 1, we see that for aluminium, which is "normal" in that it shows a positive volume change on melting, there is a positive volume change on irradiation, i.e. a positive volume change upon introduction of Frenkel pairs. In silicon on the other hand there is a negative volume change on melting and zero volume change upon introduction of Frenkel pairs. For argon ion bombardment of silicon the volume change has even been found to be negative (18). We consider these suggestive observations later.

THEORIES OF THE MELTING TRANSITION

General Remarks

During the last sixty years there have been proposed a number of melting theories, some, but not all, of which have actually attempted to give a detailed description of the atomic motions involved in the transition. In this section we review briefly the different approaches that have been applied to the problem. Because this article is chiefly concerned with defect theories and the necessity to distinguish between the different defects that could be involved in melting, we deal with these theories under separate subheadings at the end of this section.

Lindemann's theory (1) suggested that a solid should melt when the root mean square amplitude of the thermal vibrations reaches a certain value. It is not difficult to show that this quantity is given by

$$r_{rms} = \left(9h^2T \,/\, 4\pi^2Mk\Theta_D^2r_s^2 \right)^{\frac{1}{2}} \qquad (7)$$

where T is the temperature, M the atomic mass, Θ_D the Debye temperature, and r_s the Goldschmidt atomic radius. It is indeed found that for all elements having a mutually common crystal structure, r_{rms} is approximately constant. This parameter is, however, only about 20% of the nearest-neighbour distance, and it is difficult to see how such relatively mild disturbances would lead to melting. Recently McLachlan and Chamberlain (19) have reconsidered the Lindemann theory and have managed to calculate reasonable values of T_m for a number of metals, working essentially from first principles. A simple physical picture emerges if one considers the potential energy of a third atom situated between two neighbouring atoms. This potential is, of course, easily calculated if the interatomic potential is known. For low temperatures there is only one maximum in this potential, but at higher temperatures the overall thermal expansion and the increasing amplitude of vibration of individual atoms give rise to a situation in which two minima can exist. When this occurs the solid structure changes catastrophically into a disordered state in which the individual atoms are so distributed that they with equal probability occupy either of the two potential wells. The weakness of this type of cell approach is that it ignores the fact that the atoms that constitute the cell walls are themselves at the centres of other cells, and should be subjected to the same arguments. In principle this is, of course, possible, but it makes the analysis prohibitively complicated.

Born (2), who derived the general conditions for the stability of a crystal lattice, suggested that at T_m the rigidity modulus, C_{44} , should vanish. Subsequent experimental investigations have shown this to be incorrect. Varshni (20), for instance, finds that for a series of fcc metals (Pb,Al,Ag,Cu,Au, and Ir) C_{44} has decreased to 55% of its value at 0°K when the melting point is reached. One might note here that melting might still involve shearing deformations even though C_{44} is not zero at T_m. Dislocations, which could be generated by thermal activation near to T_m, (see also later), could produce shear if they had the appropriate Burgers vectors. Theories that are somewhat related to that of Born have invoked dynamic instability in certain vibrational modes. As the temperature increases the anharmonicity of the interatomic interactions leads to thermal expansion and a decrease in the strength of the coupling forces ; and hence to a lowering of the vibrational frequencies. There are various ways of expressing what then happens. Ida (21) uses the concept of vibrational elongation and shows that beyond a certain point no further energy can be fed into the vibrations, so it goes into producing disorder. Alternatively there is the possibilty that certain vibrational frequencies go imaginary beyond T_m thereby overdamping the oscillations and producing permanent displacements that give the disorder of melting.

Cell theories, of both the analytical and Monte Carlo type, have been proposed which treat both the liquid and solid phases as "lattices" with the difference in their equations of state arising because of differences in the free volume (22). Analytical approaches, such as that of Lennard-Jones and Devonshire (23), are actually capable of producing a Van der Waals loop, and T_m is then derived by employing the Maxwell equal area construction. Some of the Monte Carlo studies have simply produced parallel solid and liquid lines instead of a loop, but Hoover and Ree (24) were actually able to detect an instability point directly.

There are various crystal defects that could be candidates for the role of triggering the breakdown of the regular crystal structure. The vacancy, the interstitial and the dislocation all cause local relaxations of the lattice and might therefore cause sufficient disruption if they were present in large enough concentrations. These defects have usually been invoked individually, although Eckstein (25) has discussed the possibility of point defects and dislocations appearing together at T_m as a result of the sudden lowering of their formation energies at that temperature. Reasonable agreement with several experimental parameters was obtained by this model and Eckstein describes a melt as a paracrystalline structure with a "grain size" of about 10^{-7} cm. There has also been experimental evidence of such a situation (26). The Eckstein paper does not make specific proposals for a detailed picture of the atomic displacements and rearrangements which accompany melting, such as are given in the theories discussed in the following subsections.

Vacancies

The vacancy (or hole) theory of melting was first propounded by Frenkel (27). It was based on the well-established fact that an otherwise-perfect crystal will always contain a certain number of vacant lattice sites, the actual concentration at temperature T being given by the formula

$$n_V / N = A \exp(-E_V^F/kT) \qquad (8)$$

where n_V and N are respectively the number of vacant sites and the total number of lattice sites. A is a vibrational entropy term and E_V^F is the vacancy formation energy (strictly the enthalpy, but at atmospheric pressure the two are essentially equal). The available

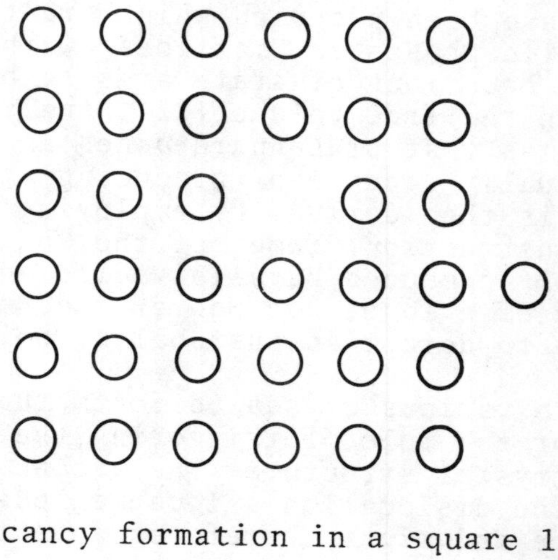

Figure 6. Vacancy formation in a square lattice.

experimental data indicate that for elements having the same crystal structure, A does not vary significantly. Moreover it is found that the ratio E_v^F / T_m is remarkably constant. Thus the concentration of vacancies at the melting point is itself fairly constant for all elements having a given crystal structure. This has led to the conclusion that vacancies play a central role in the melting process. The difficulty with this theory is that the vacancy concentration at the melting point, while being constant, is only about 10^{-3}. It is difficult to see how such a small fraction of vacant sites could actually lead to the total disruption of the solid structure. Moreover, and notwithstanding the small atomic relaxations which are known to occur near these defects, it is not possible to explain the marked differences between the solid and liquid PCF's on the basis of vacancies in such small concentrations. On the other hand it should be noted that it is possible that vacancies in some way trigger the transformation, and that the latter is then completed by other more complicated processes.

Interstitials.

The first melting theory based on interstitials was put forward by Lennard-Jones and Devonshire (23). In this theory a distinction is made between those atoms which reside on normal lattice sites (the A atoms, say) and those which have, through thermal activation, been displaced so that they lie in the interstitial positions (the B atoms). Clearly the concentration of B atoms will be increased with increasing temperature. Further-

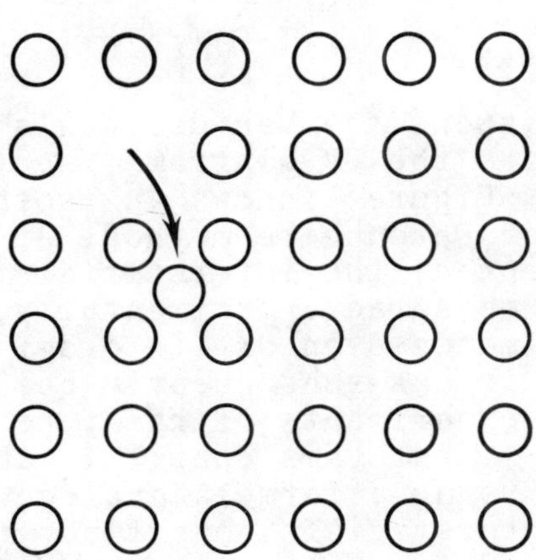

Figure 7. Interstitial formation in a square lattice.

more, the activation energy required to move an atom from an A site to a B site will depend on the local environment of that A atom; it will depend on how many of the Z atoms in the coordination shell are on A sites and how many are already on B sites. Defining the order parameter Q by

$$Q = N_A / N$$
$$1-Q = N_B / N \qquad (9)$$

where N_A and N_B are the number of atoms on A and B sites respectively, the mean energy (above the ground state) of an atom on an A site, when the instantaneous value of the order parameter is Q, will be

$$\overline{W}_A = ZW(1-Q) \qquad (10)$$

where W is the interaction energy between adjacent A and B atoms. The activation energy required to move an atom from an A site to a B site will therefore be given by

$$W_{A-B} = \Delta W = ZW(2Q-1) \qquad (11)$$

The cooperative nature of the process embodied in eqn 11 is simply the Bragg-Williams behaviour frequently encountered in order-disorder theory. Clearly $\Delta W = ZW$ for the complete order case, $Q = 1$, and ΔW falls with decreasing Q, until it becomes zero for complete disorder, at $Q = \frac{1}{2}$. It is not difficult to show that these cooperative effects give rise to a second term in the expression for the pressure, which becomes

$$P = P_{ORDERED} - ZNQ(1-Q)(dW/dV) \qquad (12)$$

and equation 12 shows the Van der Waals loop behaviour that characterizes the PV diagram (see figure 1) of a melting solid. Figure 8 shows an isotherm derived by Lennard-Jones and Devonshire method. At the time this theory was presented, the interstitial formation energy E_I^F had not been measured experimentally, and no reliable theoretical estimates even of its order of magnitude were available. It has subsequently been found (by irradiation damage experiments of the type refered to in an earlier section) that this energy is about five times larger than E_V^F. Using a formula analogous to equation 8 applicable to interstitials, one finds that the concentration of these defects even at the melting point is quite negligble. This constitutes a major draw-back to

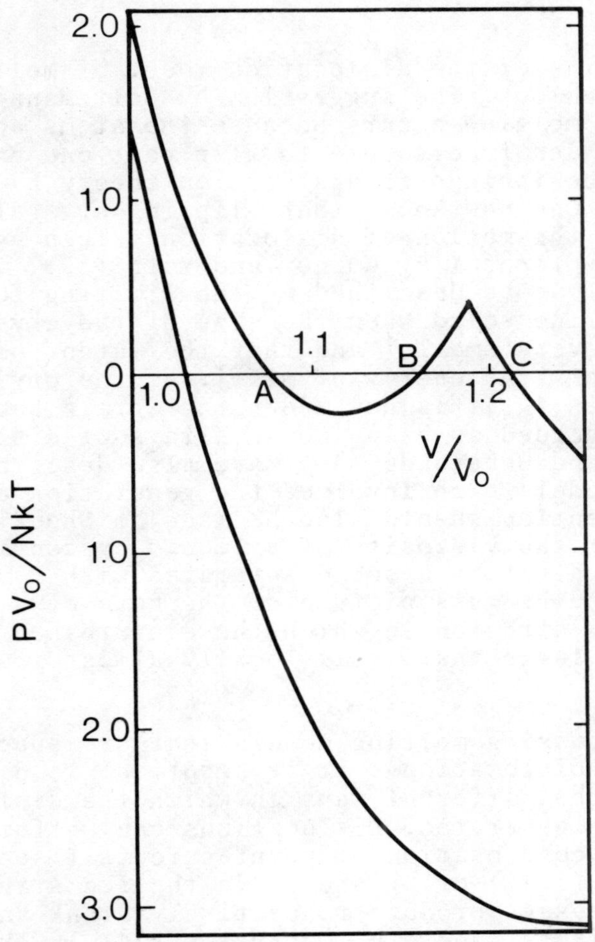

Figure 8. Isotherm for a Lennard-Jones substance derived by the theory of Lennard-Jones and Devonshire (23). The lower curve corresponds to perfect order, and the upper curve shows the result of allowing disorder to occur.

the theory. Another shortcoming of the Lennard-Jones and Devonshire approach is that it predicts a solid-liquid critical point, which for argon would be located at T=132°K and P=2400 atmospheres. No critical point has been found experimentally even up to 6000 atmospheres, and, as was discussed earlier, there are good reasons for believing that critical melting cannot occur.

Dislocations.

The origins of the dislocation model of melting are rather nebulous. The suggestion by Boltzmann (28), in 1896, that melting occurs because the atoms acquire enough energy and free volume to slip past one another, predated the beginnings of dislocation theory by over thirty years. One now knows that slip in crystals proceeds through the motion of dislocations. Then again there was the proposal by Gurney and Mott (29), in 1939, that a liquid can be described as the limiting form of a polycrystalline solid when the size of the crystallites becomes very small,(and that the latent heat of fusion is simply the energy of misfit of the crystallite boundaries). It is now known that grain boundaries can be regarded as being dense arrays of dislocations. Oldham and Ubbelohde (30) have also described a cooperative model which involved the generation of grain boundaries. Mention should also be made of Shockley's calculation of the viscosity of a liquid by considering it to be equivalent to a solid saturated with dislocations (31), and Bragg's picture of the core of a dislocation as being a region in which the disturbance to the lattice is so large that it is locally almost liquid--like(32).

When discussing melting models that are specifically based on dislocations, it is important to distinguish between the different ways in which the dislocations might be generated. Dislocations can be formed either by the condensation of point defects (see e.g. Hirsch et al, (33)) or by shear. In the fcc structure these two processes produce respectively Frank and Shockley dislocations (see e.g. Cottrell, (34)). The dynamics of these two formation processes would be quite different, because the former involves diffusion while the latter does not. Rothstein (35) proposes a melting model which involves vacancy condensation, whereas Mizushima (36), Ookawa (37), Siol (38), and Kuhlmann-Wilsdorf (39) have variously discussed melting and the liquid state in terms of dislocations that are spontaneously generated in the interior of a crystal.

A key feature of the Mizushima theory concerns the fact that dislocation energy and concentration are interdependent. The energy per unit length of a dislocation is given by

$$W_{DISLOC} = (Gb^2/4\pi)\ln(R/r_C) + W_C \qquad (13)$$

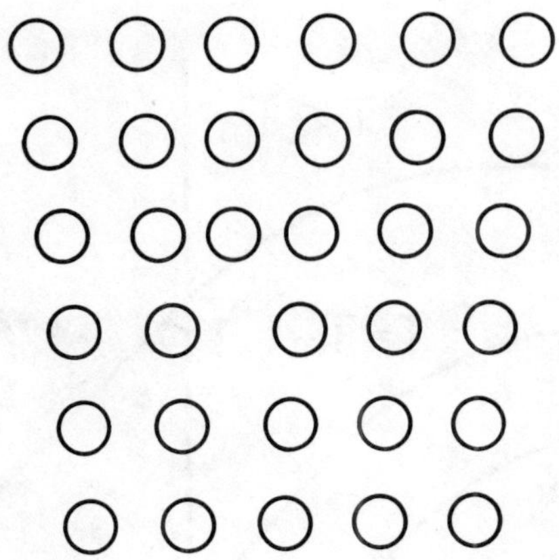

Figure 9. A dislocation in a square lattice.

where G is the shear modulus, r_c is the core radius (usually between one and three nearest-neighbour distances), and W_c is the core energy. R is the distance from the core of the dislocation to the point at which the stress falls to zero. In a single crystal containing a single dislocation at its center, R would equal the crystal radius. For an array of dislocations, however, R becomes half the distance between neighbouring dislocations. It therefore decreases with increasing dislocation concentration. The implied cooperative effect is quite similar to the one described earlier in connection with interstitials.

The theory predicts that the dependence of the free energy of an array of dislocations depends on the concentration, C, and temperature according to the following equation

$$F(C,T) = -(Gb^3/8\pi)C\ln(C(r_c/b)^2) - 5kTC \qquad (14)$$

where b is the Burgers vector. This shows that there is a critical temperature, T_m, for which $F(1,T)=F(0,T)$ and this is identified with the melting point. If T is below T_m, then $F(0,T)<F(C,T)$ for all possible values of C so thermodynamic equilibrium prevails only if C=0. For T above T_m one finds that $F(1,T)<F(C,T)$ for all possible values of C, and equilibrium is achieved when the crystal is saturated with dislocations. This behaviour is indicated in figure 10. Melting, in this model, therefore corresponds to a sudden and catastrophic increase in the concentration of dislocations.

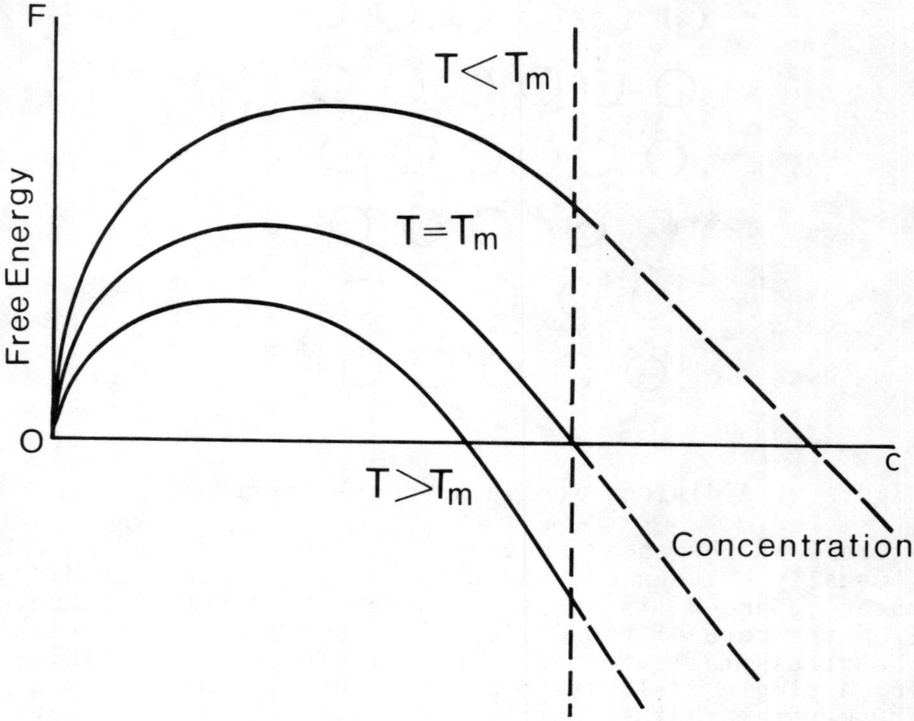

Figure 10. Variation of the free energy of a dislocation array as a function of dislocation concentration, for the three different regimes.

Of the various articles cited earlier in this sub-section, that of Siol (38) is particularly interesting because it contains specific suggestions for the actual dislocation configurations that could be generated by thermal activation. Siol proposed that the dislocations should be of the shear type on {111} planes, and that they should form closed loops. Kuhlmann-Wilsdorf (39) suggested that they would be generated in the form of dipoles. We shall see later that both predictions appear to be correct.

Before closing this section it will be appropriate to discuss an aspect of dislocation theory which has an important bearing on the mode of dislocation generation during melting. In Equation 13, the first term arises from the elastic strain field that surrounds the dislocation while the second term is the energy of the core. The latter is composed of contributions from the strain field (which, for the region lying within r_c, exceeds the elastic limit) and the "dangling bond" that is the chief characteristic of the dislocation. It has been shown, by calculations on discrete atomistic models of the type that will be described in the next section (Cotterill and Doyama (40)), that for R equal to 1000 nearest-neighbour distances, which would be the situation in a well-deformed metal, only about one quarter of the total dislocation energy would be stored in the core. This means that a considerable lowering of energy will occur if the dislocations in an array are so arranged that the long-range strain fields cancel with one another. Just how this is achieved in practice will be described in the next section.

COMPUTER SIMULATIONS OF MELTING

General Description of the Method.

The recent advent of fast electronic digital computers with large memory capacities has afforded the development of the technique of molecular dynamics (41-44). With this technique the motions of individual atoms under the influence of known forces can be studied. It is now possible to follow the temporal evolution of an assembly of atoms and to observe directly the response of such a system to mechanical and thermal stimuli. The method is based on classical mechanics and involves the solution of Newton's equations of motion for the individual atoms

$$Ma_i(t) = - \sum_j \text{grad } V(r_{ij}) \qquad (15)$$

where $a_i(t)$ is the instantaneous acceleration of the ith atom at time t, and V is the known interaction potential. The r_{ij} are the distances from the ith atom to those neighbours that are close enough to influence its motion. When the acceleration of an atom has been calculated, the position to which it will move in time interval Δt can be found from

$$r_i(t+\Delta t) = - r_i(t-\Delta t) + 2r_i(t) + a_i(t)\Delta t^2 \qquad (16)$$

The optimal time increment for Δt is usually about 2×10^{-14} sec. Since the positions of all the atoms are known as a function of time, their velocities are known at every instant, and this defines the classical temperature through

$$3nkT/2 = (M/2) \sum_{i}^{n} v_i^2 \tag{17}$$

where n is the number of atoms in the model, and v_i is the velocity of the ith atom. The temperature can be changed simply by scaling the velocities. The pressure can be calculated from the virial equation

$$P = \rho kT - (\rho/6n) \left\langle \sum_i \sum_j r_{ij} \, (\partial V_{ij}/\partial r_{ij}) \right\rangle \tag{18}$$

where ρ is the number density of the system, and where the angled brackets indicate a time average. The pressure in the system can be varied by uniformly scaling distances. The system is therefore completely defined thermodynamically, and the thermodynamic conditions can be changed at will. In the following subsections brief descriptions are given of simulations of the melting of both two and three dimensional crystals. Most of the studies were carried out on systems that were made psuedoinfinite by use of periodic boundary conditions, but a description will also be given of the results obtained with two and three dimensional microcrystals.

Two-dimensional Pseudoinfinite System.

A two-dimensional study of the molecular dynamics type has been carried out by Cotterill and Pedersen (45), using a computational cell of 440 atoms arranged in a hexagonal lattice. The

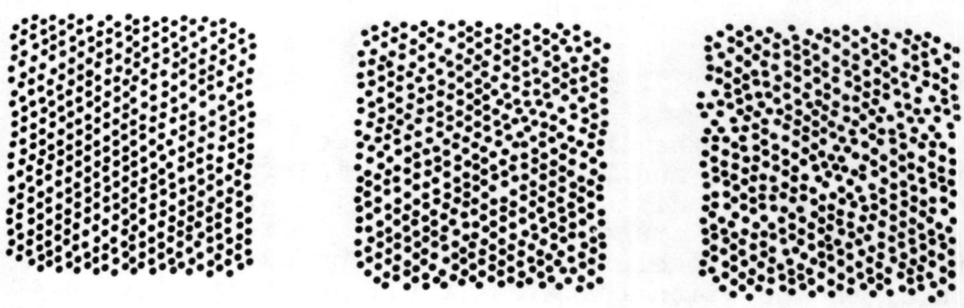

Figure 11. Three instantaneous situations during the melting of a hexagonal two-dimensional Lennard-Jones crystal (45). The time interval between the pictures is 3×10^{-12}sec in each case. Departures from perfection can best be detected by observing the pictures from a low angle along one of the three systems of close-packed atomic rows.

interactions were simulated by a Lennard-Jones potential truncated between the third and fourth neighbour shells. At each time step of the calculation the atomic positions were displayed on a television screen and photographed as single frames of a motion picture, selected frames of which are shown in figure 11. The model was heated through the melting point (approximately $80^{O}K$) by periodically scaling up the velocities. The first type of imperfection that could be discerned was the "kink" in a close-packed row, which occured when nearest-neighbours in a row vibrated in anti-phase with sufficient amplitude. As the melting process proceeded dislocations began to appear in increasing numbers, and at the terminal stage there was diffusion of single atoms,particularly at the points of high dislocation concentration. Because the dislocations are formed in an otherwise-perfect crystal, they must be formed in pairs with positive and negative Burgers vectors with respect to their common slip plane. This pair generation has important significance for the dislocation theory of melting. As can be seen from the schematic diagram of a dislocation in figure 9, the strain field of the dislocation is compressive above the slip plane and dilational below. In a close positive-negative pair of dislocations the individual strain fields will tend to cancel with each other, and this will considerably diminish the magnitude of the first term in equation 13. In fact, when the pair of dislocations has a separation distance comparable to the core radius (as indeed it must have at the moment of formation) the cancelation will be complete and the activation energy of dislocation generation will consist solely of a core component. During this two-dimensional study, the pressure was periodically monitored and the area of the model was varied to maintain the pressure constant. In this way the area change on melting was found to be 9.9% which would correspond to a change of about 15% in volume in a three-dimensional model. This is in good agreement with the entries for noble gases in Table 1.

Three-dimensional Pseudoinfinite Systems.

Apart from their geater demands on computer time and memory size, three-dimensional models are more difficult to work with than their two-dimensional counterparts. The detection and visualisation of defects, for instance, is much more of a problem in three dimensions. The extra work involved is however offset by the opportunity it provides of direct comparison with experiment, and a number of three-dimensional studies have now been carried out. Jensen et al (46) have investigated the change in the FDF caused by the systematic introduction of randomly oriented dislocations into an fcc Lennard-Jones crystal. The Burgers vectors of the dislocations were all of the perfect (a/2)<110> type, but they were chosen at random from among the twelve possible directions. Perfect dislocations would normally try to dissociate in the fcc lattice, but this was artificially inhibited in the calculations by

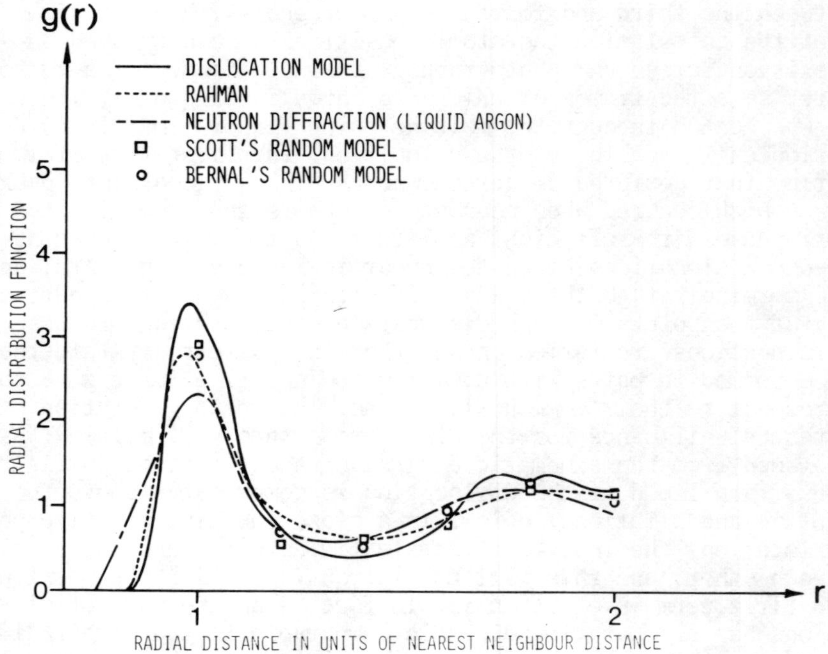

Figure 12. The pair distribution function for the liquid state derived from experiment and also from various theoretical approaches. For the dislocation model the temperature was half the absolute melting temperature.

omitting the customary periodic redefinition of neighbours. The temperature and pressure of the system were monitored and controlled in the manner described earlier. Several interesting facts emerged from these calculations. It was found that the increase in volume due to the insertion of the dislocations showed a saturation behaviour, with saturation being achieved when the dislocation density had reached a value that corresponded to the individual cores being contiguous. The saturation value of volume increase was 10.8%, which is comparable to (though slightly lower than) the entries for the noble gases in Table 1. The PDF was calculated for various values of the inserted dislocation concentration and it was found that a distinctly liquid-like PDF had already been developed by the time the dislocation density was about half its saturation value. This is shown in figure 12, where the model PDF has been compared that obtained by neutron diffraction studies of liquid Ar. The small radius behaviour of the curve for this latter case is not very reliable because of the experimental cut-off of high wave numbers. Also shown are the random hard sphere results alluded to earlier, and Rahman's molecular dynamics results (43) for liquid Ar. The temperature of the dislocation model was maintained at half the absolute melting temperature in order to prevent the dislocations from moving and thereby mutually annihilating. The good agreement with experiment and the fact that it was obtained in spite of the relatively low model temperature is important, in view of what was noted earlier. Namely, that thermal vibrations cannot alone account for the difference between the solid and liquid PDF's.

Kristensen et al (47) have shown that the model also reliably predicts the change upon melting of several thermodynamic quantities. The model latent heat, for instance, was only 1% less than the experimental value for Kr, and 5% less than the experimental value for Ar. In fact the agreement could be said to be perfect since the computational accuracy was only ±10%. The calculated melting temperature was respectively 20% and 14% higher than the Kr and Ar values, while the computational accuracy was in this case ±13%. Both configurational and vibrational contributions to the entropy were calculated, and the model value was less than both the Kr and Ar values by 18%. It was shown that an overall improvement in the fitting to experimental parameters would have been obtained had the inserted dislocations been of the Shockley partial type.

A similar Lennard-Jones model (Cotterill et al (48))(different only in that periodic boundaries were used) was employed in a dynamic simulation analogous to the two-dimensional study described in the previous subsection. As had been discussed earlier (Kristensen et al (47), Cotterill (49)), one might expect dislocations of the Shockley partial type to be generated by thermal activation in an fcc lattice. A systematic examination of the atomic arrangement in the model, at various instants as the model was taken through the melting transition, revealed that it is indeed this type of dislocation that is associated with the melting of the fcc structure. In three dimensions the dislocations must form closed loops, and these are the topological counterparts of the positive-negative pairs that are seen in two dimensions. The Shockley partial loops were detected by examining pairs of adjacent {111} planes, with separate searches for each of the three possible Shockley Burgers vectors. The latter lie in <112> directions at right angles to their respective <110> close-packed lines. The development of a dislocation can easily be detected by noting that this produces a change in the characteristic linkage sequence along the close-packed row. Figure 13 shows the situation (averaged over 50 computational cycles, i.e. about half a vibrational period) in two adjacent {111} planes at an advanced stage during the melting of the model. The three pictures correspond to the three different Shockley Burgers vectors, and the shaded areas indicate the regions surrounded by dislocation. Because the Shock-

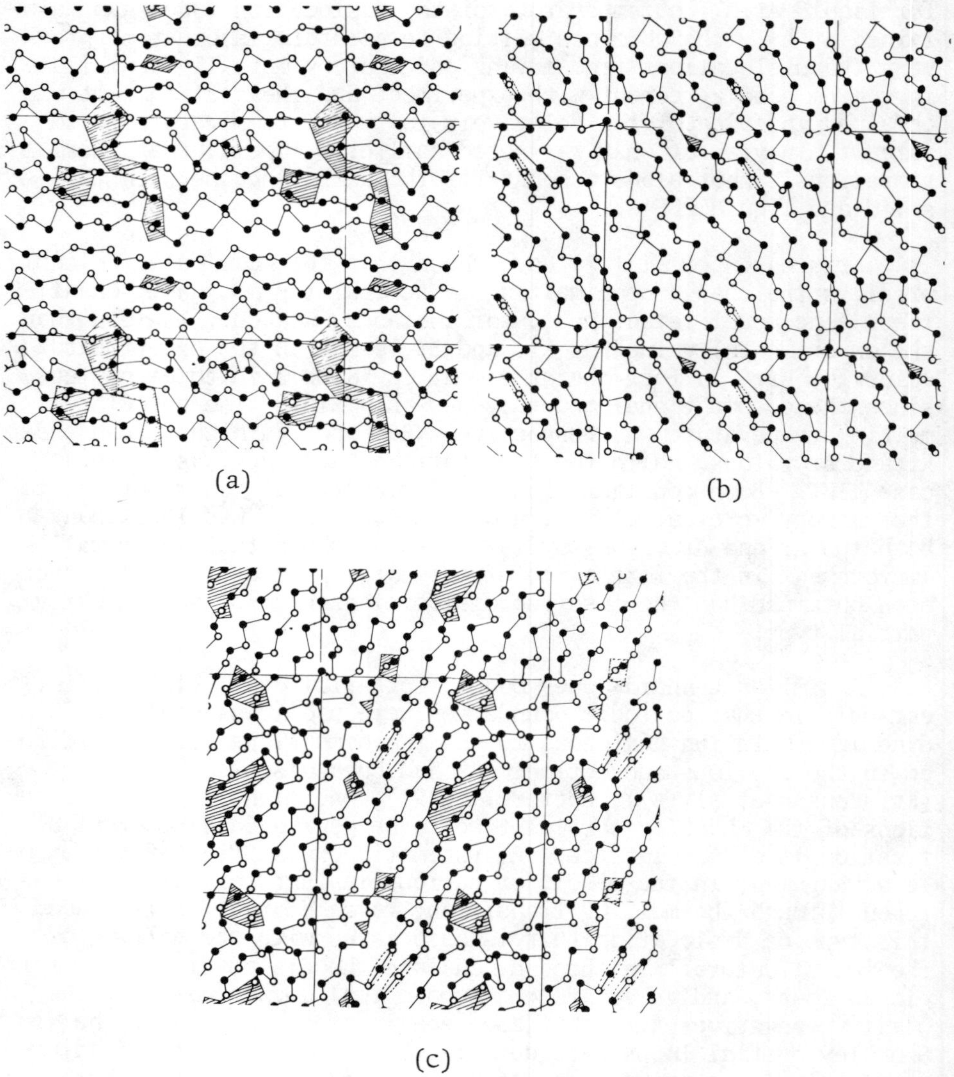

Figure 13. Detection of Shockley partial dislocations and associated regions of stacking fault (shaded), in two adjacent close-packed planes, during the melting of an fcc Lennard-Jones system. The three pictures show separately dislocations having the three possible Burgers vectors. The dotted lines indicate incipient "point dislocations".

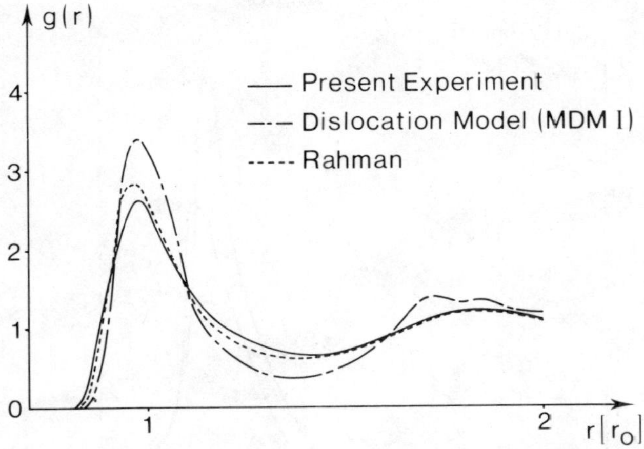

Figure 14. Pair distribution function obtained during the terminal stage of the dynamic simulation of melting of a three-dimensional Lennard-Jones system (48). MDM1 refers to the pseudostatic simulation (see figure 12). Rahman's curve was obtained during a molecular dynamics simulation of liquid Argon (43).

ley dislocation is of the partial type (i.e. because the Burgers vector is less than the minimum translation vector of the lattice) these shaded areas are stacking faults. It was found that the loops tended to adopt a "dipole" configuration, in that they elongated along predominantly <112> directions. Cross-sections normal to the directions of elongation would produce pictures analogous to the two-dimensional ones shown in Figure 11. This shows that the principle of positive-negative dislocation pair generation applies also to the three-dimensional case, and again gives the net lowering of activation energy which favors the dislocation mode of melting. It was also observed that the dislocation dipoles tended to lie in the screw orientation, which would be expected since this form has a lower energy than the edge type. It was found that the three-dimensional counterpart of the "kinks" observed in the two-dimensional model are what might be called "point dislocations" which are found when an atom in one close-packed plane moves in the Shockley Burgers vector direction while two of its three nearest neighbors in an adjacent plane move in the opposite direction. The dotted lines in Figure 13 show triplets of atoms which are just about to form such point dislocations. Figure 14 shows the PDF for this dynamical model compared with two of the curves which appeared in Figure 12 ("MDM I" refers to the dislocation model shown in Figure 12). One sees that the agreement with the Rahman result, and with the experimental curve shown in Figure 12, is now remarkably good. Two other interesting

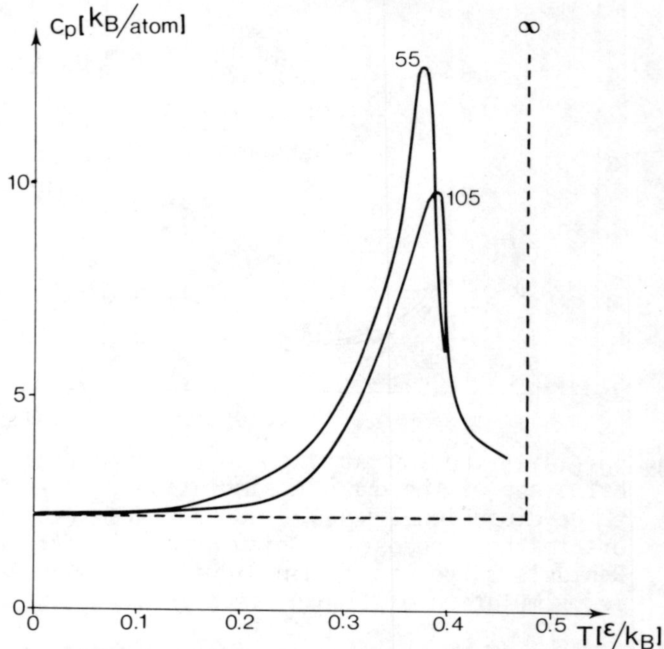

Figure 15. Specific heat curves for melting two-dimensional microcrystals, containing the number of atoms indicated.

points emerged from the dynamic simulations. By comparing the effect of time-averaging over periods less than and greater than the mean vibration period, it was established that the dipole loops "pulsate" during the early part of the transition. Individual loops are constantly being generated, growing, shrinking, and disappearing. But the overall dislocation density was found to rise at an ever-increasing rate (thus demonstrating the cooperative effect discussed in the previous section of this review), and the terminal stage corresponded to a situation in which the crystal had become saturated with dislocations.

Two- and Three-Dimensional Microcrystals

Molecular dynamics is well suited also to the study of very small systems; microcrystals and microdrops. The melting behavior of microcrystals is known to show significant departures from what is seen in infinite systems. Electron microscopic investigations have shown that the melting temperature decreases with decreasing radius (50,51) and heat capacity measurements on films of a few monolayers' thickness show that this parameter no longer shows the abrupt change that is observed in infinite systems. Figures 15 and 16 show the molecular dynamics results for two- and three-

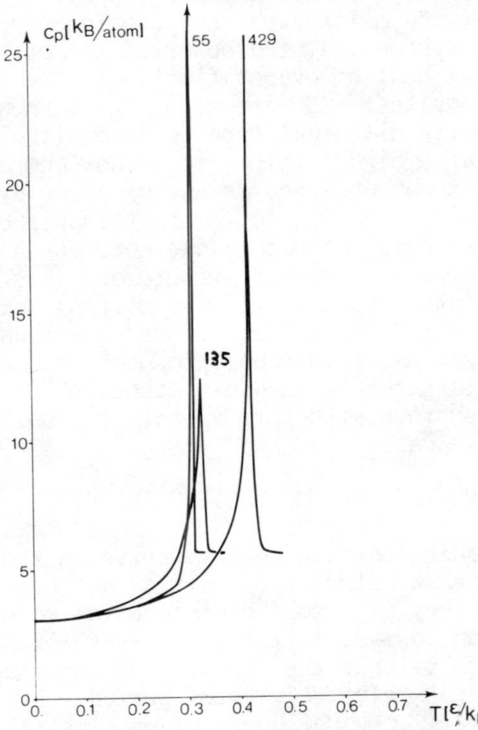

Figure 16. Specific heat curves for melting three-dimensional microcrystals, containing the number of atoms indicated.

dimensional systems respectively (52,53). Both the depression of T_M and the smearing out of the C_p curve are clearly visible. By inspection of the time dependence of the actual atomic arrangements in these models, the origin of their "anomalous" properties was found to be the relaxations that occur at the surfaces. In a microcrystal the surface atoms account for an unusually large fraction of the total system and the relaxations are no longer negligible. In fact, it is found that the surface becomes rather diffuse in structure. The pressure of the surface is found to promote defect generation, and in the two-dimensional system at least the defects are again of the dislocation type. It was found that the dislocations were generated in the core of the microcrystal where the structure was still solid-like. The melting of these small systems does not appear to proceed by the gradual solution of the solid core in a surrounding liquid layer, but rather by a relatively sharp change in the internal structure. As can be seen from both Figures 15 and Figure 16, there is another type of anomaly in these small systems. In the two-dimensional case, the 55 particle system has a higher apparent L_m. This is due to the fact that the melting and evaporation processes overlap so much in this system that the observed L_m contains a fraction of L_e (the latent heat of evaporation). In the three-dimensional systems the smallest (55 particles) again shows an anomaly, but it is of a quite different type because although the C_p curve again peaks at a higher value, it is now comparatively narrower. No evaporation was observed in any of these systems, and the anomaly was found to be due to the fact that the 55-atom system adopts the icosahedral structure and not, as in the case of the larger microcrystal, the FCC structure. This icosahedral structure is maintained right up to the melting point. The greater stability of this modified structure, accompanied by a greater binding energy, as it must be, produces both a higher melting point and a larger L_m than would otherwise be expected. Moreover the compact surface structure appears to inhibit the premelting effect in this case.

MICROSCOPIC PICTURE OF THE MELTING TRANSITION

In the preceeding section we have surveyed the results of several molecular dynamics studies of melting and have described the strong support that they give to the dislocation mechanism. This type of study does, of course, suffer from several limitations, the most acute of which is that the behavior of the model system can be followed for only a relatively brief period. Even with the most advanced computers, several hours of machine time are required for the simulation of 10^{-10} seconds of real time. The model solids in these studies are therefore forced to melt within a few tens of vibrational periods. This is achieved by slightly superheating whereas superheating is not normally observed in nature. Nevertheless, it should be noted that the probability of occurrence of the different possible mechanisms of melting would depend only on their associated activation free energies, and these are unaffected by the superheating.

A microscopic picture of the melting process clearly emerges from these studies. The driving force for melting arises, of course, from the fact that above a certain temperature the disordered state of the liquid has a lower free energy than that of the ordered crystalline solid. The transition from order to disorder is then accomplished by the sudden proliferation of dislocations, and it is completed when the dislocation density is so high that the dislocation cores are contiguous, whereupon all long-range order has disappeared. It has been shown that the introduction of these dislocations accounts for the changes in volume, entropy, and the pair distribution function, while the latent heat is simply the energy required to generate the dislocations.

There are several experimental observations which can readily be explained on the basis of the dislocation model. The change in diffusion constant (by a factor of approximately 10^4 in passing from the solid to the liquid has already been touched upon in connection with Figure 3. An important characteristic of a dislocation is that its core is relatively more open than is the corresponding perfect lattice. Dislocation cores therefore provide routes of easy atomic transport through a crystal and this is known as "pipe diffusion". Gupta (54) has recently measured the diffusion constant for pipe diffusion in Au and finds it to be 0.05 exp(- 1.16/kT) where the preexponential is in units of $cm^2\ sec^{-1}$ and the activation energy is in electron volts. This is to be compared with the value 0.09 exp(- 1.81/kT) (55) for normal self-diffusion in Au. If a liquid can be regarded as a crystal saturated with dislocations, diffusion would be expected to occur entirely by pipe diffusion. Assuming no change in the preexponential terms, these two numbers can be used in the Boltzmann factor to find the ratio of diffusion constants in the liquid and solid states. We have

$$D_L/D_S \simeq 0.05\ \exp(-\ 1.16/kT)/0.09\ \exp(-\ 1.81/kT) \qquad (19)$$

and since $T_m = 1336^{\circ}K$ for Au, this gives

$$D_L/D_S \simeq 1.53 \times 10^2 \qquad (20)$$

which is in fairly good agreement for the entries for the noble metals in Table 1.

The change in electrical resistivity upon melting can also be explained by the dislocation model. Martin and Cotterill (56) have compiled dislocation resistivity values for a number of metals and compared them with the liquid resistivities of the same metals. The ratios are remarkably constant. Moreover, if one again assumes a liquid to be equivalent to a crystal saturated with dislocations, one obtains excellent quantitative agreement with experiment so long as one assumes a reduction of the dislocation resistivity by a factor of five in the liquid state because only the cores are present. This reduction factor is of course similar to that applying to the dislocation energy when only cores are present.

We have earlier mentioned the various empirical rules that have been found to apply to melting. The origin of the Lindemann rule (1), in particular, is easy to demonstrate on the basis of the dislocation model. In the FCC structure it was shown earlier that the "point dislocation" which initiates melting is formed when an atom in one close-packed plane moves by half the Shockley Burgers vector, while two of its three nearest neighbors on an adjacent close-packed plane move by the same amount in the opposite direction. This motion requires all three atoms to move slightly out of their respective planes. The net distance moved can easily be shown to be 0.197 of the nearest-neighbor distance. This compares very well with values calculated from Equation 7 for a number of FCC metals; 0.183 (Al), 0.206 (Au), 0.193 (Ag), 0.199 (Cu). The Lindemann rule arises therefore from the simple geometrical requirement that neighboring atoms have to move with respect to each other by a certain fraction of the nearest-neighbor distance in order to surmount the energy barrier which acts against dislocation formation.

The most pronounced characteristics of liquids, compared with solids, is their inability to support shear stress and their very low viscosity. The first of these is readily understood through the dislocation model. Shearing motions will occur spontaneously in the liquid through the motion of dislocations. The model predicts both the presence of dislocations and their spontaneous motion (the latter being necessarily invoked in the proliferation mechanism). The change in viscosity upon melting is also easily explained by the model. Let us consider first two well-known expressions for viscosity (57). For the liquid state we have the Stokes-Einstein relation

$$\eta_L = kT/(3\pi d\, D_L) \tag{21}$$

where d is the diameter of an atom in the liquid and D_L is the diffusion constant. For a solid we have the Nabarro-Herring equation

$$\eta_S = R_S^2\, kT/(\alpha\, \Omega_o\, D_S) \tag{22}$$

which was derived for a model in which the applied stress causes vacancy motion, from sources at high stress to sinks at low stress. R_S is the distance that the vacancies move, and in polycrystalline specimens this is usually taken to be the grain diameter. Ω_o is the atomic volume and D_S is the diffusion coefficient in the solid state. The value of the constant α is approximately 4. In the dislocation model of melting, a liquid may be regarded as a solid with a very small grain size because, as was noted in an earlier section, grain boundaries are equivalent to dense arrays of dislocations. We will therefore attempt to apply Equation 22 to a liquid. The relationship between grain size and dislocation concentration, C_d, is simply

$$R \simeq C_d^{-\frac{1}{2}} \tag{23}$$

Moreover, we have shown earlier that in a liquid the dislocation concentration is such that cores are contiguous, so that the mean spacing between dislocations is about $2r_c$ where r_c is the core radius. This implies that

$$C_d \simeq (2r_c)^{-2} \tag{24}$$

and hence we may write

$$R_L \simeq 2r_c \tag{25}$$

Noting also that

$$\Omega_o = \beta\, d_o^3 \tag{26}$$

where d_o is the nearest-neighbor distance, and where β is a constant for a given crystal structure ($\beta = 1/\sqrt{2}$ for the FCC structure) and that

$$r_c = \gamma\, d_o \tag{27}$$

where γ is usually of the order of 1, we have finally (putting 25, 26, and 27 into 22)

$$\eta_L = 4\,\gamma^2\, kT/\alpha\beta\, d_o\, D_p \tag{28}$$

where D_p is the diffusion constant for the pipe diffusion mode, as discussed and justified earlier in this section. The similarity between Equations 21 and 28 is striking. Indeed, they differ only by a small numerical factor.

Let us close by actually calculating the ratio of the viscosities of the solid and liquid states. We use Equation 22 and we set the temperatures equal (i.e. we compare the liquid immediately above T_m with the solid immediately below T_m), obtaining

$$\eta_S/\eta_L = (R_S/R_L)^2\ (D_L/D_S) \tag{29}$$

We have already seen that $D_L/D_S \simeq 10^4$ (see Table 1), and typical C_d values for well annealed solids are about 10^4 cm^{-2}. Hence

$$R_S/R_L \simeq 10^6 \tag{30}$$

and hence

$$\eta_S/\eta_L \simeq 10^{16} \tag{31}$$

The dislocation model is therefore quite capable of explaining in a quantitative manner the most striking difference between the solid and liquid states.

REFERENCES

1. LINDEMANN, F.A., Z.Physik, 11, 609, (1910)
2. BORN, M., J.Chem.Phys., 7, 591 (1939)
3. IVANOV, V.A., MAKARENKO, J.N., and STISHOV, S.M., I.E.T.P. Letters, 12, 7 (1970)
4. MUKHERJEE, K., Phys.Lett., 8, 17 (1964)
5. VAN LIEMPT, J., Zeits.F.Phys., 96, 534 (1935)
6. NACHTRIEB, N.H., RESING, H.A., and RICE, S., J.Chem.Phys., 31, 135 (1959)
7. EGELSTAFF, P.A., An Introduction to the Liquid State, (Academic Press, London, 1967) pp 15 and 73
8. BERNAL, J.D., in Liquids: Structure, Properties, Solid Interactions edited by T.J. Hughel (Elsevier, N.Y.,1965)p.25 See also: BERNAL, J.D., and KING, S.V., Disc.Faraday Soc., 43, 60 (1967), BERNAL, J.D., and FINNEY, J.L., ibid p.62 BERNAL, J.D., KNIGHT, K.R., and CHERRY, I., Nature, 202, 852 (1964)
9. BERNAL, J.D., and FINNEY, J.L., Nature, 214, 265 (1967)
10. FINNEY, J.L., and BERNAL, J.D., Nature, 213, 1079 (1967) See also: BERNAL, J.D., and FINNEY, J.L., Nature, 215, 269 (1967)
11. ANDRADE, E.N. da C., Phil.Mag., 17, 698 (1934)
12. CARNEIRO, K., and NIELSEN, M., This Volume
13. UBBELOHDE, A.R., Melting and Crystal Structure (Clarendon Press, Oxford, 1965), Chapter 11.
14. BALL, A.F., J.Res.Nat.Bur.Stand., 45, 23 (1950)
15. ECKERT, R.E., and DRICKAMER, H.G., J.Chem.Phys.,20, 13 (1952)
16. WAGNER, H., DWORSCHAK, F., and SCHILLING, W., Phys.Rev., 2, 3856 (1970)
17. VOOK, F.L., Phys.Rev., 125, 855 (1962)
18. ANDERSSON, S., Phys.Lett., 33A, 455 (1970)
19. MCLACHLAN, D., and CHAMBERLAIN, L.L., Acta Met., 12, 571 (1964)
20. VARSHNI, Y.P., Phys.Rev.B., 2, 3952 (1970)
21. IDA, Y., Phys.Rev., 187, 915 (1969)
22. TURNBULL, D., in Liquids: Structure, Properties, Solid Interactions, edited by T.J. Hughel (Elsevier, N.Y., 1965) p.8
23. LENNARD-JONES, J.E., and DEVONSHIRE, A.F., Proc.Roy.Soc., A169, 317; A170, 464 (1939)
24. HOOVER, W.G., and REE, F.H., J.Chem.Phys., 49, 3609 (1968)
25. ECKSTEIN, B., Phys.Stat.Sol., 20, 83 (1967)
26. HOSEMANN, R., WILLMANN, G., and ROESSLER, B., Phys.Rev.A., 2243 (1972)
27. FRENKEL, J., Kinetic Theory of Liquids (Dover, New York, 1955)
28. BOLTZMANN, L., Vorlesungen über Gastheorie (Leipzig, 1896) 1, 7
29. MOTT, N.F., and GURNEY, R.W., Trans.Faraday Soc., 35, 364 (1939)

30. OLDHAM, J.W.H., and UBBELOHDE, A.R., Proc.Roy.Soc.A, 176, 50 (1940)
31. SHOCKLEY, W., in l'Etat Solide (Institute Internationale de Physique Solvay, Brussels, 1952) p.431
32. BRAGG, W.L., Symposium on Internal Stresses (Institute of Metals, London, 1947) p.221
33. HIRSCH, P.B., SILCOX, J., SMALLMAN, R.E., and WESTMACOTT, K. H., Phil.Mag., 3, 897 (1958)
34. COTTRELL, A.H., Dislocations and Plastic Flow in Crystals (Clarendon Press, 1953) p.72
35. ROTHSTEIN, J., J.Chem.Phys., 23, 218 (1955)
36. MIZUSHIMA, S., J.Phys.Soc.Japan, 15, 70 (1960)
37. OOKAWA, A., J.Phys.Soc.Japan, 15, 2191 (1960)
38. SIOL, M., Z.Physik, 164, 93 (1961)
39. KUHLMANN-WILSDORF, D., Phys.Rev., 140, A1599 (1965)
40. COTTERILL, R.M.J., and DOYAMA, M., Phys.Rev., 145, 465 (1966)
41. ALDER, B.J., and WAINWRIGHT, T., J.Chem.Phys., 31, 459 (1959)
42. GIBSON, J.B., GOLAND, A.N., MILGRAM, M., and VINEYARD, G.H., Phys.Rev., 120, 1229 (1960)
43. RAHMAN, A., Phys.Rev., 136, A405 (1964)
44. VERLET, L., Phys.Rev., 159, 98 (1967)
45. COTTERILL, R.M.J., and PEDERSEN, L.B., Sol.State Comm., 10, 439 (1972)
46. JENSEN, E.J., and KRISTENSEN, W.D., and COTTERILL, R.M.J., Phil.Mag., 27, 623 (1973)
47. KRISTENSEN, W.D., JENSEN, E.J., and COTTERILL, R.M.J., Phil. Mag. (in the press)
48. COTTERILL, R.M.J., JENSEN, E.J., KRISTENSEN, W.D., Phys.Lett., 44A, 127 (1973)
49. COTTERILL, R.M.J., High Temperature Materials Phenomena, ed. by J.G. Rasmussen (Polyteknisk Forlag, Copenhagen, 1973) p.285
50. WRONSKI, C.R.M., Brit.J.Appl.Phys., 18, 1731 (1967)
51. BLACKMAN, M., and SAMBLES, J.R., Nature, 226, 938 (1970)
52. COTTERILL, R.M.J., KRISTENSEN, W.D., MARTIN, J.W., PEDERSEN, L.B., and JENSEN, E.J., Computer Phys.Comm., 5, 28 (1973)
53. KRISTENSEN, W.D., JENSEN, E.J., and COTTERILL, R.M.J., J. Chem.Phys. (in the press)
54. GUPTA, D., Phys.Rev.B., 7, 586 (1973)
55. MAKIN, S.M., ROWE, A.H., and LE CLAIRE, A.D., Proc.Phys.Soc. (London) 70, 545 (1957)
56. MARTIN, J.W., and COTTERILL, R.M.J., (Unpublished)
57. See e.g. COTTRELL, A.H., The Mechanical Properties of Matter (Wiley, 1964)

VOLUME EXPANSION, THERMAL AND ELECTRICAL CONDUCTIVITIES AND SOUND VELOCITIES OF SODIUM AT TEMPERATURES VERY CLOSE TO ITS MELTING POINT

G. Fritsch

Physik Department E 13, TU München
8046 Garching, Germany

ABSTRACT

Data on thermodynamic properties of high purity sodium (99.99 %) with a temperature resolution of about 1 mK are reported. In addition, results on sound velocities in the [110] - and [111] - directions at a frequency of 10 MHz are presented. In a temperature region of about 40 mK below the melting point these properties show an anomalous behavior. As an interpretation we suggest that the instability limit of the solid phase is close to the thermodynamic transition temperature.

INTRODUCTION

Phase transitions of first order are characterized by jumps in the first derivatives of the thermodynamic potentials at the transition temperature. This temperature is given as a function of pressure by the equality of the molar Gibbs free energies of the two phases in equilibrium. Another feature is the existence of metastable regions beyond the thermodynamic transition temperature. There exists some controversy (1) whether the metastable region is terminated by some kind of instability of the respective phases or not as this point can never be reached experimentally. This fact is due to seed growing modes in the metastable phase.

The melting transition shows some pecularities with respect to the metastable regions. Firstly the solid crystalline state cannot be superheated to any appreciable amount, if at all, whereas secondly a large supercooling of the corresponding liquid is possible. In addition no critical point seems to exist.

To examine the first point further we used the simplest metallic substance, sodium, as its melting temperature at 97.81 ^{o}C can easily be approached with high precision. A pure lattice transition is to be expected, because the electronic system shows only very small variations upon melting.

DEFINITION OF THE MELTING TEMPERATURE

Measuring very close to the melting point, it is necessary to define a criterion which gives the onset of melting. For this purpose we used the volume expansion as a function of temperature. Solid soluble impurities lower and broaden the melting transition. This is due to diffusion of that impurity from the solid crystalline state into the liquid state, because such impurities are often more soluble in the liquid phase. It can be shown(2), using purely thermodynamic arguments that the onset of the melting transition is indicated by a linear portion of the curve volume versus temperature.

EXPERIMENTAL RESULTS

The specific volume was measured with a capillary method (3). The procedure for determining the onset of melting is illustrated in Fig. 1

Below T_{MO} there is an additional rise in volume. Its magnitude amounts to 0.4 % of the total volume at T_{MO}. This figure should be compared with the volume jump of about 2 % between crystalline and liquid state. The result of a 99.99 % pure sample is shown in Fig. 2. The effect changes only slightly with increasing impurity concentration (potassium). The rise in volume is of course coupled with an increase in the expansivity.

The thermal conductivity and the electrical resistivity were measured simultaneously on the same sample (4). The chosen temperature gradients were typically smaller than 10 mK. The onset of melting was determined from the influence of the volume change during melting on the electrical resistivity. That resistivity in-

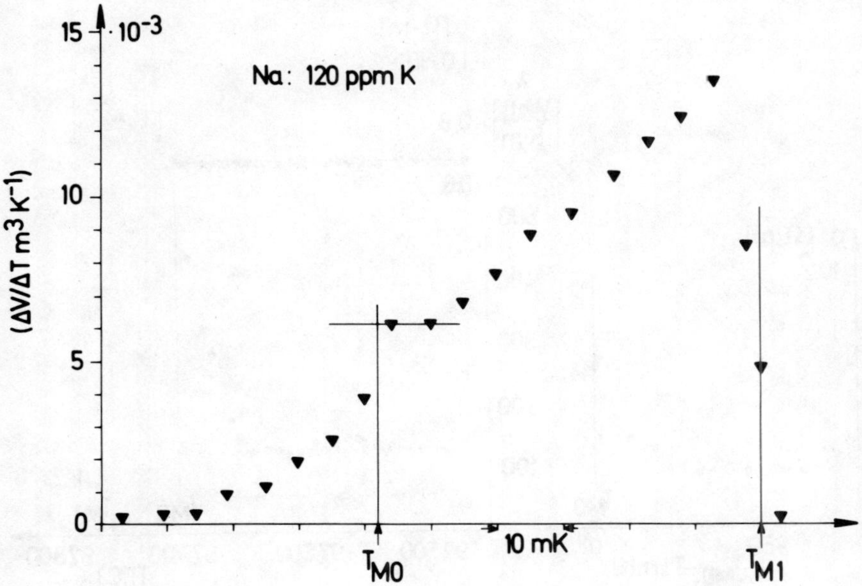

Figure 1. First derivative of the volume data with respect to temperature. The onset of melting is indicated by T_{MO}.

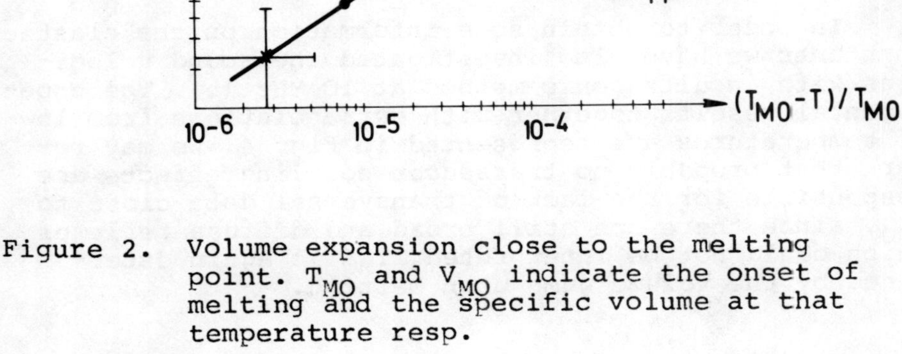

Figure 2. Volume expansion close to the melting point. T_{MO} and V_{MO} indicate the onset of melting and the specific volume at that temperature resp.

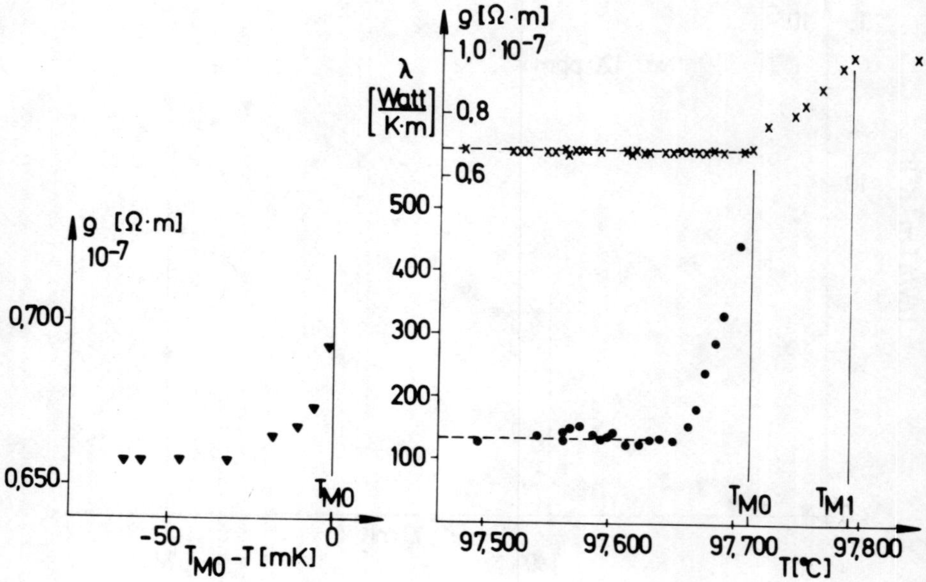

Figure 3. Thermal conductivity and electrical resistivity close to the melting transition. The melting region extends from T_{MO} to T_{M1}.

creases by a factor of about 1.5 upon melting. The thermal conductivity shows a steep rise in front of T_{MO}, whereas the electrical resistivity exhibits only a small increase (3 %). A typical result is presented in Fig. 3. It should be mentioned that during those measurements we find hints for a rise in the specific heat also.

In order to obtain some information on the elastic constants we have also investigated the sound velocities with an ultrasonic method at 10 MHz (5). The experimental results together with extrapolations from lower temperatures are represented in Fig. 4. We may remark that propably no transducer coupling effects are responsible for the lack of transversal data close to T_{MO}, since there are still broad and diffuse reflexes which could not be interpreted. T_{MO} is again determined by the volume jump upon melting.

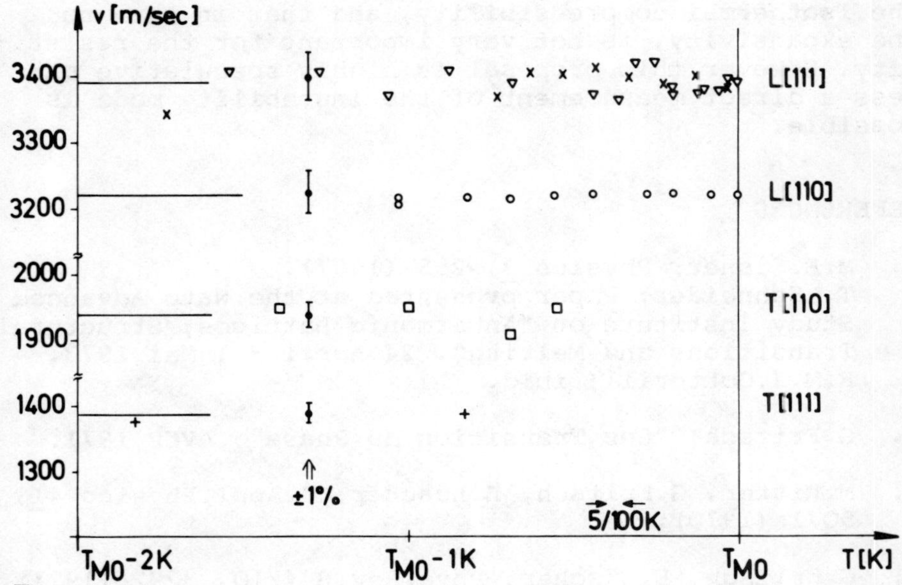

Figure 4. Sound velocities in [110] - and [111] - directions. Longitudinal data are indicated by L, transverse ones by T. The lines are extrapolations from low temperature data.

AN EXPLANATION

As there is no closed theory for the melting transition we propose as an explanation a very small metastable region for the solid crystalline phase. The observed effects should then be due to an instability, which terminates the metastable region. The physical nature of this instability is discussed elsewhere (6).

Excess volume expansion and the increase in the specific heat can be traced back to the excitation of a mode which is the origin of the instability. The large increase of the lattice thermal conductivity - the electronic part remains fairly constant due to the electrical resistivity result - might be related to a heat transport by a strongly enhanced diffusion via coupling of heat and mass currents. The low rise in the electrical resistivity can be explained by the Ziman formula (4), which allows an analysis of the resistivity in terms of the static structure function S(q). The low q region, where S(O) is proportional to the isothermal compressibility, and that in turn to the expansivity, is not very important for the resistivity. However this proposal is highly speculative unless a direct measurement of the instability mode is possible.

REFERENCES

1. M.E.Fisher; Physics 3, 255 (1967).
T. Schneider; Paper presented at the Nato Advanced Study Institute on "Anharmonic Lattices, Structural Transitions and Melting". 24 April - 1 Mai 1973.
R.M.J.Cotterill; ibid.

2. G.Fritsch; "Les Transition de Phase", AVCP 1971.

3. M.Ritter, G.Fritsch, E.Lüscher; J.Appl.Physics 41, 5071 (1970).

4. G.Fritsch, E.Lüscher; Phys.Rev.B 4/10, 3292 (1971).

5. G.Fritsch, F.Geipel, A.Prasetyo; J.Phys.Chem.Sol. to be published.

6. G.Fritsch, E.Lüscher; to be published.

LIST OF PARTICIPANTS

NAME	ADDRESS (at the time of the meeting)
ALEFELD, G.	Physik-Department der Technischen Universität München, 8046 Garching, Germany
ANDRESEN, A.F.	Institutt for Atomenergi, 2007 Kjeller, Norway
AUBRY, S.	DPHG/PSRM, BP No. 2, 91 Gif-sur-Yvette, France
AXE, J.D.	Department of Physics, Brookhaven National Laboratory, Upton, L.I., N.Y. 11973, USA
BERRE, B.	Physik-Department der Technischen Universität München, 8046 Garching, Germany
BOHLIN, L.	Department of Theoretical Physics, University of Umeå, S-901 87 Umeå, Sweden
BLUME, M.	Department of Physics, Brookhaven National Laboratory, Upton, L.I., N.Y. 11973, USA
BRUCE, A.D.	Department of Physics, University of Edinburgh, Mayfield Road, Edinburgh EH9 3JZ, U.K.
CARNEIRO, K.	Department of Physics, Risø Research Establishment, 4000 Roskilde, Denmark
COMES, R.	Université Paris-Sud, Batiment 510, 91 Orsay, France
COTTERILL, R.	Technical University of Denmark, 2800 Lyngby, Denmark

CUMMINS, H.Z.	Department of Physics, New York University, 4 Washington Place, New York, N.Y. 10003, USA
CURRAT, R.	Institut Max von Laue - Paul Langevin, Cedex 156, 38 Grenoble-Gare, France
DIETRICH, W.	Institut Max von Laue - Paul Langevin, 8046 Garching, Germany
van DIJK, C.	Reactor Centrum Nederland, Petten (NH), The Netherlands
DOLINO, G.	IBM Research Laboratory Zurich, Säumerstrasse 4, 8803 Rüschlikon, Switzerland
DOLLING, G.	Atomic Energy of Canada Ltd., Chalk River Nuclar Laboratories, Chalk River, Ontario, Canada
DORNER, B.	Institut Max von Laue - Paul Langevin, Cedex 156, 38 Grenoble-Gare, France
EIRIKSSON, V.R.	Department of Physics, University of Edinburgh, Mayfield Road, Edinburgh EH9 3JZ, U.K.
FANKHAUSER, H.R.	Department of Theoretical Physics, University of Umeå, S-901 87 Umeå, Sweden
FEDER, J.	Department of Physics, University of Oslo, Blindern, Oslo 3, Norway
FLEURY, P.	Bell Laboratories, 600 Mountain Avenue, Murray Hills, N.J. 07974, USA
FOLK, R.	Institut für Physik, Hochschule Linz, A-4045 Linz Auhof, Austria
FOSSHEIM, K.	Department of Physics, NTH, University of Trondheim, 7034 Trondheim, Norway
FRITSCH, G.	Physik-Department der Technishen Universität München, 8046 Garching, Germany
GRØNVOLD, F.	Institute of Chemistry, University of Oslo, Blindern, Oslo 3, Norway
HATTA, I.	Department of Chemical Engineering, Imperial College, Prince Consort Road, London S.W.7, U.K.

HELLER, P. IBM Research Laboratory Zurich, Säumerstrasse 4, 8803 Rüschlikon, Switzerland

HEWAT, A.W. Materials Physics Division, Bldg. 521, AERE, Harwell, Didcot, Berks., U.K.

HÜLLER, A. Institut für Festkörperforschung KFA, Postfach 365, 517 Jülich 1, Germany

JENSEN, E.J. Department of Structural Properties of Materials, The Technical University of Denmark, 2800 Lyngby, Denmark

JØSSANG, T. Department of Physics, University of Oslo, Blindern, Oslo 3, Norway

KLEIN, R. Fachbereich Physik, Universität Konstanz, Postfach 733, 775 Konstanz, Germany

KRAGLER, R. Institut für Theoretische Physik der Universität Frankfurt/M, Robert-Mayer-Strasse 8-10, 6 Frankfurt/Main 1, Germany

KRISTENSEN, W.D. Department of Structural Properties of Materials, The Technical University of Denmark, 2800 Lyngby, Denmark

LAMBERT, M. Université Paris-Sud, Batiment 510, 91 Orsay, France

LECHNER, R. Institut Max von Laue - Paul Langevin, Cedex 156, 38 Grenoble-Gare, France

LEKKERKERKER, H. Vrije Universiteit Brussel, Adolphe Buyllaan 105, 1050 Brussels, Belgium

LOCKWOOD, D.J. Department of Physics, University of Edinburgh, Mayfield Road, Edinburgh EH9 3JZ, U.K.

MEIER, P.E. IBM Research Laboratory Zurich, Säumerstrasse 4, 8803 Rüschlikon, Switzerland

MICHEL, K.H. Universiteit Antwerpen, Fort VI Straat, 2610 Wilrijk, Beglium

MÜLLER, K.A. IBM Research Laboratory Zurich, Säumerstrasse 8803 Rüschlikon, Switzerland

NICOLL, J.F. MIT, Room 13-2142, Cambridge, Mass. 02139, USA

OTNES, K. Institutt for Atomenergi, 2007 Kjeller, Norway

PRESS, W. Institut für Festkörperforschung KFA, Postfach 365, 517 Jülich 1, Germany

PRÆSTGAARD, E. H.C. Ørsted Institutet, Universitetsparken 5, 2100 Copenhagen Ø, Denmark

PYNN, R. Institutt for Atomenergi, 2007 Kjeller, Norway

RANNESTAD, A. Scientific Affairs Division, NATO, 1110 Brussels, Belgium

SAMUELSEN, E.J. Institutt for Atomenergi, 2007 Kjeller, Norway

SCHNEIDER, T. IBM Research Laboratory Zurich, Säumerstrasse 4, 8803 Rüschlikon, Switzerland

SCHLENKER, C. Groupe des Transitions de Phases, CNRS, BP No. 166, 38042 Grenoble-Cedex, France

SCHRÖDER, G. Institut für Theoretische Physik der Universität Frankfurt/M, Robert-Mayer-Strasse 8-10, 6 Frankfurt/Main 1, Germany

SCHWABL, F. Institut für Festkörperforschung KFA, Postfach 365, 517 Jülich 1, Germany

SHAPIRO, S. Department of Physics, Brookhaven National Laboratory, Upton, L.I., N.Y. 11973, USA

SJÖLANDER, A. Institute for Theoretical Physics, Fack, S-402 20, Gothenburg 5, Sweden

SRINIVASAN, G. Institute for Theoretical Physics, Fack, S-402 20, Gothenburg 5, Sweden

STEIGMEIER, E.F. Laboratories RCA, Badenerstrasse 169, 8048 Zürich, Switzerland

STEINSVOLL, O. Institutt for Atomenergi, 2007 Kjeller, Norway

STILLER, H. Institut für Festkörperforschung KFA, Postfach 365, 517 Jülich 1, Germany

STIRLING, W.G. Institut für Festkörperforschung KFA, Postfach 365, 517 Jülich 1, Germany

STØLAN, B. Department of Physics, NTH, University of Trondheim, 7034 Trondheim, Norway

SVARE, I. Department of Physics, NTH, University of Trondheim, 7034 Trondheim, Norway

SVENSSON, E. Section for Solid State Physics, AB Atomenergi, Studsvik, Fack, S-611 01 Nyköping, Sweden

TAYLOR, W. Department of Physics, University of Edinburgh, Mayfield Road, Edinburgh EH9 3JZ, U.K.

THOMAS, H. Institut für Theoretische Physik der Universität Frankfurt/M., Robert-Mayer-Strasse 8-10, 6 Frankfurt/Main 1, Germany

TORRIE, B.H. Department of Physics, University of Edinburgh, Mayfield Road, Edinburgh EH9 3JZ, U.K.

TUTHILL, G.F. MIT, Room 13-2126, Cambridge, Mass. 02139, USA

WALL, L.S. Department of Physics, Colorado State University, Fort Collins, Colorado 80521, USA

YELON, W.B. Institut Max von Laue _ Paul Langevin, Cedex 156, 38 Grenoble-Gare, France

VALENTA, M.W. Ludwig Boltzmann Institut, Kopernikusgasse 15, A-1060 Vienna, Austria

ØSTGAARD, E. Department of Physics, NLHT, University of Trondheim, 7000 Trondheim